MEDIA PLANNING AND PRODUCTION

Published by
Charles E. Merrill Publishing Company
A Bell & Howell Company
Columbus, Ohio 43216

This book was set in Univers.
Production Coordination: Martha Morss
Text Design: Cynthia Brunk
Cover Design: Tony Faiola
Cover Photo: Robert Maust Photography

Library of Congress Catalog Card Number: 83-062598
International Standard Book Number: 0-675-20113-6
Printed in the United States of America
1 2 3 4 5 6 7—90 89 88 87 86 85 84

PREFACE

Today's instructors certainly do not themselves produce all the media they use to support their teaching. It takes a commitment of time, equipment, and interest to edit videotape, create sophisticated diazo transparencies, or write a computer program. Yet the majority of training centers and schools now have videotape equipment, overhead projectors, and microcomputers. While a large percentage of the software used by instructors in these machines is commercially produced, most educators also want to design and teach with materials of their own. They want to produce and use items they created themselves in addition to knowing how to include in their lesson materials produced by others. It is for these educators that this text is written.

Teachers have a natural interest in making media and often prefer to create materials because they are more personalized. These materials may not be fancy but are preferred because they are nicely tailored to fit a teacher's students, subject area, and style. Unfortunately, "home production" by teachers often suffers in its early stages of growth because procedures appear to be too complicated. There seem to be too many instructions to remember, and when it was explained in college classes everything seemed so technical.

The media text used to present production, or the classroom instruction provided to explain techniques, may have been at fault. Instructors and texts frequently begin with the technical end of things—the audiovisual nuts and bolts of a technique—and by the time the instructional problem is addressed students have lost interest. This book attempts to correct this problem by approaching the use of media from the trainer's or instructor's point of view. We assume that those who use this book already know something about teaching, even though their experiences may have been in a summer recreation program, Scouts, church school, or as a tutor to a younger brother or sister. We also assume that these new instructors-to-be have used some media already, since tape players, television, and photographic equipment are commonly available to everyone. While many examples used in *Media Planning and Production* apply to elementary and secondary schools, there are a number of references made to college

CONTENTS

MEDIA PLANNING AND PRODUCTION

GOALS

□ Learn about ways that verbal instruction can be used as the primary teaching method.
□ Gain an overview of the process of instructional development.

OBJECTIVES

□ List methods that can be used to help students remember information.
□ Describe the characteristics of set induction and closure.
□ Outline effective methods for handling questions and answers in teaching.
□ Describe several uses of the chalkboard, as well as the unique characteristics and correct techniques for use.
□ Distinguish among audiovisual aids, educational media, and instructional technology.

Good teaching is like good salesmanship. Neither accomplishes its purpose by mere recitation of dull facts. The best-sold customer is one who gets excited about a commodity or service and sees in it some benefit to personal welfare. The best-taught student is convinced that the subject matter provides personal benefit. Since both teaching and selling rely a great deal on verbal skills, the use of your voice is an appropriate place to begin a text like this one.

You can see from our goals and objectives that this chapter concentrates on verbal skills that are related to teaching, with the production of simple visuals included where appropriate.

HOW DO WE TEACH?

When you think of teaching, do you have just one method in mind? Actually, teaching can be achieved in several ways.

Teaching by Talking

The way in which an instructor presents verbal information provides the foundation for other instructional events during a lesson. Of course, visual elements of the lesson are also important, and a great deal of teaching is done from a chalkboard. These two media—your voice and a chalkboard—provide a good starting point for a book such as this. As an illustration of methods for producing an instructional strategy that involves these media, consider the following example:

Suppose it is Tuesday morning in your social studies class and you are attempting to interest your students in American history of the 1800s. You have worked your way up to the Civil War. To compare the resources of both the Union and the Confederacy just as the war began, you refer to differences in population, number of factories, value

1 ARTICULATING INSTRUCTION

of manufactured goods, miles of railroads, and farm acreage. Your instructional strategy might be the often-used technique of asking questions and calling on students who raise their hands. The class proceeds this way during most of the period, as you move systematically through the lesson. Toward the end of the class you may provide some time for students to read or study on their own; you make an assignment for next time, and the class is over.

Let's pause and think back over this example. Think as a student in the class might think. Students daydream a little. They may not concentrate on the questions unless the teacher calls on them. Their minds wander. After all, what would *you* do, as a student, during teaching like this? Classroom interaction is almost entirely verbal. There is little for you to look at, read, or write.

Teaching by Participation

Even though teaching by talking is perfectly natural and can be quite effective, it involves a style that requires little activity on the part of students. Moving teaching and learning out of the realm of a spectator sport and into the arena of active involvement is the subject of this book. We have known for some time that active involvement—"direct purposeful experience," as Edgar Dale puts it—produces the biggest payoff. Nearly forty years ago Dale worked out a *Cone of Experience*[1] to describe various levels of learning; his idea is more fully explained in a later chapter. The development of a system for understanding instructional strategies has been a long-standing concern. In the 1600s, Johann Amos Comenius, another educator, listed thirteen instructional principles that make good sense today. These principles were:

1. Instructional method should follow the order of nature. Content should be related to the developmental stage of each learner.

2. Instruction should begin at infancy and should be designed for the age, interest, and capacity of each learner.

3. Whatever is taught should have practical application to life and should have value for the learner.

4. Subject matter should be organized according to its difficulty. Instruction should proceed by the inductive process from the simple to the complex.

5. A graduated series of textbooks and illustrative materials should be correlated with instruction.

6. Sequence is important. For example, it is irrational to teach a foreign language before the mother language has been learned.

7. General principles should be explained and examples given before rules are learned; nothing should be memorized until it is understood.

8. Reading and writing should be taught together; subjects should be correlated whenever possible.

9. Learning is to be approached through the senses; actual objects and things should be studied and associated with words.

[1]E. Dale, *Audiovisual Methods in Teaching* (Hinsdale, Ill.: Dryden Press, 1946).

a kind of "method acting." Sure enough, ratings of the trained teachers improved significantly. Apparently teachers who are effective with the spoken word are judged to be effective overall. Of course, knowledge of subject matter, ways of dealing with students, organization, and other characteristics are important too. Verbal skill has been recognized as important for some time, and by using portable videotape equipment a procedure was developed in the early 1960s to provide help to teachers desiring to improve themselves. Termed microteaching, students teach, critique the taped performance, and then reteach under controlled conditions.

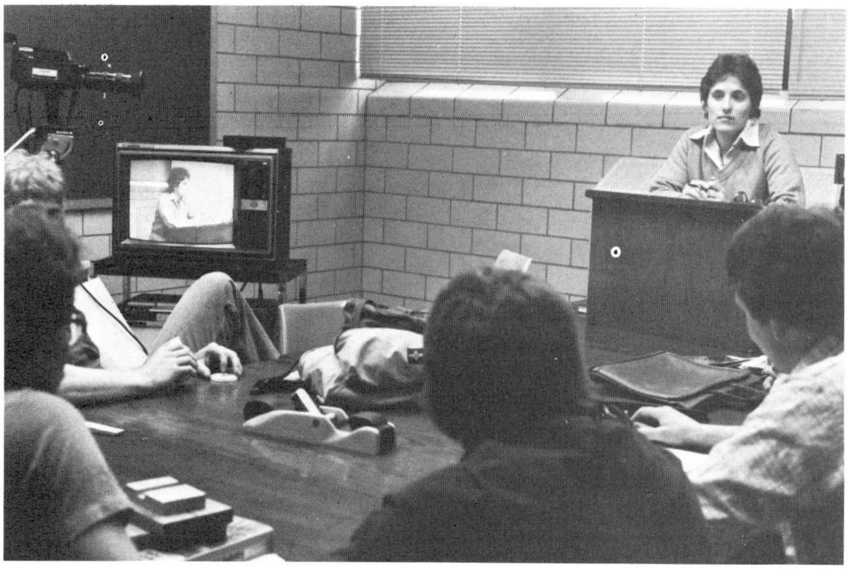

Microteaching is also called "mirror teaching" because students can see themselves on videotape after they have presented a lesson.

Microteaching derives its name from the fact that a single concept is taught to a small group (5 to 10) for a short time (around 5 minutes). Because the process is designed to improve verbal delivery, the procedures used are important to us here.

SOME SKILLS USED IN MICROTEACHING

Some suggestions for beginning a class are:

☐ Use an example:
"I remember a case where . . ."
"About three years ago there was a . . ."

☐ Make a promise:
"One of the advantages of learning about this . . ."
"People in this field tell me that . . ."

☐ Assess readiness:
"With the background you now have, we are ready to . . ."
"I sense from your informal feedback that we now can . . ."

☐ State objectives:
"By the end of class today, you should be able to . . ."
"After you have completed the assignment, you should be able to . . ."

☐ Give directions:
"Take notes on . . ."
"Read . . ."
"Use this apparatus . . ."
"Work these problems . . ."

Set Induction

This is the process of opening class activities or "setting the stage" for effective teaching and learning. Techniques include:

Readiness Formal and/or informal pretesting to indicate if students have a foundation for new concepts. Pretesting helps avoid wasted time from teaching above or below students' abilities.

Motivation Demonstration of applications of new concepts, examples from everyday life, or the promise of new skills that can be learned.

Behavioral Objectives Specific instructions on the attainment of goals, stated in terms of student performance, with words such as *". . . list, describe, explain, identify. . . ."* Objectives should be clarified at the beginning of periods of study rather than after testing.

Directions Assignments for study, including the provision for needed resources and deadline dates.

Discussion Questions or comments initiated by students. Expressed in early stages of instruction, this method often points out misinterpretations, confusion, or vital information accidentally omitted in other aspects of establishing set.

Closure

This is a summary of newly learned information, recognition of general student achievement, and illustrations of ways to use new knowledge in the future. Closure of one lesson often leads naturally into Set Induction for the next.

Summary Teachers should ask students to summarize new information and relate it to prior knowledge. Application, analysis, synthesis, and evaluation of new ideas aid long-term memory and use. Summary can be formal or informal.

Demonstration Students can demonstrate achievement by giving reports, using apparatus, or by showing appropriate behaviors. Emphasis should be on demonstration of what students *do* know rather than on deficiencies. Every demonstration experience is viewed as an opportunity for continued learning and reinforcement of attained skills.

Appraisal Compare student levels of achievement to acceptable levels of achievement. Help students understand their own levels of proficiency in comparison to their personal goals. If those goals include the necessity to compete with others it is reasonable that students be informed of their comparative standing. Restatement of original behavioral objectives is appropriate as a basis for appraisal.

Some suggestions for ending a class are:

☐ Summarize:
"These are the three points we've covered today . . ."
"Let's review what we've talked about."
☐ Look ahead:
"Our topic today relates to the next topic because . . ."
"You can see that our current topics are naturally leading to . . ."
☐ Evaluate:
"I know you may have some questions. Write a question on a card and I'll answer it next time."
"Raise your hand if you feel pretty comfortable with the areas we've been discussing."
☐ Make an assignment:
"With this background you should be able to . . ."
"Next time we will be . . . , so you should . . ."

Questioning

To handle questions and answers:

Questions
☐ Call on individuals:
"Sally, here's a question for you."
"Bill, do you remember . . ."
☐ Be nonthreatening:
"What do you think about . . ."
"What's your opinion on . . ."
"Can you think of a situation in which . . ."

Answers
☐ Repeat the answer:
"Apparently you mean . . ."
"OK, so your view is . . ."
☐ Shop around for answers:
"Alright. Now, Sam, what do you think?"
"Is that OK, or should it be . . ."
☐ Use nonverbal feedback:
(Smile and nod; frown and scratch your jaw.)
☐ Take a vote:
"How many of you agree?"
"How many of you disagree?"
"How many of you don't know?"

Models can be used to introduce a topic in a realistic way.

Chalkboard Production Ideas

Verbal Information

☐ Solve math problems (or have students copy their work onto the chalkboard).
☐ List chemical equations or formulas.
☐ List important terms or ideas.
☐ Write a famous quotation every day or each week in one corner of the board.
☐ Outline the lecture, discussion, or lesson plan at the beginning of your class.

Nonverbal Information

☐ Graphs, charts, histograms
☐ Drawings or models that show relationships

Combination

☐ Diagrams with labels.
☐ Geometry and algebra problems that need nonverbal drawings.

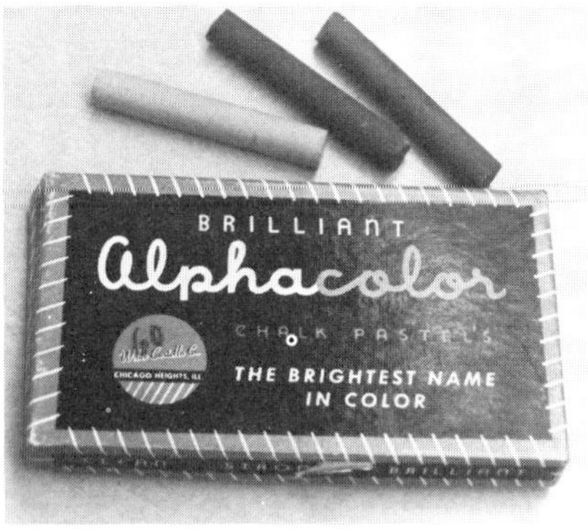

Colored chalk can be used to separate ideas and increase understanding by emphasizing parts of a chalkboard display.

Some of the uses you make of the chalkboard may seem perfectly natural, and might not require a great deal of advance planning. For example, you may list important points that emerge from a class discussion, or sketch a diagram or model as you are explaining an idea.

TRY THIS

List some uses of the chalkboard for your subject area.

Although it is a simple medium, the chalkboard is sometimes misused. You may remember a teacher's writing that was too small, messy, or difficult to read. You may have tried taking notes from a teacher who kept writing and erasing rapidly. Some of the correct techniques for producing chalkboard instruction are listed here:

One common problem when using the chalkboard is talking to the board rather than the class.

PRODUCE

Correct Techniques for Using a Chalkboard

☐ Print, unless your writing is especially easy to read.
☐ Verify that your words are large enough to be seen easily from the back of the room.
☐ Use colored chalk to separate ideas or add emphasis.
☐ Do not try to talk and write at the same time. When you write on the board stop talking and concentrate on what you are writing.
☐ Maintain eye contact with the class, not with the board. Avoid looking at the board during discussion, but use a long pointer instead so that you can stand aside, using peripheral vision to see the board.
☐ Write short sentences or phrases, but avoid abbreviations; students taking notes will have difficulty later in deciphering your shorthand.
☐ Hold the chalk like a knife and pretend you are slicing tomatoes on the kitchen counter. Use firm downward strokes, and rotate the chalk as you use it so that lines will be uniform as the chalk wears.

The Chalkboard as a Unique Medium

Earlier in this chapter we mentioned that each type of educational medium should be chosen for what it can do best, or for what it can do that no other medium can. As you begin to accumulate ideas for teaching with media, consider these important characteristics of the chalkboard:

- ☐ Materials can be big, bold, and colorful.
- ☐ Unplanned information can be listed as needed during class.
- ☐ "Down time" is no worry; it requires little maintenance, no electricity or operating instructions, and cost for materials is very low.
- ☐ Availability is nearly universal; you can count on always having one in a classroom. (Portable boards are also available for use in unusual locations. You can even make your own chalkboard by using special paint on masonite, hardboard, or plywood.)
- ☐ Long-term notices and announcements can be recorded on the board and kept there, since the board is permanently located in the classroom.

TRY THIS

Make your own chalkboard with special slate paint on masonite. Most paint stores or lumber yards have this paint.

There is no doubt that the use of a chalkboard creates a much richer instructional setting than does a presentation that is strictly verbal. You might be tempted to place the chalkboard in the category of "instructional aid" because of its rather mild, accessory role in teaching.

Because of the high esteem that live teaching enjoys, it is natural to assume that machines or devices are merely aids. However, when any device is used—even a simple one such as the chalkboard—a new form of teaching results. With the chalkboard and other technology as examples, consider the comparison between audiovisual aids and educational media.

AUDIOVISUAL AIDS OR EDUCATIONAL MEDIA?

What do a plumber, a violinist, and a teacher all have in common? For one thing, they possess talents and expertise that set them apart from others. They earn money with their knowledge. They have acquired formal training and continue to perfect their skills through practice.

Beyond these characteristics, each person needs tools, instruments, or devices to accomplish a task. Without a violin the violinist cannot make music, and the violin is helpless without someone to play it. A teacher and educational media become a combination of talent and tools that represents a new system. In the case of a teacher who uses a slide set during instruction, or a film, videotape, or transparency, the medium is no more an aid to the teacher than the teacher is an aid to the medium. Each—taken alone—is less effective than the new combination. When teachers design instruction that involves media they are engaged in *instructional development.* The product that results is called an *instructional system.* During production, while media are selected, a basic rule must be followed: *Media should be selected for their potential to implement a behavioral objective.*

Instructional systems can be very simple or quite complex, but each contains a set of fundamental components. They are:

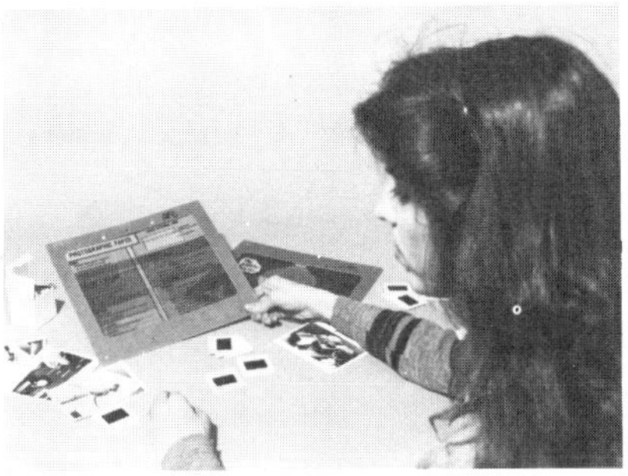

A teacher is an artist with tools, like other artists.

might be equally effective. As you design instruction with media rejoice that there is usually more than one correct, proper, and effective way to create instruction.

TRY THIS

Select a topic in your subject area and suggest two media that would be appropriate for use in teaching the idea.

A FINAL WORD

In this chapter we have begun a study of media production that starts with ideas you should find familiar. Your voice and a chalkboard are appropriate media for many instructional settings. To perfect your skills, try producing some videotapes that document your teaching style, as suggested in this chapter. With a firm foundation in these basic skills you will be in a good position to begin the process of making and using a wide range of media.

LOOKING AHEAD

In this first chapter we have characterized good teaching—the interaction between the instructor and the student. Instruction can take place with relatively few media. However, if you want to expand your teaching repertoire to include the sights and sounds of your subject, it is necessary to rely on a wider variety of materials and techniques.

While it is tempting to pick up a camera or a video recorder and begin production, it makes better sense to examine commercially prepared materials first. They may be expensive, but they may also be well done and fit your requirements. Some software can be modified using the production techniques we will talk about in later sections of this book. Minor changes in format or sequence can adapt software for your needs. Commercial materials also suggest many ideas for projects you might want to produce yourself, or have your own students work on.

The next chapter in this section suggests ways of identifying commercially prepared materials and producing instruction from them.

REVIEW QUESTIONS

1. List a topic in your subject area that would lend itself to instruction that is largely verbal.
2. Choose one of Comenius' thirteen instructional principles and describe how it is applicable to teaching today.
3. Invent a mnemonic device for use in teaching.
4. What impact does nonverbal communication have on the teaching and learning process?
5. List some common uses of the chalkboard.
6. What is instructional development, and what are its components?

ANSWERS TO REVIEW QUESTIONS

1. Answers will vary, of course, but certain topics in language arts, where the spoken word is being studied, are generally handled by having students listen and then react. Recordings of Robert Frost reading his own poetry are examples.
2. One of the principles—No. 4—has been important for a long time. Educators in nearly all subject areas recognize the value of proceeding from the simple to the complex. K–6 programs in mathematics follow this principle.
3. Several mnemonic devices are mentioned in the chapter. In science, for example, to remember the classification system (kingdom, phylum, class, order, etc.), it is helpful to remember "Kings play chess or fry good steak." The first letter of each word helps in the recall of the levels of classification.
4. It is recognized by many people that smiles, frowns, arms folded across the chest, tone of voice, and other aspects of communication convey feelings and impressions that go beyond the actual content of a lesson.
5. Some common uses of the chalkboard include listing terms for discussion, making simple sketches to show relationships, and diagramming a model that represents how ideas relate to each other.
6. Instructional development is the systematic planning of instruction, including pre- and posttesting, determination of objectives, planning and producing media, delivery of instruction, and follow-up evaluation. The concept is applicable to virtually every type of teaching.

REFERENCES

Bennett, Neville, and McNamara, David. *Focus on Teaching* (London and New York: Longman's, 1979).

Cooper, James M., et al. *Classroom Teaching Skills: A Handbook* (Lexington, Mass.: D. C. Heath, 1977).

Cooper, James M., et al. *Classroom Teaching Skills: A Workbook* (Lexington, Mass.: D. C. Heath, 1977).

Dunkin, Michael J., and Biddle, Bruce J. *The Study of Teaching* (New York: Holt, Rinehart and Winston, 1974).

Flanders, Ned A. *Analyzing Teacher Behavior* (Reading, Mass.: Addison-Wesley, 1970).

House, Ernest R., and Lapan, Stephen D. *Survival in the Classroom* (Abridged) (Boston: Allyn and Bacon, 1978).

Hoyle, Eric. *The Role of the Teacher* (New York: Routledge and Kegan Paul, 1969).

Hunkins, Francis P. *Involving Students in Questioning* (Boston: Allyn and Bacon, 1976).

Kim, Eugene C., and Kellough, Richard D. *A Resource Guide for Secondary School Teaching* (New York: Macmillan, 1974).

Levin, Tamar, and Long, Ruth. *Effective Instruction* (Alexandria, Va.: Association for Supervision and Curriculum Development, 1981).

Maggs, Margaret Martin. *The Classroom Survival Book: A Practical Manual for Teachers* New Viewpoints—(New York: Franklin Watts, 1981).

McKeachie, Wilbert J. *Teaching Tips: A Guidebook for the Beginning College Teacher,* 7th ed. (New York: D. C. Heath, 1978).

Saettler, Paul. *A History of Instructional Technology* (New York: McGraw-Hill, 1968).

Waterman, Floyd, et al. *Designing Short-Term Instructional Programs* (Washington, D.C.: Association of Teacher Educators, 1979).

Westbury, I., and Bellack, A. A. *Research into Classroom Processes* (New York: Teachers College Press, 1971).

School district media centers usually provide expensive and one-of-a-kind services to all teachers in the system.

One of the first school district media centers was authorized in 1904, shortly after the St. Louis Exhibition (commemorating the 100th anniversary of the Louisiana Purchase). Many of the materials featured at the Exhibition were purchased by the St. Louis Board of Education and, on April 11, 1905, a museum was opened in an old school building. A horse and wagon were used to make weekly deliveries to schools in the area.

TRY THIS

Check with a local school district office of central administration to determine availability of media that teachers can use. Get a catalog of software and make a list of items you could use in your subject area.

Area or Regional Media Centers

In the past two decades, regional, multicounty media centers for public schools have developed throughout the country. Establishment of these services was spurred in part by the infusion of federal funds, and began at a time when the long tradition of county school systems was fading. Regional media centers, or regional educational service agencies, were designed to offer a wide range of services in addition to media production and instructional development. Some of the features of regional media programs include filmstrip production (from teacher-made slides), video- and audiotape duplicating, graphics production and laminating, open-house work sessions, in-service programs in media production and utilization, and consultation in instructional development. Regional media centers usually also offer large and comprehensive collections of curriculum materials and professional books, journals, and nonprint software. Regional trade and professional associations often have collections of instructional materials for lending.

TRY THIS

Examine a district or building-level catalog of media services to determine what types of media production *you* could carry out there. Also note the types of production the *agency* will do for you.

Media Centers on the State Level

Since each state has an agency that coordinates educational programs, it is natural that various collections of teaching materials might be found there.

Media to support bilingual education, assistance for children with special needs, and programs for talented and gifted children often come from a state department of training or education. Consultants in these agencies may become focal points for the collections. Although media are rarely produced by state agencies and the software housed by them are usually in specialized collections that are not widely circulated, teachers are encouraged to make contact with appropriate consultants who work at the state level to determine what kinds of support they do provide.

Media Centers at the National Level

Nearly every national organization (National Science Teacher's Association, National Council of Teachers of English, American Manufacturer's Association) publishes booklets, references, and other items that can be used for teaching. Several national professional associations in the media fields also have many publications that are of use to teachers and media specialists. Among the groups are the Association for Educational Communication and Technology (AECT), the National Audiovisual Association (NAVA), the American Library Association (ALA), and the American Association of School Librarians (AASL).

MEDIA COLLECTIONS TO MEET SPECIAL NEEDS

One example of media to meet specific needs is collected by the National Information Center for Special Education Materials (NICSEM) in California. Established in 1966 and using a computerized information base, the Center assists special educators in meeting the requirements of Public Law 94-142. This law states that an Individualized Education Program must be developed for each handicapped child in special or mainstreamed classrooms.

The index published by NISCEM contains information on more than 10,000 materials. Other publications offer help in locating and purchasing materials for talented and gifted students, vocational-technical education, and career education. A professional library maintained by an educational agency can help in locating these special indexes.

FINDING READY-MADE MEDIA

Several hundred publishers and producers manufacture hundreds of thousands of different types of software for education and training. You can imagine the confusion you would face in locating a specific item without some sort of index or reference system. The following sections describe several means for identifying commercially produced software.

El-Hi Texts in Print

This is a yearly book of several hundred pages, containing over 36,000 currently available texts for use in elementary and secondary schools. No evaluations are attempted; this is merely a reference for locating the materials. Author, title, publisher, copyright, price, and other important descriptive information are listed.

El-Hi Textbooks in Print is published yearly. It lists all textbooks available for purchase, and is indexed by subject and topic.

NICEM Indexes

The National Information Center for Educational Media (NICEM) publishes a number of volumes that list nonprint software. Organized by topics and subject areas, the NICEM system provides only title, publisher, and minimal descriptive information. Materials are not evaluated, and no prices are included. To use this system you must have a current catalog from a publisher to determine price, description, and other details. A single NICEM Index in the series may list all media of a certain type, such as "16mm FILMS" or "FILMSTRIPS." Several single volumes, however, list all media for a single content area, such as "PSYCHOLOGY" or "BLACK HISTORY."

NICEM Indexes list instructional materials by type or by subject.

Scope and Sequence Charts

Large charts, produced by publishers, permit comparison within a series of texts, such as K–6 reading series, a mathematics program, or

Scope and sequence charts show how textbook series should be used, and how different texts in the series relate to one another.

Many instructional materials can be obtained free by requesting them on school stationery.

The Educators Progress Service, Inc. catalogs list free teaching materials in both subject area and media-type volumes.

a junior high science program. The charts often contain goals and objectives, a description of tests and activities, and a matrix of grade levels and major concepts on which the series is based.

Curriculum Guides

Available in several forms, curriculum guides are published by many school systems to help teachers plan instruction. Frequently a school will fail to print extra copies of its guides, making it difficult to build a collection from a variety of schools. There are attempts at the national level to produce collections of curriculum guides on microfiche, so that a wide variety of school systems and subject areas is represented. One of these, the Fearon-Pitman Curriculum Laboratory, lists several thousand guides on fiche. Abstracts and a description of each guide make it possible to scan the information quickly, before viewing details on the microfiche.

Free and Inexpensive Materials

Many useful teaching materials are provided free of charge by several thousand industries in the United States. Embassies of foreign countries also publish a great deal of free material about their countries and cultures.

One such service, the Educator's Progress Service, lists films, pamphlets, filmstrips, and other media from several thousand companies, industries, and associations. Teachers can obtain this material by requesting it. A brief description of each item is included in this service, which consists of over a dozen volumes. Some volumes are titled by subject area (FREE MATERIALS FOR SOCIAL STUDIES, SCIENCE, AND GUIDANCE), and others are titled by type of medium (FREE FILMSTRIPS, FILMS).

Professional Literature

For strategies on how to teach a subject at a selected grade level, there are many methods books that can provide assistance. Some contain lesson plans, sources of materials for teaching through simulations and games, plans for construction of bulletin boards, and suggestions for using community resources. In addition to methods, there are books with sample lesson plans, test items, performance objectives, ideas for planning new facilities, and suggestions for individualizing instruction.

TRY THIS

Locate some software, using the suggested references, that would be of value in teaching in your content area.

USING SOFTWARE TO INSTRUCT

It would be ideal if each school, classroom, or teacher had specific software for each subject and grade level, keyed to each objective. While this situation is rare, there are often occasions in which the software may be a little sparse, but adequate. Usually the media on hand will treat a given subject in a broad and general way, providing the teacher with the opportunity to round out the instruction with locally produced materials.

One example of the broad treatment sometimes found in commercial software can be illustrated by considering a 16mm film. While the content may be applicable to your class, and fit the general area of the subject you are teaching, you may find yourself wishing that certain portions of the film could be changed. Some scenes, some of the language used, and often the organization of topics may be inappropriate. In another case, suppose you are using a transparency that is part of a larger set, purchased to support the instructional program. Certain terms may be included that you would like to omit; some parts of the diagrams may contain too much detail, or not enough. Again you find yourself with the urge to redesign the work to fit your particular teaching situation.

For example, you might find a filmstrip that provides a suitable, general treatment of the subject, yet lacks specific detail in one or two places. You might be able to create a small collection of slides to expand certain concepts. Then, using a filmstrip projector *and* a slide projector you could intermingle the commercial images with your own.

Begin to formulate ideas for your own teaching material. A chart in a text, for example, might suggest a bulletin board you want to make. A graph could be redesigned and used as a transparency. The description of a process could be worked up as computer-assisted instruction for a microcomputer. In each case your creativity might be triggered by media that are already on hand.

TRY THIS

Select a transparency, filmstrip, or other software and redesign it to personalize it for your own teaching.

Many excellent instruction books and manuals give specific examples of how to teach concepts in different ways.

READY-MADE MEDIA VERSUS LOCALLY PRODUCED MATERIALS

There was a time, not long ago, when the only ways to obtain a chocolate cake were to buy one from a bakery or make one yourself, from scratch. When cake mixes came along it became possible to obtain a finished product merely by adding some water and baking the mixture. Where teaching materials are concerned should you buy them ready-made or try the do-it-yourself route? We cannot recommend that you always attempt to make your own media, nor can we advise you never to do so. Every teacher must make the decision after weighing several factors.

Factors to Consider

The teacher who is interested in producing "homemade" media should carefully consider the following elements before actively undertaking the project.

Interest and Desire Your attitude about production may be the single most significant factor. For some, the creative process of discovering a new way to explain a concept, using a personal media "invention," is very rewarding. It involves a craft, a knack with graphics, and some mechanical aptitude. The same qualities motivate the do-it-yourselfer to insulate a house, build a bookcase, tune an engine, or paint a picture. If you have a strong desire to accomplish a task, time, money, and skill rarely interfere. You simply take the time

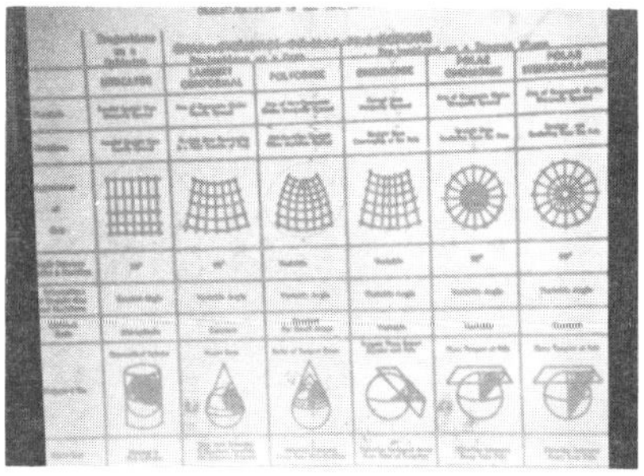

This transparency, made from a commercially printed handout, must be redesigned to include less information and larger type.

and spend the money to get the job done. And while skill—or lack of it—affects the final product, you will enjoy the feeling of accomplishment. Somehow this compensates for a lack of finesse and may justify your use of hand-made materials that lack a truly professional touch.

Time It is natural to take time for activities we enjoy. You rarely complain about lack of time, for example, if you enjoy being a member of a bowling team.

On the other hand, if you dislike pulling weeds and hoeing the tomatoes you may claim you ". . . don't have *time* to garden." The time required to finish a task is also related to efficiency. If you find you enjoy media production, but if it takes three hours to shoot twenty slides on a copy stand, the slow rate at which you work can make it impractical.

Lest you be too quick to rationalize by saying ". . . my time is too valuable; I can't afford to do that myself . . . ," consider what you would spend your valuable time doing instead. Media production may be a form of relaxation and recreation—activities you would not expect to be paid for. Furthermore, chances are that you would not be earning money if you were not making media anyway—so beware of that method of rationalizing.

Money This is a more objective consideration, since you can make a direct comparison between costs of do-it-yourself media products and those you might purchase. In almost every case the actual out-of-pocket cost is less if you produce the media yourself. The reason is simple. Media production is a labor-intensive endeavor. Much of the cost of a slide set, videotape, transparency, or other medium is tied up in salaries, royalties, artist's fees, and other expenses generated by people involved in the project. Only a small fraction is needed to pay for the film, transparency plastic, raw tape stock, or photographic print paper.

Equipment This turns out to be an important consideration for schools and other agencies that produce media. Initial cost of the equipment is only part of the story; there are also maintenance cost and depreciation. Since much of the production equipment is sophisticated, it is possible to spend thousands of dollars on the equipment for a modern and comprehensive media center.

However, a great deal of equipment may already be on hand. Nearly every school has a camera (to support the school newspaper and yearbook) and videotape recorder (for use in the athletic programs in evaluating competitive team sports). Audiotape recorders are widespread in schools today, and thermal copiers and spirit duplicators are standard equipment in the principal's office, if not in the school media center. Microcomputer sales to schools are increasing rapidly and, with the growth of community cable TV systems, a number of schools are finding themselves in the television business. Thus a sizeable pool of equipment for your own media projects may be available to support some fairly complex production.

Skill The greater your talent and expertise, the more effective the media production. The suggestions in this book are intended to increase skill in selecting, producing, and using media. While these abilities increase with experience, you can expect some outside help as well. Media specialists and instructional developers are trained to provide specialized services and advice for instructors who want to increase the use of media. Other sources of ideas include your fellow instructors, in-service meetings, workshops, and short-term courses.

In addition, the designers and manufacturers of equipment have always been sensitive to the needs of their clients. Easy-to-use cameras that automatically adjust to lighting conditions, tape recorders with automatic volume controls, and self-threading projectors are a few examples of strides made in the media industry to provide equipment that is nearly foolproof.

STUDENT PRODUCTION—A PRACTICAL SOLUTION

As an instructor, you have two fundamental methods for obtaining media for teaching. You can purchase them, or you can make them yourself.

With the availability of media production centers, media specialists, and the rich variety of commercially available material, it is natural to ask if teachers need to become producers of their own materials. The answer is that teachers may not be the primary source of media production. The actual producers can be students. We suggest that students become involved in production, as an alternative to ". . . read-a-book . . . ," ". . . write-a-paper . . . ," "take-a-test. . . ." In other words, consider that a student prepares a slide set, a couple of transparencies, a videotape, or a film in the form of a report or project. There are a number of benefits of student media production, including:

- ☐ A large pool of time and talent can be tapped.
- ☐ Students learn the content thoroughly if they script and produce media packages.
- ☐ Students acquire media skills.
- ☐ The role of the teacher is to advise and assist in planning, with students carrying out the production.
- ☐ Relevant teaching material for various forms of instruction is produced for use by the teacher in subsequent years.

Ideas for Low-Cost Production

Some effective sources, inexpensive and close at hand, provide a rich supply of teaching materials. While some require a small amount of media production, others can be used in their original condition.

Picture File Collect pictures from magazines and newspapers and file them in categories for future use in slide sets, bulletin boards, or as study prints. If your community has a magazine exchange at the public library you can take advantage of the opportunity to enlarge your photo file.

Free Materials Collect free materials from business, industry, publishers, and other sources. The brochures, pamphlets, charts, and other types of materials often contain diagrams, drawings, photos, and other visual materials that can be used as is, or modified and reworked into teaching materials that are more suitable for your purposes and adapted to your subject area.

TRY THIS

Select some desirable free materials and send for them to begin your teaching file. Ask your instructor for advice; you may need to use school letterheads and a letter that explains you are in a teacher education or training program. Mention that you are beginning a teaching file for use in your chosen subject area.

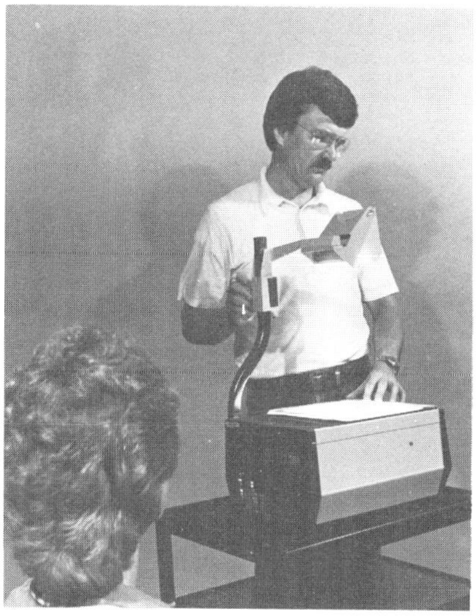

After teachers demonstrate media production, the students can give class reports using the media.

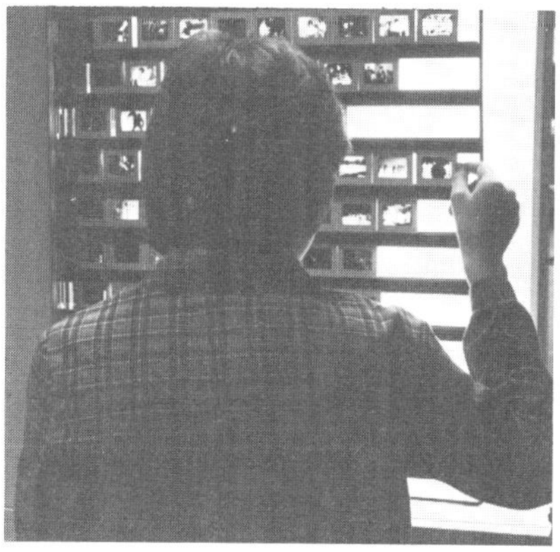

In addition to photograph or picture files, it is a good idea to keep a record of your slides for use in putting together new presentations.

With permission, you can often produce presentations during field trips.

EPIE reports give in-depth evaluations of many educational products.

Pictures from Field Trips, Museums Using a simple camera and high-speed film, produce photo essays (either slides or prints) for instruction. If certain views, scenes, or situations are not available from live circumstances, you can add photos where necessary by copying from your picture file. For example, visit a store to photograph products or displays that you could use in teaching.

Audiotapes Collect sounds appropriate for your subject area from radio broadcasts, conventions, personal interviews, television, disk recordings, or other sources.

Videotapes Create videotapes live, during field trips, special events, or visits to historic places. Or record selected broadcasts off the air, as long as copyright is not violated.

Teleconference Phone Calls Audiotape long-distance phone calls in which particularly significant people or situations may be involved. Be sure to obtain permission first, however.

Microcomputer Programs Become a member of a local "users group" in order to learn programming techniques and to share software that you and others have developed.

EVALUATING PRODUCED MEDIA

When it comes to deciding whether or not a particular kind of software is good, a highly personal decision must be made. Things that work for one person may not be at all satisfactory for someone else. As the evaluation criteria included here indicate, each medium should be examined from a slightly different point of view. Cost, suitability for your students, complexity, and quality are among the important aspects of evaluation.

A type of "consumer report" service—the Educational Products Information Exchange (EPIE)—provides evaluations of hardware and software. EPIE publishes monthly newsletters and periodic in-depth reports to help in decisions about the type of teaching material in which to invest. You should use the same care in selecting instructional resources that you would use in buying a new car or washing machine.

TRY THIS

Evaluate some selected software using the checklists on the next page.

MATERIALS EVALUATION PLAN

What kind of material is covered by this questionaire?

What specific material is covered by this plan--title, etc.?

Summarize the contents of the material below.

Circle YES, NO, or NA for not applicable.

Content

1. Does the material include information useful for the purpose for which it was chosen? YES NO NA
2. Is the level of the material appropriate for the classes involved? YES NO NA
3. Is the material current--modern machines, dress, etc.? YES NO NA
4. Has any important information been omitted? YES NO NA

Sex Bias

5. Are the sexes treated equally in visual effects? YES NO NA
   If no, which gets less emphasis? __females __males
6. Are the sexes treated equally in the audio or copy? YES NO NA
   If no, which gets less emphasis? __females __males
7. Are sex-free occupational titles used? YES NO NA
8. Are gender-free pronouns used? YES NO NA

Racial Bias--Omission

9. Have any of the racial groups shown below been omitted from the visual effects? YES NO NA
   If yes, which group(s) is omitted?
   ___American Indians
   ___Asian-Americans
   ___Blacks
   ___Hispanics
   ___Whites
10. Are all of the groups listed below found in the audio or copy material? YES NO NA
   If no, which group(s) is omitted?
   ___American Indians
   ___Asian-Americans
   ___Blacks
   ___Hispanics
   ___Whites
11. Is any ethnic group presented unfavorably? YES NO NA
   If yes, which group(s)? ____________________
   ____

Stereotyping

12. Is any group stereotyped? YES NO NA
   If yes, which group(s)?
   ___females
   ___males
   ___American Indians
   ___Asian-Americans
   ___Blacks
   ___Hispanics
   ___Whites
   ___Others--List them ____________________
   ____
   ____

Additional Comments:

Prepared for: Curriculum and Instruction 506x
By: Gary Donlin

MATERIALS EVALUATION FORM

NAME
DATE

(1) TITLE AND AUTHOR				
(2) TOPICS COVERED				
(3) COURSE AND SUBJECT				
(4) TYPE OF MATERIAL (CIRCLE ONE)	TEXT	STUDY GUIDE	LAB MANUAL	TAPE
	FILM	FILM STRIP	SLIDE SET	VIDEOTAPE
(5) WHERE AVAILABLE		(6) TIME REQ'D TO ORDER		
(7) LENGTH OF MATERIAL		(8) SUPPLEMENTAL MATERIALS		
(9) USAGE (CIRCLE ONE)	IN-CLASS	STUDENT REF.	TEACHER REF.	
(10) SUGGESTIONS ON USING MATERIAL (WHEN, HOW)				
(11) STUDENT'S REACTION				
(12) OVERALL EVALUATION				

(13) RATING SCALE ON SUBJECT CONTENT		1	2	3	4	5	
a) PREREQUISITE KNOWLEDGE	ADVANCED						ELEMENTARY
b) DIVERSITY OF TOPICS	BROAD						NARROW
c) EXPLANATIONS	THOROUGH						INCOMPLETE
	CLEAR						UNCLEAR
	CONCISE						WORDY
d) ORGANIZATION	WELL ORGANIZED						DISORGANIZED
	CONSISTENT						INCONSISTENT
e) TERMINOLOGY	DEFINED						UNDEFINED
	ADVANCED						ELEMENTARY
f) VISUALS	MANY						FEW
	CLEAR						UNCLEAR
	COMPLEX						ELEMENTARY
	INNOVATIVE						OLD HAT

(14) CHECKLIST ON SEXUAL BIAS AND STEREOTYPES

	MALE	FEMALE	ASEXUAL	BALANCED
PRONOUNS				
NARRATOR				
VISUALS				
STEREOTYPES				
ROLES: AUTHORITY				
ROLES: SUBORDINATE				
ROLES: INDEPENDENT				
ROLES: NURTURING				

COMMENTS:

(15) CHECKLIST ON ETHNIC BIAS AND STEREOTYPES

a) GROUPS REPRESENTED (CIRCLE)

ASIAN	EUR.	AFRO-A.	HISP.	NAT. A.	JEW.	WASP

b) HOW PRESENTED (CIRCLE)

NARRATOR	PICTURES	TEXT. REF.

c) STEREOTYPES? YES- NO

d) BIAS IN LANGUAGE? YES- NO

e) COMMENTS:

EVALUATION OF NON-PRINT MATERIAL

Title ____________________
Distributer ____________________
Subject ____________________
Grade Level ____________________

Black and White ________
In Color ________
Content: Ex. ____ Good ____
Fair ____ Poor ____

To evaluate sex bias, mark M for male, F for female. Mark both if it applies. Leave blank if not observed.
To evaluate sex stereotypes, mark Y for yes, N for no, and leave blank if not observed.
To evaluate races depicted, mark W for White, B for Black, A for Asian American, H for Hispanic, NA for Native Americans, and O for Others.

Catagory	Bias M	Bias F	Stereotypes Y	Stereotypes N	Race W	Race B	Race A	Race H	Race NA	Race O
1. Title										
2. Narration										
3. Audio										
4. Visuals										
5. Main Characters										
6. Supporting Characters										
7. Home Settings										
8. Occupational Settings										
9. Community Settings										
10. School Settings										
11. Recreational Settings										
12. Independent Role										
13. Nurturing Role										
14. Aggressive Role										
15. Subordinate Role										

ADDITIONAL COMMENTS:

COMPLETE EVALUATION: According to this criteria, this non-print material is:

____ **balanced** ____ mixed ____ **racially acceptable**
____ **male biased** ____ stereotyped ____ **racially unacceptable**
____ **female biased** ____ expanded

I do/do not recommend this film.

AVOIDING BIAS IN MEDIA YOU PRODUCE

Not too long ago it was common practice to refer to all people as "mankind," to speak of students as "him," and to pay little more than sympathetic attention to minority groups in our culture. Some of this feeling—especially the inaccuracies in treating minorities—may have stemmed from the "melting pot" philosophy that prevailed in the United States for such a long time. Many people had come to believe that various ethnic groups had been assimilated into the country to form a new group called "Americans" and, in doing so, these groups needed to forsake their heritage. The current feeling is that the "melting pot" idea is inaccurate.

Social studies materials have become a particularly significant target for locating bias, along with basic reading texts and other language arts software. The movement has gradually made an impact on our popular media, such as newspapers, magazines, and television programs.

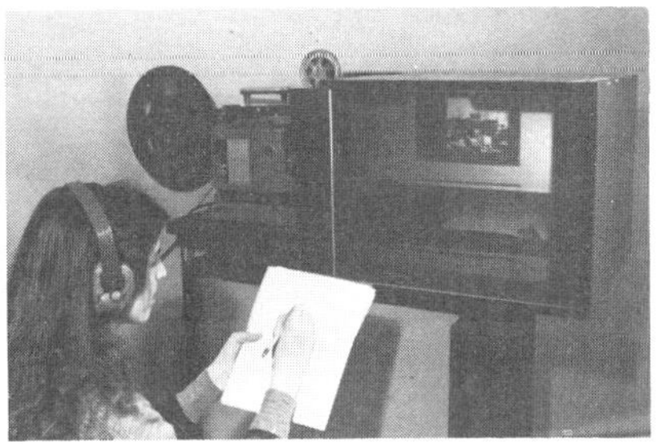

All films should be reviewed before use in class.

TRY THIS

Collect examples of racism, sexism, cultural bias, or stereotyping of occupations from magazines or newspapers. Look at recent publications as well as back issues. If possible, you could create a slide set from pictures taken from a variety of sources. You can also make a list of television shows that have been created to make the general public more aware of the contributions of certain cultural groups to ". . . the American way of life."

Various studies that cite inaccuracies in teaching material may be of help to those who wish to produce their own media. The Council for Interracial Books for Children has made a detailed analysis of U.S. history texts to determine where stereotypes, distortions, and omissions exist. Observations from this study fall into three categories:

- ☐ *Perspective* Too often the information is presented *from the perspective* of the dominant groups. Students are sometimes taught that America is a white country and the only prime movers in history have been white males.
- ☐ *Methods of Inclusion* Third-world people are often included in a text if they can be categorized as "greats," "contributors," or "protesters." The isolated examples that these stereotypes provide ignore the broader base of support furnished by large numbers of lesser-known members of a culture.
- ☐ *Underlying Assumptions* Certain generalizations have been made that cloud the more specific motives, reasons, and explanations that account for the contribution of various groups. Among these assumptions are: (1) U.S. society is a true democracy; (2) sexism and racism are not linked to economic exploitation; (3) individual bigots, not the culture that fosters bigotry, are to blame for inequalities.

If bias-free media are to be produced these guidelines should be considered:

- ☐ Numerical balance between the sexes, with each gender in an equally dominant role, depicted in nontraditional occupations and varied levels of occupational status and responsibility.
- ☐ Both genders depicted in expanded societal roles, and in attire appropriate for the role.

☐ Illustrations of a variety of body types and some evidence of handicaps.
☐ Gender-free language (plural pronouns, gender-free nouns).
☐ No reference to title or physical appearance unless there is a purpose in doing so.
☐ Expression of a range of emotions regardless of gender.

TRY THIS

Using the six criteria above, as well as additional considerations that you may want to include, choose an example of instructional material in your subject area and evaluate it. Note specific strengths and weaknesses by citing quotations from the material you evaluate.

A FINAL WORD

So far, in these two chapters, we have suggested ways by which you can sharpen your existing skills in teaching with media, as well as indicating how you can get ideas from commercial materials. While your instructional style may improve a great deal if you can use the ideas presented so far, you will find it is often satisfying to produce your own software. Actually, that is the topic and theme of this book: media production. As we move through the text, the techniques and suggestions for production will become more and more dominant.

LOOKING AHEAD

While we move into the area of designing and producing your own software, we must take a moment to consider the fundamentals of visual literacy on which future production is based. Generally accepted rules, such as elements and principles of design, are a part of visual literacy. So are a communication model and Dale's *Cone of Experience*. Think of visual literacy as you might think of any other sort of literacy. It is a sensitivity to a way of thinking; it is a point of view; it is a perspective. But all of this must be based on some cognitive information. That foundation is the essence of Chapter 3.

REVIEW QUESTIONS

1. List the resources you would examine to find:
   ☐ Free materials
   ☐ Ready-made transparencies
   ☐ Texts in your subject area
   ☐ Curriculum guides
2. What is meant by "bias in instructional materials"?
3. List the hierarchy of levels of resource collections found in an educational system.
4. List the steps to be followed in preparing for use of commercial software, and the steps to follow after its use.

ANSWERS TO REVIEW QUESTIONS

1. Free materials—Educator's Progress Service
   Ready-made transparencies—NICEM Indexes
   Texts in your subject area—*El-Hi Texts in Print*
   Curriculum guides—*The Curriculum Development Library* (Fearon-Pitman) or locally produced curriculum guides from selected school systems
2. Bias is opinion not based on reason—a slanted, prejudiced, or stereotyped point of view that may be found in instructional materials. Bias can take the form of inaccuracies about vocational roles of men and women, cultural or racial roles, or other relationships in a society.
3. The lowest and smallest level is the classroom collection. Moving up from there: school building, school district, regional or area collection, state, and finally national collection of materials.
4. Prepare by: Preview, design a worksheet, and explain to class what the material is about. After use: Discuss questions on the worksheet, use a quiz if you wish, and consider using the material again to further reinforce the content.

REFERENCES

Educator's Guide to Free Materials (Randolph, Wisc.: Educator's Progress Service, annual publication).

El-Hi Textbooks in Print (New York: R. R. Bowker, 1981).

Index to Educational Materials (University of Southern California, Los Angeles: National Information Center for Educational Media (NICEM), 1980).

Promoting Sex Equity in the Classroom: A Resource for Teachers (Iowa State University, Ames: Vocational Education Council, Department of Home Economics Education, College of Education and Des Moines: State of Iowa Department of Public Instruction, 1979).

Stereotypes, Distortions, and Omissions in U.S. History Textbooks New York: The Council on Interracial Books for Children; a Division of the Racism and Sexism Resource Center for Educators, 1977).

GOAL

☐ Understand the need for visualized teaching in order to promote effective learning.

OBJECTIVES

☐ Describe the process of reception, perception, and recognition.
☐ Describe how experiences can be sequenced from real to abstract, and the relationship this sequence has to learning.
☐ List and explain the elements of the process of communication between individuals.
☐ List and explain the use of the elements and principles of design for creating literate visuals.
☐ Create a visual sentence.
☐ Organize visuals to direct a viewer toward some idea.
☐ Organize visuals to tell a story or convey a message chronologically.
☐ Incorporate research results into the process of producing visuals.

In 1972, a team of press correspondents was being briefed on the Pioneer 10 and 11 space missions at the TRW Systems, Inc., offices in Southern California. The ultrasophisticated spacecraft these newspaper reporters were viewing were scheduled for launch later that year. They would go toward the outer planets in the Solar System, and then into the depths of space where they would hurtle outward for millions of years. Eric Burgess, of the *Christian Science Monitor,* commented that he saw the Pioneer spacecraft as the first emissaries of humankind to other civilizations of the Universe. He wondered what thoughts would cross the mind of some intelligent life form that intercepted a Pioneer craft in hundreds of thousands of years. What would the finder know of Earth and of the intelligent species that sent the Pioneer on its mission?

Based on Burgess' comments, and those of others, it was decided that a message about Earth should accompany each spacecraft when it was launched. The communication was planned as a visual one. After considerable discussion and planning, it was decided that a 6 × 9 inch (15.25 × 22.8 cm) plaque would be attached to the side of the Pioneer spacecraft. Its message was of man, woman, Earth, and the Solar System in visual language. It was thought that a visual message would be more universal than any other form of communication, and that this message would be the one most likely to be understood by any life form that might read it millions of miles and years from the earth.

Visuals may not really be a universal language, and a picture may not really be worth a thousand words as a 1930s audiovisual company once advertised, but pictures are an effective, important, and precise means of communication. In this chapter we will discuss the language of pictures and how we can use visuals to communicate. Specifically, this chapter will discuss:

3 COMMUNICATION AND VISUAL LITERACY

☐ How people perceive their environment through their senses.
☐ How people communicate with one another.
☐ Techniques used to create visuals with meaning.
☐ Ways to sequence visuals to tell a story, or to convey a complex idea.
☐ Research that explains how visual media can and should be used to promote learning.

UNDERSTANDING COMMUNICATION

In order to understand communication we must first discuss the "simple" communication process that occurs when an individual interacts with the environment. The techniques important during this basic process are then applied to the more complicated interactions that occur when individuals attempt to communicate with one another.

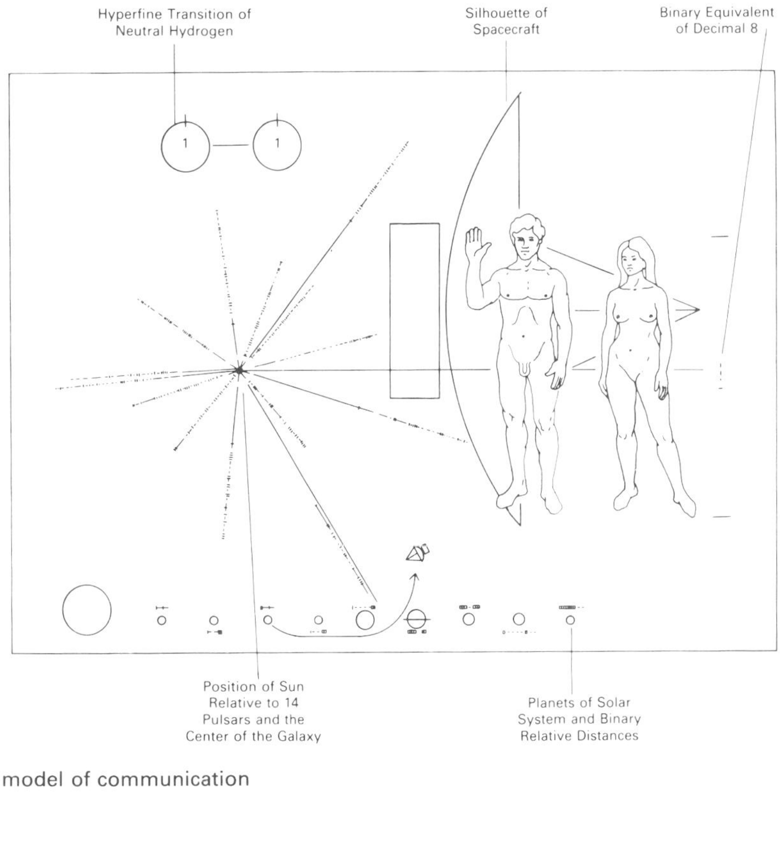

A model of communication

Five Senses

Sight
Sound
Touch
Taste
Smell

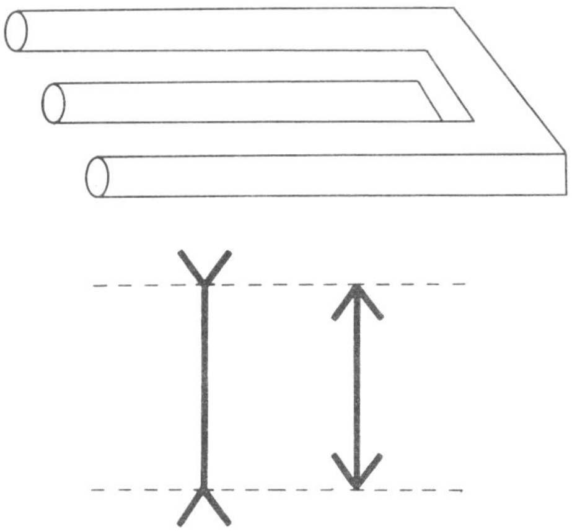

The senses can be fooled easily, as shown by these optical illusions.

Simple Communication—Reception, Perception, and Recognition

Simple communication is the interaction of the individual and the environment, and is possible because our senses react to transmitted stimuli. The eyes respond to light waves, the ears to sound waves, and the nerve endings of the nose and mouth to odors and tastes. The nerves in the skin are sensitive to pressure and temperature. These "conduits for knowledge" allow the individual to receive information from the environment.

Perception occurs when stimuli are received by the brain. The senses feed data to the brain which, in turn, acts on or reacts to them. Perception is often referred to as the process whereby individuals become aware of the world and themselves. The process is actually one of reception and then perception. We receive impulses from our nerve endings. Perception is the understanding of this information. It involves intuition and is a cognitive process.

Recognition, the last step in simple communication, occurs as perceptual judgments become familiar. What we perceive today we recognize tomorrow. Perceptions are experiences. They build upon one another so that the individual constructs a foundation of experience that enables higher-level understanding.

Children provide excellent subjects for studying the processes of perception and learning. They are continuously acquiring new information. They are constantly perceiving new objects in the environment, and building their levels of experience.

In 1946, Edgar Dale reported on an article by Edward Vernon in the *British Weekly* where he asked young children to respond to the question "What are the loveliest things you know, persons not counted?" Some of the replies he received were:

- ☐ The smell of rain
- ☐ Cool wind on a hot day
- ☐ The feel of clean clothes
- ☐ Climbing uphill and looking down
- ☐ Rain on your cheeks
- ☐ Our dog's eyes
- ☐ The crunch of dry leaves when you walk through them
- ☐ The feel of running
- ☐ The taste of strawberries
- ☐ The feel of a dive
- ☐ The smell of new-mown hay

Stages of Perceptual Development

1. Sensation—The detection and awareness of change by senses.
2. Figure Perception—The awareness of an entity—a real object.
3. Symbol Perception—The awareness of figures and the ability to distinguish one from another. The ability to understand and use symbols and to recognize them.
4. Perception of Meaning—Interpretive ability. The ability to understand the significance of forms, patterns, or events, and the ability to assign personal significance to them.
5. Perceptive Performance—The ability to sensitively and accurately observe and make complex decisions where many factors are involved, and the ability to change ongoing behavior in response to feedback received.

Source: M. R. Moore, "The Perceptual-Motor Domain and a Proposed Taxonomy of Perception," *AV Communication Review* 4 (1970): 379–413.

TRY THIS

The Importance of the Senses to Children

Ask a group of children under eight years of age to list ten things that they feel are the loveliest they know, except for people. When all the statements are written, collect them and identify the senses to which they seem to refer. Make a priority listing of what children consider to be the most important senses.

Children are acutely aware of their senses. They use them continuously to acquire new skills and ideas. Dale used this observation about the senses, in part, to propose a way to organize types of sensory experience from abstract to realistic. His *Cone of Experience* consists of twelve levels of experience. The twelve levels of experience in Dale's *Cone of Experience* can be subdivided into three categories according to the amount of involvement required of the learner. The most abstract level, at the top of the cone, is the verbal symbol—the written word. The least abstract experiences are those described at the bottom of the cone, ones Dale called direct, purposeful experiences.

According to Dale, the basis for all effective learning is the direct, purposeful experience. The most realistic experience is the one individually perceived by the learner. These are the events in which a learner participates directly.

In order to learn, a student must have a foundation of experience. Abstract forms of experience produce learning only when an individual can draw upon and relate to more realistic experiences. When totally new ideas are presented to students it is important that they be given a chance to participate directly in the learning experience.

Simple communication, where the individual interacts with the environment, begins with a *reception* of stimuli through one of the five senses. We smell a rose, see its color, and feel the sharpness of its thorns. This information is perceived and translated into *perceptions* by the brain. These perceptions, then, make it possible for a person to *recognize* a rose in the future. Because we have had direct, purposeful experience with roses we are able to recognize pic-

Groupings of Experience

Doing
1. Direct experience
2. Contrived experience
3. Dramatization

Observing
4. Demonstrations
5. Field trips
6. Exhibits
7. Television
8. Films
9. Still pictures
10. Radio

Symbolizing
11. Visual symbols
12. Verbal symbols

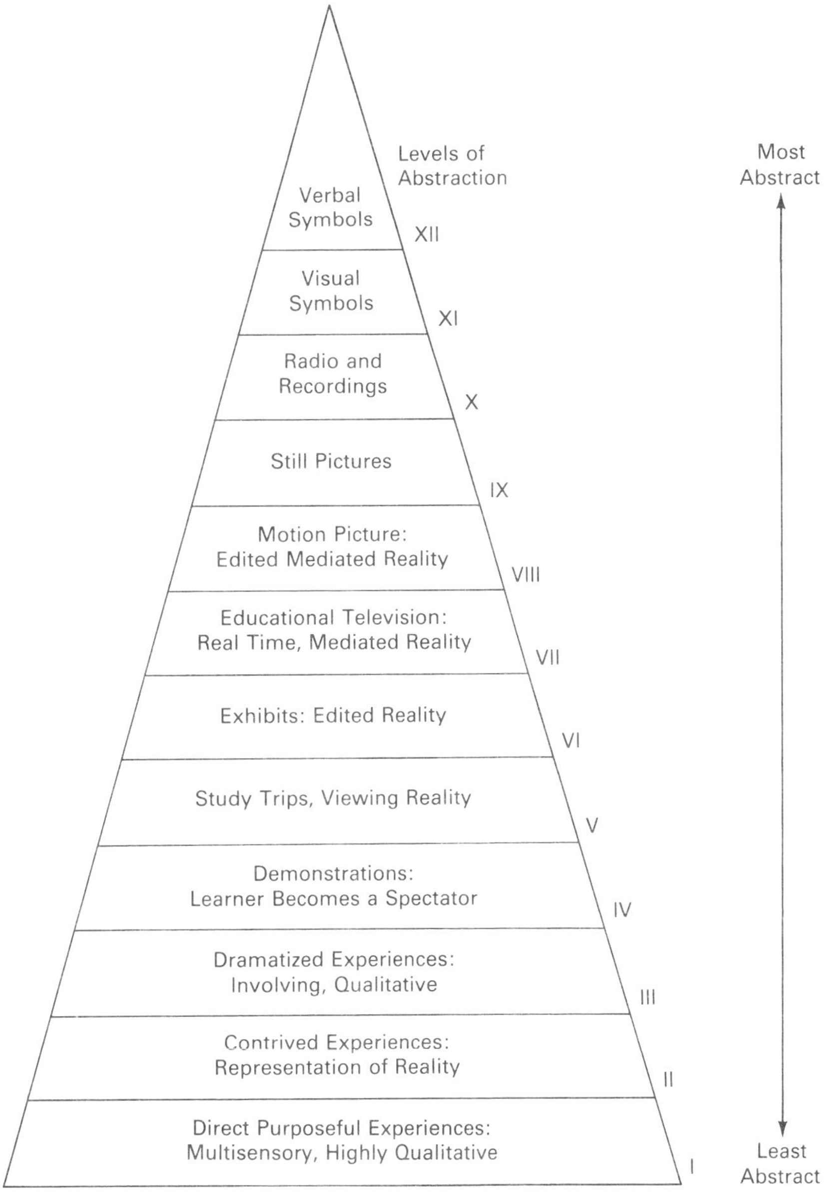

The *Cone of Experience*

Common Experiences

1. Language
2. Culture
3. Education
4. Vocation
5. Social group
6. Attitude
7. Environment

tures of roses in a magazine. Abstract learning is possible because realistic learning has occurred.

Communication between Individuals and Groups

Eric Burgess and his reporter colleagues were not proposing merely the processing of information when they planned the Pioneer spacecraft communication plaque. They were not concerned with the simple interaction of a thinking individual with the environment; rather, they wanted to communicate ideas. They wanted to tell someone, or something, about Earth and its inhabitants.

Communication is usually considered to be a process involving at least two individuals: one who has a message, or idea, and one who does not. The communication process is one of transmitting the message from the first person, the sender, to the second, the receiver.

In 1949, Claude Shannon and Warren Weaver wrote a book titled *The Mathematical Theory of Communication.* The basis for this book was a model to explain what happens when we communicate.

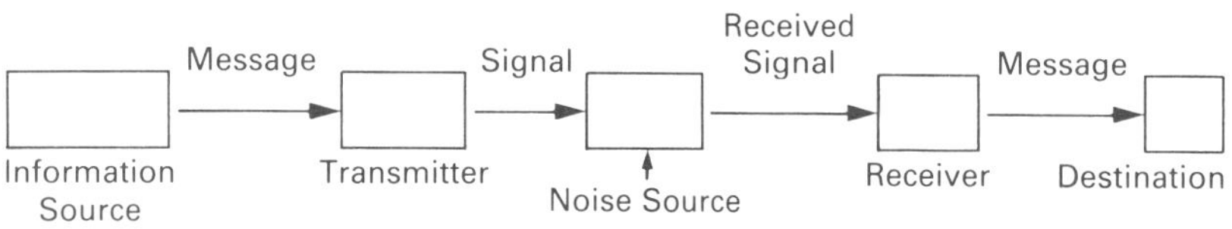

Shannon-Weaver Communication Model

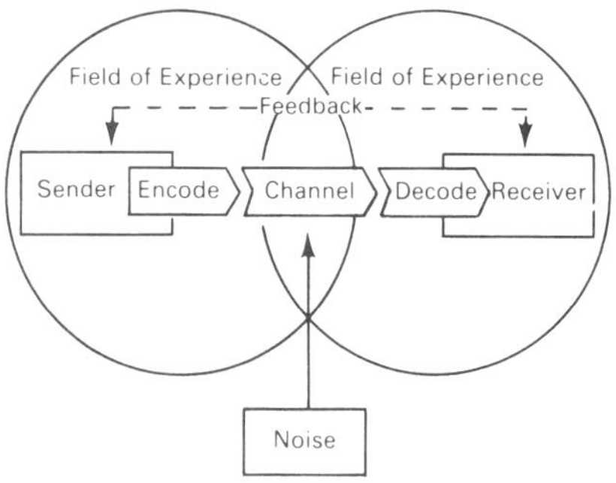

A model of communication

While the original Shannon-Weaver model was linear, other theorists have added the concepts of "Feedback" and "Fields of Experience" to more accurately describe what happens during communication. The addition of feedback makes communication a cyclical process that is dynamic, or capable of change.

Fields of Experience relate directly to Dale's *Cone of Experience.* The more varied and profound the types of experience an individual has, the more receptive that person will be to communication attempts. A young child has difficulty understanding anything except direct commands or statements. Older children and adults become adept at "reading between the lines" and inferring messages because they have a broader base of experience. It is also important to notice that the fields of experience for the sender and receiver overlap. The greater the overlap, the more likely it is that communication will be successful. Two individuals with a considerable degree of similarity in their backgrounds have a much easier time understanding one another than those from different cultures, social backgrounds, or age groups. However, it is thought that an individual can perceive and recognize some messages even though common experiences are few. Alien life forms that may someday intercept the Pioneer spacecraft will have little in common with the message designers. It is thought that they will be able to perceive visual images and recognize the binary number and hydrogen atom message clues provided them because it is hoped that these clues in the communication will be familiar to them.

It is important to remember that the greater the shared experiences and similarities between sender and receiver, the more likely and the more complete will be the understanding.

The *source,* or sender of a message, is the individual or group transmitting the message to the receiver. The task of the source is to prepare a message that informs or influences the receiver toward the object of the message. The more the teacher knows about the students in the class, their levels of language development, their attitudes, their family backgrounds, and their previous experiences, the more likely it is that a successful message will be constructed.

The *message* is the idea the sender wishes to convey to the receiver. This idea is coded in some transmittable form, usually involving symbols such as words or pictures. Symbols serve as clues to the meaning of a message. It is during coding and decoding of messages that many problems in the delivery of ideas occur. Generally, the more realistic the symbols used to communicate an idea, the more successful will be the message transmission process. The receiver of the message must be able to easily, quickly, and accurately decode the message into the idea originally coded by the sender. The fewer times a message must be coded, the more likely it is that it will be received and understood.

TRY THIS

Send a Message

A game often used by youth groups called "Pass the Story" is a good example of how difficult communication is. Try it and find out for yourself.

1. Think up a short idea no longer than a sentence or two. Quotes work well. Write it down.
2. Arrange a group of five or more in a circle. Whisper the quote to the first person.
3. Have the first person whisper the quote to the next person and so on, around the circle.
4. Have the last person write down what was whispered.
5. Compare the original quote to what the last person wrote. Effective communication is difficult, isn't it?

The *channel* is considered to be the vehicle for carrying the message. The channel is also called the medium. There are two categories of channels—sensory and technological. Sensory channels are those involving only the five senses. A teacher talks to the class. Dinner speakers use gestures to visualize ideas. Lovers touch to show their feelings. Sensory channels are generally quite limited in their capabilities for communication. Teachers can make their voices heard only over short distances, and gestures convey only limited meanings. Touching, tasting, and smelling are limited both in variability and because of the need for close proximity of the sender and receiver. Sight is the most variable of the five senses and can be used over the greatest distance. However, the eyes only receive messages, just as the voice is only used to send messages.

Technological media include all nonsensory techniques for carrying a message. Radio carrier waves can be used to send audio and video signals over vast distances. Words and pictures, written and drawn on paper, can be sent through the mails, included in a book, or printed in a magazine. Electrical signals can be used to send sound and image information through wires. Ingenious advertisers have found that chemicals embedded in a page of a magazine can be used to simulate the smells and tastes of perfumes or foods. Technological media greatly expand and extend the message delivery capabilities of the teacher and trainer.

Noise refers to anything that impairs the efficiency or delivery of the message. This interference can occur because of the improper coding or decoding of messages. Noise occurs because extraneous information such as static is picked up during the delivery of the message. Noise could also be any event that conflicts with the message for the attention of the receiver.

Examples of Noise in Teaching

- ☐ Static on the TV
- ☐ Complex language used by the teacher
- ☐ Distractions outside the classroom
- ☐ Improper channel used by the teacher to communicate to students
- ☐ Garbled soundtrack on 16mm film
- ☐ Hot or cold classroom
- ☐ Proximity to lunch time

Johann Gutenberg (1400–1468), founder of modern printing

The Gutenberg press, based on the contemporary wine press. The lever-operated screw lowered the platen to press the paper against the type. (Reprinted from George A. Kubler, *A New History of Stereotyping,* by permission of the Certified Dry Mat Corporation).

In a classroom, noise might influence the communication process because the teacher uses words and pictures with which the students are unfamiliar, because the soundtrack of the film is garbled and muffled, or because other children are loudly engaged in band practice in the auditorium just down the hall. All are noise and all impede the delivery of the message from the sender to the receiver. One concern of the message sender is to minimize the influence of noise on the communication process. This can be done by repeating the message several times, by sending it over several channels, or by asking for feedback from the receiver.

The *receiver* is the target of the communication process. Often the sender and receiver alternate their roles as communication occurs. If communication is successful, after it is completed the student should have the same idea that the teacher had before communication began.

Feedback allows for some degree of control over the communication process. Through feedback the sender can obtain information about how the receiver reacts to the message. In the classroom, feedback is usually direct. Students ask and answer questions, exhibit facial expressions, or act bored. All this information is feedback to the teacher that can be used to direct and improve subsequent attempts at communicating. Teachers obtain additional feedback information when they give tests. While not as direct and immediate, this feedback is also used to evaluate classroom communication.

Summary

Communication is a process of message delivery. A student is able to learn because messages can be received through one or more of the senses, perceived by the brain, and recognized when compared to previous experiences. Simple communication involves one person interacting with the environment. Complex communication, such as that which occurs in a classroom between teacher and students, involves two or more persons. Teachers must encode and transmit ideas to their students. Since sight is by far the most variable and complex of the senses it can be used to communicate in the largest number of situations in the classroom. The process of creating and using visuals to communicate ideas is the theme of the next section of this chapter.

IMPROVING COMMUNICATION—VISUAL LITERACY

Literacy is a goal of teachers everywhere. Since Johann Gutenberg's invention of movable type and the printing press it has been possible for masses of individuals to learn to read and to write. However, in spite of the capabilities afforded by the printing press, the process of developing literacy has been a difficult one.

It has taken centuries for literacy to become widespread and, until recently, literacy was largely restricted to the skills of writing and reading. However, just as the printing press made the written word universally available, inventions such as photography, television, and motion pictures have made technological visual communication an everyday occurrence. In many cases the accurate understanding of pictures has proved to be more difficult than the creation of the visuals themselves.

Visual literacy is a concept directed toward the ability of an individual to understand the meaning of a visual, and to communicate visually in an understandable way. Literacy makes possible visual communication which, in turn, is made possible by visual technology.

Visual Literacy		**Visual Communication**		**Visual Technology**
an individual skill—the ability to understand visuals	↔	the process of transferring ideas between individuals, visually	↔	the techniques and materials used to make the transfer of visual ideas possible

The cross and helmet have meaning for us, but would they to someone from India? From E. Dale, *Audiovisual Methods in Teaching.* Used by permission; William Reuter, *The New Literacy* (Toronto: University of Toronto Press, 1971).

Visual literacy is a skill that can be used to create visuals that convey meaning, or that an individual uses to understand visual messages developed by others. Some say visuals are anything we look at, including words, as well as drawings, cartoons, or pictures. Words can convey meaning, even when used alone, just by the way they are written. Consider the impact and meaning of the way the words below are written. Relate the design of the word to the meaning you feel. For example, would the word *beauty* convey the same meaning if it were written in the same lettering style as the word *chaos*?

Used by permission. Design by William Reuter from *The New Literacy* by Donald Gordon (Toronto: University of Toronto Press, 1971).

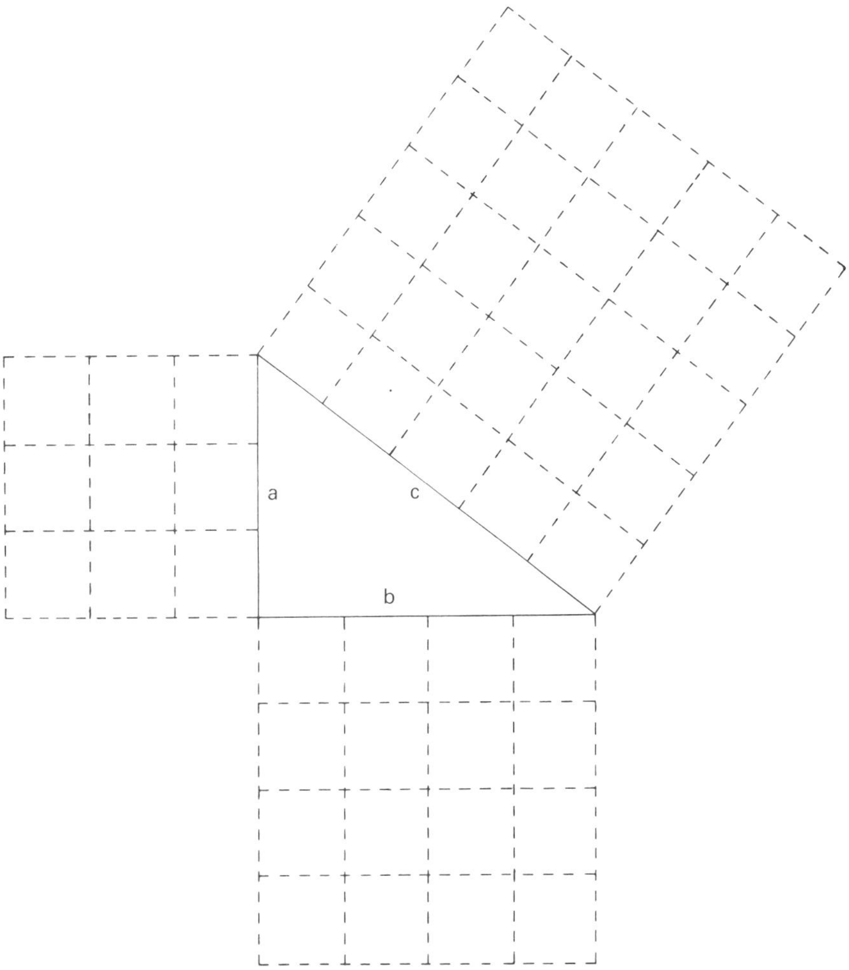

Pythagorean Theorem
$c^2 = a^2 + b^2$

Which is easier to understand?

Line

Shape

Space

Texture

Generally, when we talk about visual literacy we are not talking about words. Rather, we are talking about pictures that convey meaning through design. Here is an example of pictures and words used to convey an idea.

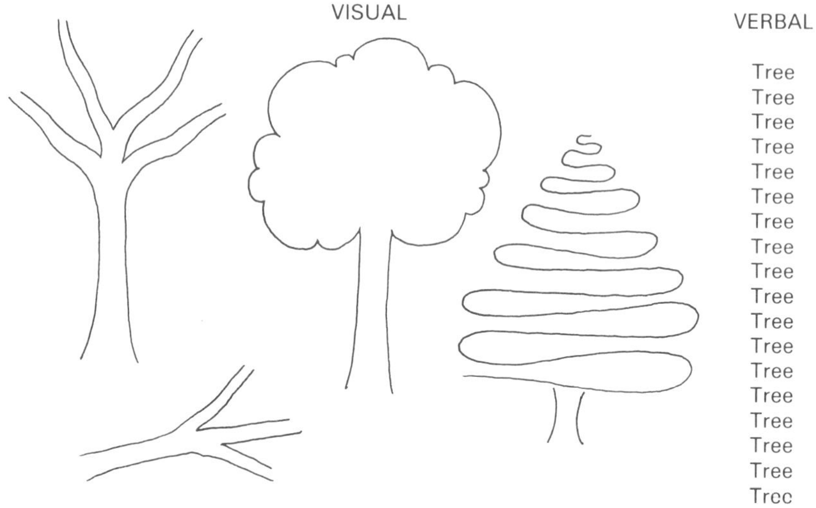

Designing Literate Visuals

The five elements of design and the six principles of design are basic ingredients of pictures and are used to design visuals that are literate.

Elements of Design The building blocks for creating visuals are (1) line, (2) shape, (3) space, (4) texture, and (5) value/color.

Line is generally considered to be one-dimensional. It has length but not width. An important function of line is to portray direction. Another is to represent objects. Line is used to define the outer edges of shape.

Shape is used to symbolize objects, or to show large or small spaces. Shapes have two dimensions: height and width.

Space may be referred to as either positive or negative. The outline of an object in a visual signifies its positive space. The most common negative space is the background portion of a visual. For example, when a bulletin board is first put up there is a large area that is not covered by letters, pictures, or other information. This is negative space. Positive space is the remaining portion of the display.

Texture is the perceived or actual roughness or smoothness of a surface. It is used to define a shape or space.

Value and color are similar elements. Value is the degree of lightness or darkness of a surface. In simple drawings value is accomplished through shading. It is used to define the elements of shape, line, texture, and color. Value shows changes in space, and is often used to create the illusion of volume or solidity in an object.

Color is related to value, and is used to visualize an object realistically, or to vividly differentiate it from some other object. Color is an important technique to use in the creation of visuals. Colors generally have hue, value, and intensity.

1. *Hue* is used to describe a specific color, such as red, green, or blue.
2. *Value* is how light or dark a color appears. Among colors, yellow has the highest value.
3. *Intensity* is used to describe the strength of a color, such as bright yellow, or dull red. Intensity is determined by the purity of a color. If a color is mixed with others it is less intense.

The color wheel is an effective tool for understanding the relationships between colors. All colors are arranged so that those highest in value are at the top of the color wheel and those lowest at the bottom.

1. The *primary colors* in pigments are red, yellow, and blue (precisely, magenta, yellow, and cyan).
2. The *secondary colors* are made from mixtures of the primaries that lie next to them on the color wheel (orange, green, violet).
3. The *intermediate colors* are produced when a secondary color is mixed with an adjacent primary. Intermediates are named for both parents (e.g., yellow-green).
4. *Tertiary colors* are formed by mixing secondaries together. These colors do not possess names such as the others, but are referred to by their appearance, such as "reddish-brown" which is produced when orange and violet are mixed.

Colors can be mixed in other ways also. Achromatic colors (white, grays, and black) can be mixed with chromatic colors.

1. *Tints* result when white is mixed with a color. White and red give pink.
2. *Tones* are produced when gray and a color are mixed.
3. *Shades* are mixtures of a color and black.

Complementary colors are those directly opposite one another on the color wheel. Complementary colors provide the greatest amount of contrast to one another. A yellow caption on a violet background would jump right out at the viewer. *Analogous colors* sit next to each other on the color wheel. In visual designs where great contrasts are not desired, it is a good idea to use analogous colors to promote the feeling of unity or rhythm.

When visuals are designed, the elements are used to portray the symbols or objects used to communicate the idea of the picture. Realism is promoted when many elements are used.

Principles of Design The guidelines for using the elements of design when visuals are created are called the principles of design. There are six principles: (1) balance, (2) center of interest, (3) emphasis, (4) unity, (5) contrast, and (6) rhythm.

Balance is the sense of equilibrium in a visual. There are two kinds of balance: formal and informal. Generally, a visual should be balanced from left to right and top to bottom. Formal balance means that objects of equal size and importance are placed at equivalent distances from the center of the visual.

Shading

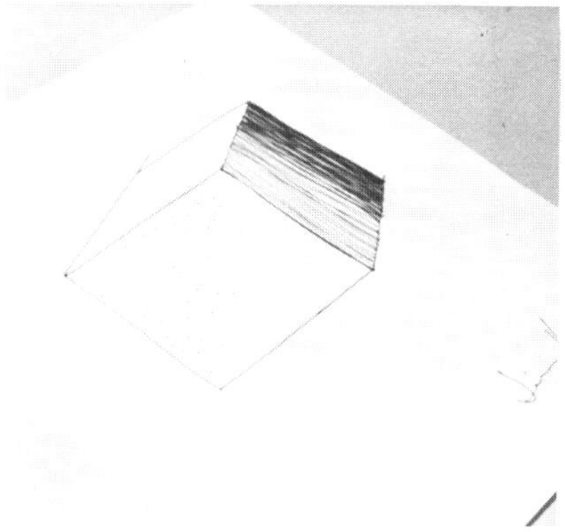

Finished product

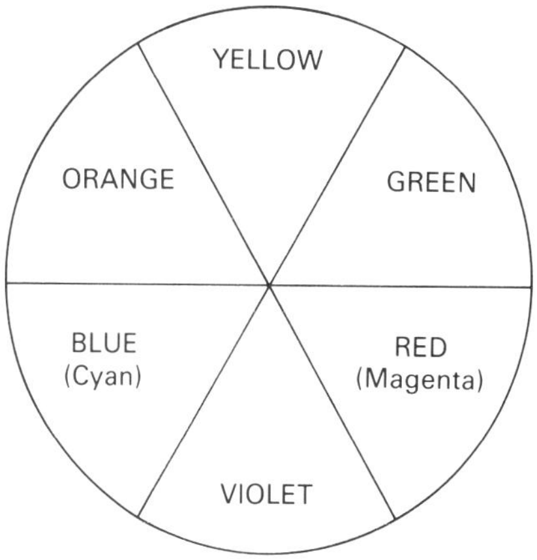

The color wheel for pigments

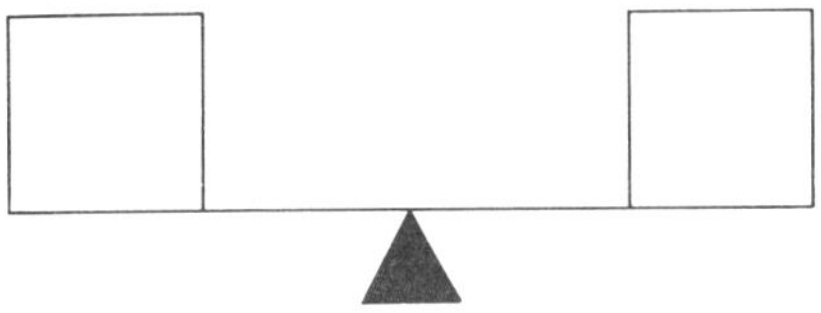

Formal balance

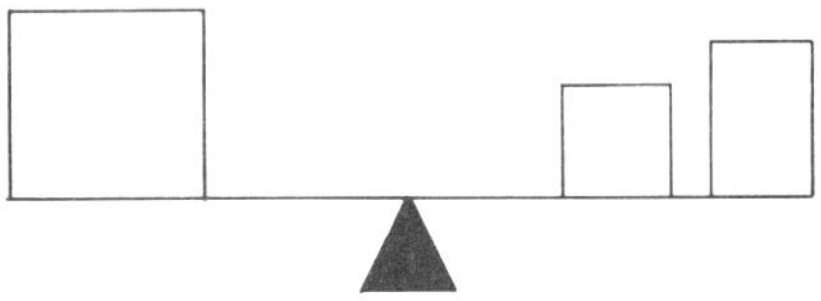

Informal balance

Informally balanced visuals are often more interesting to create and to view. The concept of equilibrium is also important for informally balanced visuals. Careful planning is important when informally balanced pictures are designed because they are difficult to create successfully. Several small objects can be used to balance a large object. Words can be used to balance pictures. Small, highly colored objects can balance larger, duller items.

The *center of interest* is the visual focal point and directly relates to the purpose of the picture. It is the primary idea which the visual intends to convey. The center of interest usually should not be at the center of the picture. To locate the center of interest, the "Rule of Thirds" should be applied. Divide your visual space into thirds horizontally and vertically. The center of interest can be at any of the intersections of these lines. Unless other cues are provided, such as arrows, the eye will usually seek out the top left center of interest and move clockwise through the visual.

Formally balanced display

Emphasis is closely related to center of interest. Often it is a good idea to emphasize the key object in a visual so that it is obvious to the viewer that it is the center of interest for the picture. There are several techniques for adding emphasis to a visual.

1. Use pointers, such as arrows.
2. Use color to highlight the center of interest.
3. Use large objects at the center of interest.
4. Place objects to direct attention to the center of interest. Students in a picture who are looking at a microcomputer emphasize it as the center of interest.
5. Use a different shape for the center of interest. A rectangular picture would stand out in a display where all other shapes were ovals.
6. Use the Rule of Thirds to place the critical object(s) of the visual at the naturally emphasized position.
7. Use more of the elements of design to create the center of interest of the visual than for less important objects.

Unity means that a visual should hold together to convey its meaning or the total effect is one of disorder. If several pictures are used in a display, they should all convey one meaning, and should not seem to be scattered haphazardly across the background.

Overlapping is a simple technique for promoting unity. Trees overlap buildings and each other. Houses overlap shrubs and people. The total message conveyed is of a unified picture, rather than disorganized objects. Overlapping pictures on a bulletin board can convey the same meaning. A large, single background also promotes the idea of unity.

Another commonly used technique to promote unity is to place a border or outline around major elements of a display. Repetition of shapes, forms, or even objects can also promote unity.

Informally balanced display

To locate a spot for the center of interest, use the Rule of Thirds.

How people look at pictures

Emphasis has to do with making the key item stand out.

Arrows or pointers are effective.

A contrasting value can emphasize an area.

The placement of an item can cause it to be emphasized.

Size can be used.

Note: One of the best ways to achieve emphasis is with color.

Emphasis

Unity involves "oneness" or a "tying together."

A "border" can be used to achieve unity.

Another technique is to use lines to unify.

Perhaps the most effective method is to overlap a common shape.

Unity

Contrast refers to the characteristics of an object that cause it to stand out. Contrast is closely related to emphasis. Most often contrast is achieved by the use of light- or dark-valued objects. Complementary colors give the most contrast (reds and greens, oranges and blues, violets and yellows), and can be used to create visual interest and stimulation. Shapes, forms, and textures can be used to create contrasts and to make an object stand out or recede in a visual.

Rhythm comes from repetition with variety, and is used to draw a viewer through the various objects in a visual. A row of houses in a picture can present a sense of rhythm. Each is different enough to avoid monotony, but alike enough to create a sense of flow. Sometimes the rhythm of a picture helps to tell the story of the picture by leading the viewer's eyes in some way.

It is important to remember that not all photographs, displays, or paintings should contain all the elements or principles of design. The use of more elements tends to promote realism and interest; use of the correct principles promotes understanding. The elements and principles are the basic tools for use in the creation of literate visuals.

TRY THIS

Elements and Principles of Design

Find an advertisement in a magazine that has many of the elements and principles of design. Mount the ad on a sheet of construction paper. Then label each element and principle in the margin and identify where it is demonstrated in the advertisement.

The important parts of a display can be made to stand out (contrast) mainly through the use of color, size, value, and shape.

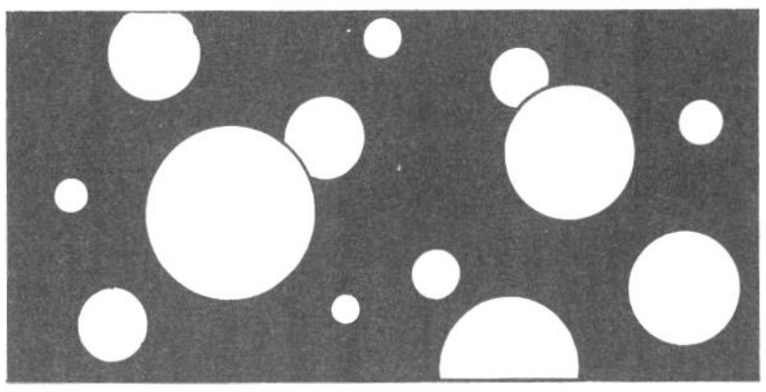

When light and dark values are used, contrast results.

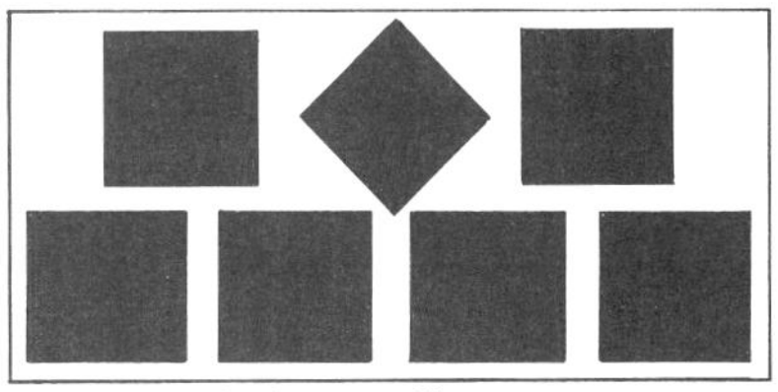

An unusual shape will contrast.

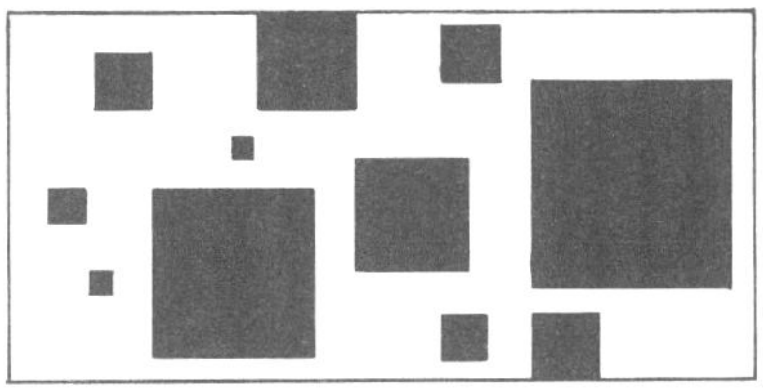

Differences in size can create contrast.

When contrast is lacking, a visual lacks "punch."

Contrast

Rhythm results when an element is repeated in some systematic manner.

Understanding Literacy in Visuals

So far we have looked at single visuals and how they can be created to convey ideas. However, in education we seldom tell a story with one picture. Rather, we use several in a series. When designing visuals to communicate an idea there are several procedures the teacher should consider.

Visual sentences are collections of pictures that convey an idea in the same way as a verbal sentence. Visual sentences have subjects, verbs (actions), and objects.

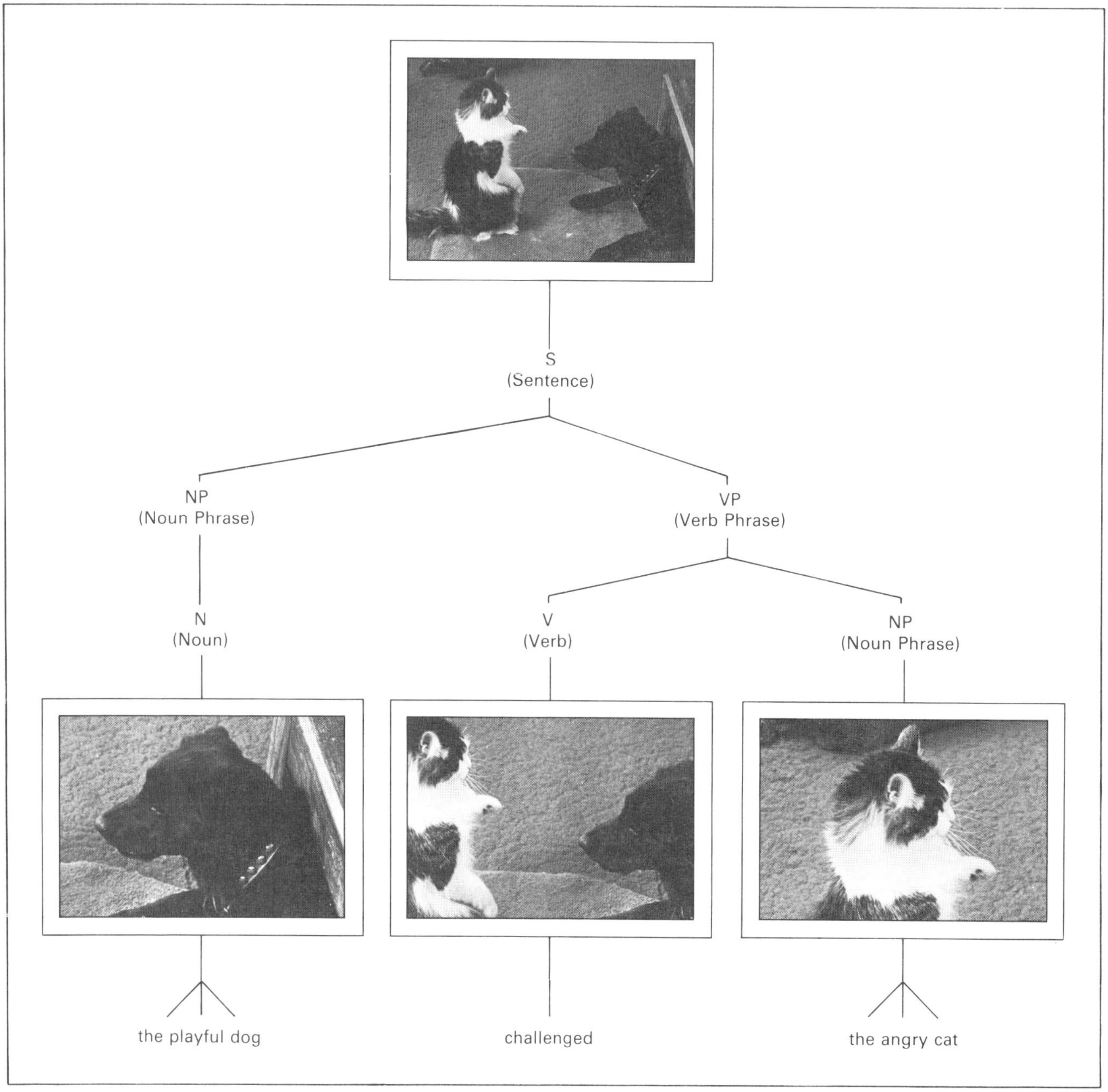

The visual sentence

TRY THIS

Producing a Visual Sentence

Using a Polaroid camera, shoot a picture that has a subject, some action, and an object. Identify these elements of the picture. It might also be fun to have students in a class try to "read" each others' visual sentences to see how understandable the pictures are and how perceptive the students are.

Two other techniques for arranging visuals to convey a message are *directed sequences* and *chronological sequences.* A series of directed pictures attempts to overtly point out or call attention to some idea. Directing attention to something or someone using a series of four pictures called a Long Shot, a Medium Shot, a Close Up, and an Extreme Close Up (LS, MS, CU, XCU) is a sequence commonly used by film makers. It is often used to orient viewers to action.

Undirected sequence

Long Shot

Medium Shot

Close Up

Extreme Close Up

Directed sequence

Chronological ordering of visuals involves the display of images portraying an event occurring over time. The pictures are arranged in the order in which they occurred or in which they should occur. Chronological sequencing is often used when processes such as mounting a picture or setting up a video-cassette recorder are visualized.

Chronological sequence

A chronological sequence using sketches (used by permission of *Popular Mechanics*).

Visually literate individuals can create and understand visual messages. The elements of design are the basic tools to be used to create visuals. The principles of design give guidance for arranging elements correctly and when several visuals are used to convey a complex idea. Pictures can be arranged like a sentence with subject elements, action elements, and object elements. At other times, visuals can be arranged in a directed way to focus a viewer's attention on the desired object. Chronologically arranged visuals show a sequence as it occurs or should occur over time. All are attempts to communicate using visuals in a literate way. Those who create visuals should apply these techniques to produce effective ones that literate viewers can understand.

TRY THIS

Tell a Story with Pictures

Using one of the commercially available photo discovery sets, a series of pictures cut from a magazine, or a collection of photographs produced with a Polaroid camera, tell a story with pictures only. Have friends explain what they think the story or sequence means, and compare these impressions to your original idea.

IMPROVING VISUAL COMMUNICATION—RESEARCH

In the last forty years there has been a great deal of effort by the educational media research community to evaluate the impact of visual instruction. Generally, this effort focused on the production, organization, and utilization of visual media when learning cognitive skills was the goal of instruction. Recently, research on the use of visual media to form and change attitudes has been analyzed and reported in the literature. Application of research-generated information can help the developer of visual messages be more literate.

Research Results on Media Production Techniques

The data gathered from research indicate several trends.

Quality Inexpensive, locally produced, even crude visuals have been found to be at least equal in effectiveness to polished, professionally produced materials, all other things being equal.

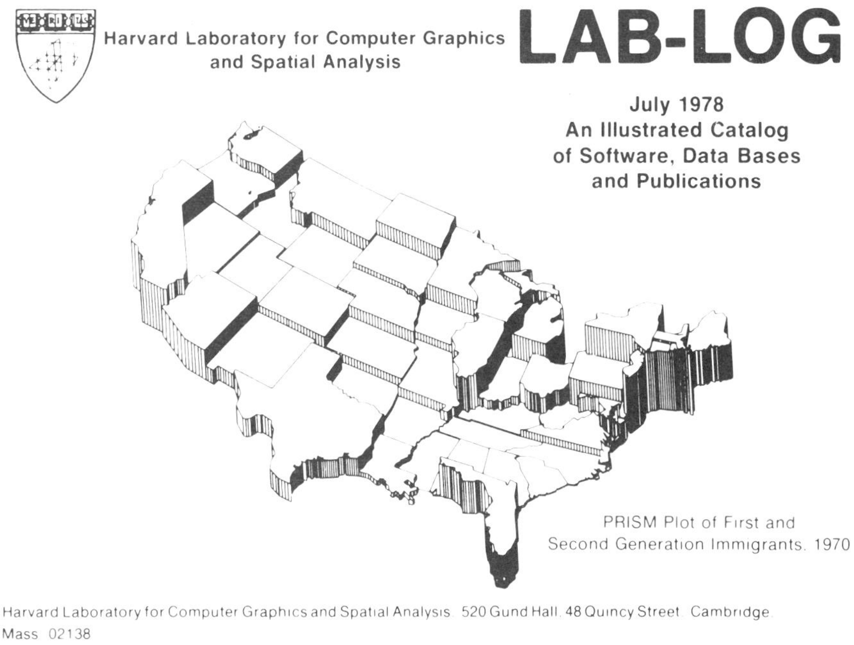

Prism plot showing numbers of second-generation immigrants in a visually literate way.

Color Traditionally, color has not been found to increase learning when used in visuals designed for teaching. However, in recently conducted studies, color has been shown to be more closely related to learning from visuals. Perhaps today's students are becoming less tolerant of black and white visuals because color is so widely used.

Color is critical for learning when it is directly related to objectives. If you were to be held responsible for knowing what the primary and secondary colors of the color wheel look like, then this book would have to have a colored picture of the color wheel rather than a black and white circle with only the names of the colors listed. Another example might be the identification of poisonous snakes. Also, younger children prefer color more than older students.

Special Effects Fancy production techniques used in films and videotapes, such as fades, dissolves, and wipes, do not seem to improve learning.

Music When cognitive objectives are of primary importance, the use of musical backgrounds to accompany a visual presentation does not seem to enhance learning. However, appropriate music, such as an inspirational march or a sad love song, is related to attitude formation and change.

Viewing Angle When skills or techniques, such as fixing a bicycle, are being shown in a visual, it seems to aid learning if the picture shows the action the way the learners would see it if they were actually completing the task (subjective rather than objective view).

Printed Titles Captions, titles, and verbal information can improve the effectiveness of visuals.

Motivators Any technique used to hold the attention of the viewer, such as color, drama, humor, or questions, seems to improve learning.

Complexity Generally, only relevant information should be depicted in visuals. Simple, complete narration, repetition of ideas, and elimination of irrelevant information improve learning from visuals. Children younger than about seven prefer realistic, simple, flat pictures of familiar objects. Older children tend to prefer more complex visuals.

Narration For films, videotapes, and slide/filmstrip presentations, the addition of a verbal narration tends to improve learning. However, it is essential that not too much information is presented in the narration or the visuals, since two senses will be processing information at the same time. Simple narratives appear to be best.

About 160 words (or less) per minute is a general standard for narratives. Using visuals interspersed with verbal or written narrative information is effective. Words are effective for directing the attention of viewers, especially those who are not visually literate.

Embellishments When figures, colors, and other irrelevant embellishments are added to visuals just to make them more attractive, learning from those visuals tends to decrease. Some embellishments, such as cartoon characters, that are a critical component of a visual can increase interest and attention and help students learn.

Research Results on the Content of Visuals

Research has also focused on the content of visuals, with the following results.

Introductions Telling the learner visually or verbally what a presentation is going to include improves comprehension.

Organization Titles and commentary that provide an obvious sequence for a presentation are desirable. Subordinate skills should be presented before more complex ones are shown.

Repetition Repeating key ideas, either identically or in a varied form, increases learning.

Reinforcement Any technique that increases the probability that a learner will remember and be able to reproduce what was presented is beneficial to learning. Visual and verbal mnemonics are helpful in aiding memory.

Rate A visual presentation should be slow enough to permit learners to adequately grasp the content. Be careful that you do not present only bare essentials in order to reduce the rate at which concepts are presented.

Errors In presentations showing techniques or skills, including the kinds of mistakes that can occur will often improve performance. However, extreme care must be taken so that the errors do not overshadow the correct processes.

Attention Directing The use of arrows, pointers, underlining, directed narration, or color to call the viewer's attention to key ideas aids learning.

Complexity Generally, a visual or series of visuals that show only the key elements of an idea, and that eliminate extraneous information, are more effective than those with a great deal of irrelevant detail.

Audio Visuals can be effective in presenting ideas by themselves, but understanding is increased in most learners if both visuals and narrative information are presented.

Research Results on the Utilization of Visuals

In attempting to determine the best ways in which to utilize visuals, researchers found the following principles to be true.

Student Involvement Learning increases substantially if viewers participate actively and mentally with the visual material.

Knowledge of Results If students are asked to participate during a visual presentation, they should be given feedback on the success or impact of these activities.

Pace If learners are asked to participate during a visual presentation, they should be given adequate time to complete the required actions. Stopping a presentation, such as a film, is one alternative.

Viewing When projected media are used, the viewing room should not be totally dark, the image should fill the screen horizontally, and no one should sit closer than two image widths or further than six to eight image widths from the screen.

Research Results on Attitude Formation and Change

Finally, research data on the effectiveness of visuals in relation to attitudes show the following.

Realism Learners' attitudes are more likely to be influenced by instruction that is realistic, relevant, and technically stimulating, especially when the attitude area is new to them.

New Information Persuasion is most likely to occur when visual messages present new information about the topic to the viewer.

Credibility A persuasive message should seem credible to the viewer. This can be accomplished by logical organization, by presenting both sides to an argument, and by being realistic. Using actors who are positively valued or representative of the intended audience will increase the credibility of the message and promote attitude change.

Active Involvement If individuals are actively involved in the planning, production, or delivery of a persuasive visual message, it is very likely that their attitudes toward the message will be influenced.

Discussions Learners who participate in guided postinstruction discussions and critiques are likely to develop favorable attitudes toward the content of the presentation and discussion.

Emotional Involvement Learners who experience a purposeful emotional involvement or arousal during a persuasive presentation are likely to have their attitudes influenced. This can be accomplished by producing fear, worry, or concern in the viewer, as long as viable solutions to the problems that produced the concern are presented. Attractive, interesting actors also have an impact on the attitudes of viewers.

A FINAL WORD

In this chapter we have described simple communication involving the individual interacting with the environment. More complex communication, where two or more individuals are involved in the exchange of ideas, was also presented. Since the visual sense is an important technique for receiving, perceiving, and responding to stimuli, it is important that teachers should know more about it in order to become more successful communicators. The development of visual literacy in teachers and students contributes to learning. Understanding the visual sentence and the sequencing of pictures contributes to the development of visually effective materials, as does consideration of the results of research on visual instruction. The purpose of this chapter was to call attention to the importance of visual literacy and to propose activities and techniques that will contribute to the development of this skill in teachers and students.

LOOKING AHEAD

The next chapter marks the beginning of Part B of this text. Beginning with goals and objectives, the unit moves through the use of some of the simpler media, such as prepared software for computers, overhead projectors, the production of duplicated worksheets and other handouts, and bulletin boards.

Before launching into production of these media, it is necessary to consider the rationale for using *any* instructional media: What do you expect students to *do* after they have been in your class? Their terminal behavior should be expressed in very specific terms. For this reason the first chapter of Part B begins with the production of goals and objectives.

REVIEW QUESTIONS

A. Communication

1. Describe the three stages of simple communication.
2. Abstract learning is possible because an individual can draw upon and relate to previously acquired realistic learning. TRUE/FALSE
3. According to Dale, which is the most realistic experience, a field trip or a film?
4. What has been added to the Shannon-Weaver Model of Communication to make it dynamic and capable of change, rather than linear?
5. Is communication possible without overlapping fields of experience?
6. Define *message.*
7. List several technological channels of communication.

B. Visual Literacy

1. Visual literacy is an individual skill that can be learned. TRUE/FALSE
2. What element of design is used to define the outer edges of shape?
3. Define *texture.*
4. Another name for color is __________.
5. If you mixed red and blue you would get __________.
6. Colors that contrast the most are opposite one another on the color wheel and are called __________.
7. Define *balance.*
8. The technique used to find the center of interest in a visual is __________.
9. List three techniques for sequencing visuals to communicate an idea.

C. Research

1. In a slide set showing the varieties of roses, color is important. TRUE/FALSE
2. If a student is viewing a film on bicycle repair it is a good idea to stop the motion picture periodically to let the student try an outlined procedure. TRUE/FALSE
3. A neatly done, hand-lettered sketch of a paramecium is just as effective as a full-color picture in a textbook if major body part recognition is the objective. TRUE/FALSE
4. In a photo set showing how to throw a football, it is a bad idea to show improper techniques or common mistakes. TRUE/FALSE
5. In a videotape showing several events occurring at the same time on a battlefield, a special effect such as a fade should be used as a transition between scenes. TRUE/FALSE
6. Pictures showing how to operate a microwave oven should probably be shot over the shoulder of the actor. TRUE/FALSE
7. Totally nonverbal presentations are most effective at teaching skills. TRUE/FALSE
8. Showing a presentation more than once will not increase learning. TRUE/FALSE

9. It is possible to scare students into changing their attitudes about a subject such as smoking by showing them a film with only frightening scenes depicted. TRUE/FALSE
10. If students assist in producing a slide presentation on the need for soil conservation, it is likely that their attitudes toward this topic will be influenced. TRUE/FALSE

ANSWERS TO REVIEW QUESTIONS

A. Communication
   1. *Reception* of stimuli through the nerve endings, *perception* of these stimuli by the brain, and *recognition* of the stimuli when they are compared to previous experience
   2. True
   3. Field trip
   4. Feedback
   5. Probably not. There must be something in common between the sender and receiver. What if alien life forms that find the Pioneer spacecraft do not have senses?
   6. The idea the sender wants to transmit to the receiver.
   7. Television, film, books, slides, transparencies, drawings, radio

B. Visual Literacy
   1. True
   2. Line
   3. The actual perceived roughness or smoothness of a surface
   4. Hue
   5. Violet
   6. Complementary colors
   7. The sense of equilibrium in a visual
   8. The Rule of Thirds
   9. Visual sentence, directed sequence, chronological sequence

C. Research
   1. True
   2. True
   3. True
   4. False
   5. False
   6. True
   7. False
   8. False
   9. False
   10. True

REFERENCES

A. Communication and Perception

Dale, E. *Audiovisual Methods in Teaching* (Hinsdale, Ill.: Dryden Press, 1946).

Geschwind, Norman. "Language and the Brain," *Scientific American* 226, no. 4 (April 1972): 76–83.

Haber, Ralph N. "How We Remember What We See," *Scientific American* 222, no. 5 (May 1970): 104–112.

Hilgard, E. R., and Bower, G. H. *Theories of Learning* (Englewood Cliffs, N.J.: Prentice-Hall, 1975).

Shannon, C. E., and Weaver, W. *The Mathematical Theory of Communication* (Urbana: University of Illinois Press, 1963).

Toch, H., and MacLean, M. S., Jr. "Perception, Communication and Educational Research: A Transactional View," *AV Communication Review* 10, no. 5 (1962): 66–68.

Turnbull, A. T., and Baird, R. N. *The Graphics of Communication* (New York: Holt, Rinehart and Winston, 1975).

B. Visual Literacy

Arnheim, R., *Visual Thinking* (Berkeley: University of California Press, 1969).

Debes, J. L. "Some Foundations for Visual Literacy," *Audiovisual Instruction* 13 (November 1968): 961–964.

Debes, J. L., and Williams, C. M. "The Power of Visuals," *Instructor* 84, no. 4 (December 1974): 31.

Dondis, Donis A. *A Primer of Visual Literacy* (Cambridge, Mass.: MIT Press, 1973).

Elements of Visual Literacy, Kodak Pamphlet No. AT-25 (Rochester, N.Y.: Eastman Kodak, 1968).

The Elephants of Visual Literacy, Kodak Pamphlet No. T-90-1-2 (Rochester, N.Y.: Eastman Kodak, 1971).

Fransecky, R. B., and Debes, J. L. *Visual Literacy: A Way to Learn, A Way to Teach* (Washington, D.C.: Association for Educational Communication and Technology, 1972).

Williams, C. M., and Debes, J. L., eds. *Proceedings of the First International Conference on Visual Literacy* (New York: Pitman, 1970).

Williams, C., and Debes, J. *Visual Literacy* (New York: Pitman, 1970).

C. Research

Ball, J., and Byrnes, F. C. *Research, Principles and Practices in Visual Communication* (Washington, D.C.: Association for Educational Communication and Technology, 1960).

Dwyer, F. M. *Strategies for Improving Learning* (State College, Pa.: Learning Services, 1978).

Fleming, M., and Levie, W. H. *Instructional Message Design* (Englewood Cliffs, N.J.: Educational Technology Publications, 1978).

Hartman, F. R. "Single and Multiple Channel Communication," *AV Communication Review* (November–December 1961): 235–262.

Hoban, C. F., and Van Ormer, E. B. *Instructional Film Research, 1918–1950* (Port Washington, N.Y.: U.S. Naval Special Devices Center, 1950).

Levie, W. H., and Dickie, K. E. "An Analysis and Application of Media," *Second Handbook of Research on Teaching* (Chicago: Rand McNally, 1973).

May, M. A., and Lumsdaine, A. A. *Learning from Films* (New Haven, Conn.: Yale University Press, 1958).

Plach, J. J., and Shick, J. "The Effects of Color on Human Behavior," *Journal of the Association for the Study of Perception* 9, no. 1 (1974): 4–15.

Platt, J. M. *Visual Literacy: What the Research Says to the Teacher* (Washington, D.C.: Association for Educational Communication and Technology, 1975).

Simonson, M. R. "Designing Instruction for Attitudinal Outcomes," *Journal of Instructional Development* 2, no. 3 (1979): 15–19.

Travers, R. M. *Research and Theory Related to Audio-visual Information Transmission.* (Kalamazoo: Western Michigan University, 1967).

SOURCES FOR VISUAL LITERACY MATERIALS

Association for Educational Communications and Technology, *Photo Discovery Sets.* Washington, D.C. 20036.

Curtis, Ron. *Media Now* (50 learning-activity, cross-media packages for the junior and senior high curriculum) (Red Oak, Iowa: Southwest Iowa Learning Resources Center, 1973).

Debes, J. L., et al. *How Does a Picture Mean?* (filmstrip and manual) (Washington, D.C.: Association for Educational Communication and Technology, 1968).

Debes, J. L., et al. *Making Sense Visually* (filmstrip and manual) (Washington, D.C.: Association for Educational Communication and Technology, 1968).

Debes, J. L., et al. *Rhetoric of the Movie* (filmstrip and manual) (Washington, D.C.: Association for Educational Communication and Technology, 1968).

Fransecky, R. B., and Debes, J. L. *Visual Literacy: A Way to Learn, A Way to Teach* (Washington, D.C.: Association for Educational Communication and Technology, 1972).

Shaw, Judith. *Learning to See and Understand: Developing Visual Literacy* (sound-160 slides) (White Plains, N.Y.: Center for the Humanities, 1974).

PRODUCING MEDIA FROM GOALS AND OBJECTIVES

B

Media are produced for a reason, with a plan. The most significant method for organizing instruction is based on goals and objectives. Part B emphasizes the production of the more widely used media that may be used as stand-alone elements in instruction. The chapters here suggest ways of making "simple" media, such as duplicated worksheets, transparencies, and bulletin boards. This section may be thought of as a transition between the already familiar content in Part A and more complicated techniques in Parts C and D.

Microcomputers are becoming increasingly important instructional resources. However, since much of the available computer instruction was not produced by people who know the process of teaching and training, it is important to give careful attention to your goals and objectives as you select computer software. The production of instruction involving computers is outlined in Chapter 5, along with explanations of ways to modify existing programs to suit your specific needs.

Handouts, worksheets, displays, and—to some degree—transparencies are used by teachers and trainers a great deal. It would be easy to describe these techniques as the "simple" media. However, because they are inexpensive and are widely used they remain very important to the teacher or trainer. The final three chapters of this section of the text present production techniques for these media—handouts, displays, and transparencies.

Part B provides a foundation for the more elaborate production techniques that follow in later chapters. By considering the production of teaching materials that are less complicated and more quickly produced, it becomes possible to provide an important basis for the remaining portion of this text.

GOALS

☐ Gain an understanding of the use of goals and objectives.
☐ Encounter several theories that relate to learning.

OBJECTIVES

☐ Define and provide examples of goals and objectives.
☐ Write goals and objectives for selected content.
☐ List the characteristics of several schemes for organizing goals, concepts, and objectives.
☐ Classify objectives using various strategies.
☐ Choose media to accomplish objectives in the domains of learning.

One of these days you may have to take a test over this chapter. As you begin to prepare for it you will naturally attempt to organize the content into important areas and clusters. Certain details, examples, or processes may be ignored because you feel they are insignificant. Other topics may receive a great deal of your attention because you think they are particularly meaningful. Only when you take the exam will you find out if you studied the correct material. It might turn out that the teacher had a quite different idea about what was important—and *that* was what you should have studied.

Perhaps some guesswork about tests is helpful as you study; after all, it is important to learn *what* to learn as well as *how.* The problem is that some lost motion and wasted energy may be the result of the confusion of trying to decide what to study for a test. If you know ahead of time what you should be able to do *after* the lesson, your learning can be focused all along. The same holds true for teaching. If you determine ahead of time what you expect of students your instruction achieves a focus. This chapter will provide the background information on goals and objectives that you will need for designing instructional materials.

GOALS AND OBJECTIVES

These two terms are sometimes used interchangeably but they are not the same. Each is used to specify different types of outcomes. A *goal* is a desired outcome of instruction and may be based on one or more concepts. Goals are broad in scope and they are nonbehavioral. On the other hand, an *objective* (more specifically a *behavioral objective* or a *performance objective*) specifies, in behavioral terms, what a person should do as a result of instruction. Objectives are written to specify a student behavior that can be observed and evaluated. Some of the characteristics of goals and objectives are shown in the chart on the next page.

Obviously, goals and objectives are related; it is not possible to have one without the other. Goals may be reached by pursuing individual objectives. You might think of goals and objectives by comparing them to a map. If your goal is to travel from Chicago to Bisbee, Arizona, you might wish to look at a map of the United States or perhaps a map of the central United States. You could get an indication of overall travel time and general

4

GOALS AND OBJECTIVES

Objectives	Goals
Are measurable, observable psychomotor actions	May not be measurable or observable
Use action verbs such as *list, describe, explain, discuss*	Use inactive verbs such as *appreciate, be aware of, know*
Can be evaluated on a short term; minutes or days after the instruction	May require a long-term evaluation; months or years

direction. Specific highways traveled would be comparable to objectives. Toward the end of your trip, whether you go from Albuquerque to Socorro and then to Lordsburg, or from Socorro to Springerville and then Globe is a minor decision in view of your major goal. Minor, yet specific instructional decisions can become objectives. Major directions are goals.

How to Write Goals

You may obtain a slightly different set of instructions for goal writing from each person with whom you talk. You can also see in the examples in this chapter that goals are written in different forms. Regardless of these variations, these general guidelines can be used in most cases:

PRODUCE

Guidelines for Writing Goals

- ☐ Think of a broad, fundamental area of understanding that you hope your students will come in contact with in your course.
- ☐ Write a sentence that includes that understanding, beginning with, "Students should have an understanding (be aware of, have knowledge of, etc.) of. . . ."
- ☐ Assume that it may take weeks, months, or perhaps years for students to completely achieve the goal. You may only begin work on certain goals in your content area when you teach; students will continue pursuing the goals in future work.

Some Goals from the National Education Association

- ☐ Ability to think
- ☐ Maintaining good health
- ☐ Worthy home membership
- ☐ Vocational competence
- ☐ Effective citizenship
- ☐ Worthy use of leisure
- ☐ Development of ethical character

Source: Educational Policies Commission, National Education Association, *The Central Purpose of American Education* (Washington, D.C.)

Some Goals from a Public High School

- ☐ To stimulate personal growth
- ☐ To provide a foundation for continued intellectual growth
- ☐ To help students achieve feelings of security and personal worth

Some Goals from an Industrial Training Program

- ☐ To improve a feeling for the importance of positive public relations
- ☐ To provide an understanding of the health benefits program
- ☐ To appreciate the problems of other departments in the company

TRY THIS

Write a list of four or five goal statements for the course you teach or are preparing to teach. Pick one subject area and list the most important goals for that subject.

How to Write Objectives

If there is some minor disagreement between educators on goal writing, there are also several views on how to write objectives. To begin with, it is important to point out that the objectives we have in mind are specific *behavioral objectives,* also called *performance objectives.* Objectives of this sort are not necessarily long. Rather, the trick in formulating them is to include (1) information or experience a student is given to (2) perform at a (3) specific level of expertise.

As a teacher, you must look for clues to tell when a student has learned something. Examinations provide these clues, as do personal observation, a student's oral response, a term paper, a report, or other work. Objectives merely indicate *before* the instruction what you will be looking for *after* the instruction. They also indicate the nature of the instruction, and how you will measure the impact of your teaching.

Students, as well as the teacher, need clues to identify when a skill or idea is learned. Objectives make this identification intentional instead of accidental, so that students know what they should learn instead of having to guess what the teacher wants.

Formulas for writing objectives may have some minor variations, but generally these steps can be followed:

PRODUCE

Guidelines for Writing Objectives

- ☐ Think of posttest questions you feel would adequately measure skills that are related to one of your goals.
- ☐ Decide what information or experience a student would need to be able to exhibit the skill.
- ☐ Determine what behavior a student should exhibit as evidence the skill was achieved.
- ☐ Specify criteria you will use to measure the degree of skill (time to perform, degree of accuracy).
- ☐ Include this information in the form of an objective such as:

Given

(the instructional experience)

the student will

(performance or behavior)

according to

(criteria or standard of achievement expertise).

Verbs Used in Objectives

The verb used in an objective is important because it specifies a behavior. Use appropriate verbs from the list below:

Behavioral Verbs

☐ Define	☐ Discuss
☐ Identify	☐ Produce
☐ Explain	☐ Solve

Nonbehavioral Verbs (do not use for objectives)

- ☐ Appreciate
- ☐ Be aware of . . .
- ☐ Have a knowledge of . . .
- ☐ Gain a feeling for . . .

Sample Objectives

☐ Given classroom instruction and three pictures of basic geometric forms, two of which are identical, the concept of visual discrimination (identifying differences visually) will be demonstrated when students will correctly identify the dissimilar form for at least five of six such sets.

☐ Given classroom instruction, and a series of black and white photographs of lung tissue, the student will identify lungworm pneumonia, and list at least two characteristics of each type upon which the choice is based with 100 percent accuracy.

Many commercially prepared instructional packages have comprehensive study guides that include goals and objectives.

TRY THIS

Write a behavioral objective for a topic in your subject area of teaching preparation that includes the (1) instructional experience, (2) performance expected of the student, and (3) standard of achievement expected. Use the format provided above.

HOW GOALS, OBJECTIVES, AND CONCEPTS ARE RELATED

Now that we have looked at goals and objectives, it is appropriate to consider concepts, which are broad, general statements that summarize fundamental laws or ideas. Concepts are used to organize and explain the huge number of facts and ideas in any subject area. They provide a framework for understanding seemingly unrelated events. Here are some examples of concepts:

☐ Air has weight.
☐ In electricity, opposite charges attract, like charges repel.
☐ "You can't teach an old dog new tricks."
☐ A checks-and-balances system exists between the three branches of government.

Generalizations such as these can be subject to a number of criticisms. Some ideas we call concepts are in reality stereotypes or forms of bias and prejudice. Other concepts may have so many exceptions that only a few cases can be found where the concept is applicable. These illustrations are provided to bring out the idea that concepts are an attempt to generalize and state fundamental laws or theories. Many are not accurate in all instances.

Certain concepts are difficult to summarize in a single phrase. The concept of a major scale in music is an example. Suppose you are taking piano lessons. You are learning to play scales. The C scale. The F scale. The B scale. In each case your behavioral objective is to play a scale in ten seconds with no mistakes, using either the right or left hand.

The trouble is, you have to *memorize* each note in each scale. The scales have little relationship to each other.

On the other hand, if you knew the concept involved, you could play any scale. If you knew that all notes were a whole step apart except the 3rd and 4th, and the 7th and 8th, you would have the *concept* of major scales. Then you could start on any note and play any major scale. The philosophy behind teaching by concept is that specifics may change with each new situation but, if students have a fundamental rule or pattern for solving classes of problems, they are ahead of the game in solving problems at present and in the future as well.

By achieving several *objectives,* a student can reinforce understanding of a *concept.* Achieving a knowledge of concepts moves one toward established educational *goals.* You may discover exceptions to these relationships, but an overview of the pattern is what is important here. As you organize these three elements you are organizing your instruction, and making possible the appropriate choices of media for teaching.

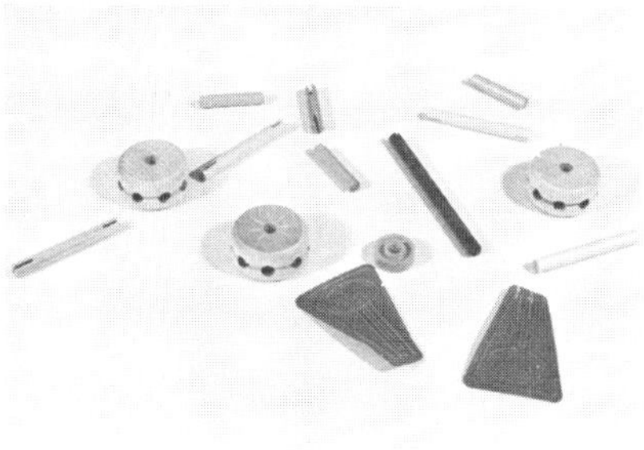

Like these Tinker Toys, ideas can be presented that do not seem to be related.

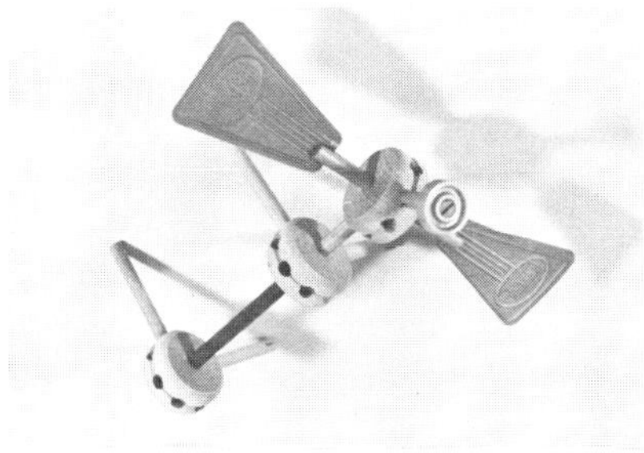

Concepts are unified structures that are synthesized from a number of related ideas, just as a child puts together the pieces of a Tinker Toy set.

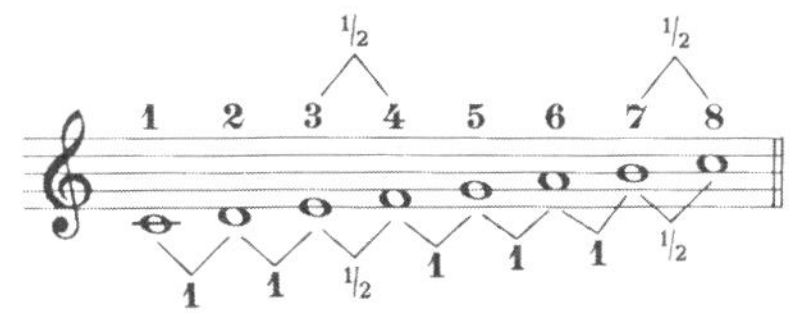

The C scale.

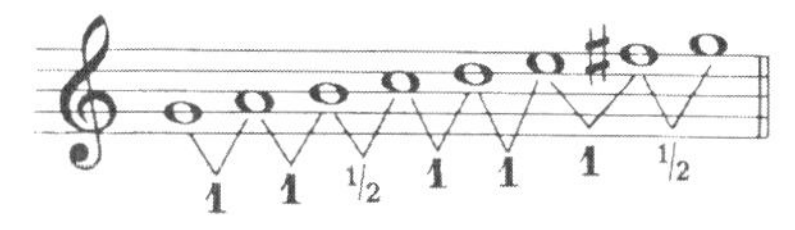

The G major scale.

SCHEMES FOR ORGANIZING GOALS, CONCEPTS, AND OBJECTIVES

As educators, psychologists, and human behaviorists have worked together over the years they have tried to determine how people learn. Is there some sort of fundamental theory to learning? What are the concepts that are agreed on? Do all people learn the same way? Does learning style vary with the subject being learned, or does a person learn geometry in a manner similar to learning automobile repair?

As you might suspect, the answers to these questions are partly yes and partly no. It depends—on learning conditions, prior knowledge, inborn intelligence, the time of day, the nature of the teacher, and so forth. It is possible, then, to come up with any single theory of learning with which all may agree?

Apparently the answer is no, because several different learning theories are currently popular. If we ask a slightly more narrow question it might be this: "Is there a single theory to explain how ideas can be organized to make learning the most efficient?" Again the answer is no, but that should not discourage you. It turns out that several important schemes have been developed to explain how instructional events, objectives, goals, and concepts can be organized to maximize learning.

Domains of Learning

Shortly after World War II, a group of psychologists began to formulate a classification scheme for various types of learning. They identified three major components that can be found in any sort of learning. The portion of learning that involves facts is called the cognitive domain; feelings and attitudes are in the affective domain; physical manipulation and performance are in the psychomotor domain. Each of the domains is further divided into a hierarchy of stages—each one leading to the next—that makes possible the detailed classification of the components of learning. Tables 4.1 through 4.3, listing levels of cognitive, affective, and psychomotor ability, will help you understand how types of learning can be categorized.

Conditions of Learning

In 1965, the psychologist Robert Gagné published a book called *The Conditions of Learning* that set forth a stepwise scheme to describe the learning process. The eight steps range from simple to complex, and may be considered as a hierarchy. For example, you would not be able to operate at the level of multiple discrimination unless you had skill in verbal-associate learning.

The Discovery Approach

Worked out in the late 1950s by psychologist Jerome Bruner, this principle is based on the assumption that people possess a natural talent for solving problems—a kind of "problem-solving intuition." Bruner feels that problem solving is based on a fundamental formula or method that is independent of the *type* of problem or the nature of the person looking for the solution. In other words, a

Gagné's Eight Types of Learning

1. *Signal learning* A simple conditioning in which a stimulus provokes a specific response. Stopping your car for a red light and the memories evoked by the sight of your old high school are examples.
2. *Stimulus-response learning* The development of specific stimuli that cause voluntary responses. Teaching a dog to "shake hands" by rewarding the dog with food for each correct response assists in the formation of an S-R bond, for example.
3. *Chaining* The connecting of two or more S-R sequences. In the case of the dog it would be possible to teach it to bark or sit up after shaking hands—or teach a chained sequence of shake hands—bark—sit up.
4. *Verbal association* A subvariety of chaining in which the emphasis is more directly on verbal processes and less on behavioral skills. Examples include the process involved in translating from one language to another through the use of cognates and certain prefixes and suffixes.
5. *Multiple discrimination* The use of many chains, all somewhat parallel but perhaps not clearly connected to each other. Learning to identify all of the new models of automobiles each year is an example. After a few years it is possible to have memorized the looks, characteristics, and other features of all of the Fords, Datsuns, and Chevys. Each sequence represents the use of chaining, but when cars are considered as a whole it is necessary to make use of a set of multiple discriminations to keep the information sorted out.
6. *Concept learning* The production of generalizations that can be formed, based on the use of chaining and multiple discrimination. The concept *middle,* meaning ". . . an object between two others . . ." is an example.
7. *Principle learning* The acquiring of ideas that have broad application and that may embody several concepts. Some examples of principles are ". . . gases expand when heated . . . ," and "singular nouns require singular verbs. . . ."
8. *Problem solving* The combination of principles in new ways that may, in turn, lead to a newer principle. Examples include the ability to solve story problems in algebra in which two unknowns are involved, and attempting to find the leak in a roof.

small child makes observations, forms hypotheses, and conducts experiments to solve a problem using techniques nearly identical to those a physicist would employ. If we allow students to discover principles and laws on their own we merely are allowing learning to occur along natural lines. Supposedly, the discovery method more completely captures student attention and promotes effective learning. It also takes considerably more classroom time.

Many of the skills we develop during childhood play are directly related to adult activities.

Lego blocks to a child may be similar to bricks to a bricklayer.

TABLE 4.1 Levels of cognitive ability

					6 *Evaluation* Ability to judge the value of ideas, procedures, methods, etc. using appropriate criteria
				5 *Synthesis* Ability to put together parts into a new, unified whole	
			4 *Analysis* Ability to break down material into its parts or to determine relations among the parts and the ways they are organized	Requires analysis	Requires analysis
		3 *Application* Ability to use generalizations in new and concrete situations	Requires application	Requires application	Requires application
	2 *Comprehension* Ability to apprehend what is being communicated and make use of the idea without relating it to other ideas or materials or seeing fullest meaning	Requires comprehension	Requires comprehension	Requires comprehension	Requires comprehension
1 *Knowledge* Ability to recall, to bring to mind the appropriate material	Requires knowledge	Requires knowledge	Requires knowledge	Requires knowledge	Requires knowledge

Source: Benjamin Bloom et al., *Taxonomy of Educational Objectives; Handbook I: Cognitive Domain* (New York: David McKay Co., 1956).

TABLE 4.2 Levels of affective ability

1 *Receiving*	2 *Responding*	3 *Valuing*	4 *Organization*	5 *Characterizations*
The learner is sensitized to the existence of stimuli. He is willing to receive or attend to them	He is not just willing to attend but actively attending. Learning becomes involved	The learner believes a behavior has worth—motivated by commitment	The learner recognizes values in a system where interrelationships are evident	Learner acts consistently in accordance with the values that he holds
Awareness	Acquiescence in responding	Acceptance of value	Concentration of a value	Generalized set
Willingness to receive	Willingness to respond	Preference for a value	Organization of a value system	Characterization
Controlled or selected attention	Satisfaction in response	Commitment or conviction		

Source: David Krathwohl et al., *Taxonomy of Educational Objectives; Handbook II: Affective Domain* (New York: David McKay Co., 1964).

TABLE 4.3 Levels of psychomotor ability

1 *Perception*	2 *Set*	3 *Guided Response*	4 *Mechanism*	5 *Complex Overt Response*
				Action, performed without hesitation, leading to automatic performance
			Habitual response	Requires mechanism
		Overt action by limitation and/or trial and error under supervision	Requires guided response	Requires guided response
	Mental, physical, or emotional readiness	Requires set	Requires set	Requires set
Become aware through sense organs. Recognize cues, make choices, and relate to actions	Requires perception	Requires perception	Requires perception	Requires perception

Source: Elizabeth Simpson, *The Classification of Objectives: Psychomotor Domain,* Research Project No. OE–5–85–104, University of Illinois, Urbana, 1966.

Advance Organizers

An additional method to make learning more effective and efficient has been offered by David Ausubel. He feels that organizers, clues, and a preview of what is to come should precede main learning events. Such organizers provide clues, set the context, and furnish necessary background. An early warning system like this can assist students who are headed into complex areas of subject matter.

Examples of advance organizers are questions for the student to answer after the lesson, a worksheet that provides the outline of a presentation, and a bulletin board that introduces a new topic the day before it is covered in class.

TRY THIS

Work out an advance organizer or two for use in your subject area. Select a specific topic; then determine what clues would help prepare a student to learn about the idea. State each clue (advance organizer) in a sentence or two.

Stimulus-Response-Reinforcement Learning Patterns

A great deal of attention has been given to this explanation of the learning process, which is based on the assumption that we tend to repeat behaviors that are positively rewarded. The *stimulus* (a question or a problem, for example) evokes a *response* (an answer, an action) that may be positively *reinforced* (a smile, a nod, some money). If this sequence happens often enough the behavior will be repeated—in other words, learned. The principle has wide application in human endeavors, and we all can think of examples in which our behavior patterns have been built up through experience with certain stimuli that have positive payoff. Conversely, we tend to extinguish those behaviors that may have negative consequences.

Gestalt Learning

Using the German word for "... a unified whole, a configuration, pattern, or organized field ...," this concept is stated as "... the whole is greater than the sum of its parts." Gestalt is sometimes offered as an explanation for those sudden flashes of inspiration or insight that result in the understanding of a complex set of events. One example that is often cited involves the problem of bonding of hydrogen and carbon atoms in the benzene molecule. The chemist Kekulé, working in the 1800s, discovered that six carbon atoms and six hydrogens were hooked together in the molecule but he was unable to work out the structure. Carbon atoms form four bonds—either to other carbons or to hydrogens. And hydrogen atoms each can form a single bond. His first approach was to imagine a chain of carbon-to-carbon atoms, with hydrogens along the edges. But you will find—as Kekulé did—that a six-carbon, six-hydrogen molecule is impossible to form in this way. One night, as the story goes, he had a dream about six monkeys in a circle, each holding one hydrogen atom. Every other monkey used two hands to hold onto the adjacent partner in the circle. Insight such as this is sometimes cited as an example of Gestalt learning.

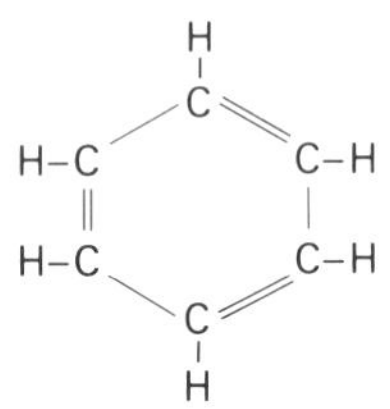

Kekulé's Molecule.

TRY THIS

Think of a concept, a fact, an idea, a process—or any new skill you have acquired recently. Then review the schemes discussed on the preceding pages and attempt to classify the type of learning that occurred.

USING A TAXONOMY TO ORGANIZE OBJECTIVES

A taxonomy is a device for organizing things. Plant and animal taxonomies have been used for years, and yet are still being modified as new relationships are discovered. Stamp and coin collectors use modified taxonomies to organize their collections. Taxonomies not only make it possible to pigeonhole information, they establish levels. Broad levels are subdivided into subsections.

As you begin your collection of objectives you will find helpful the Taxonomy of Educational Objectives presented earlier in this chapter. As already explained, learning can be partitioned into three domains: cognitive, affective, and psychomotor. Each of the three is further divided into layers or levels of sophistication. To achieve balance in your teaching it is important to strive for objectives in all three domains. In addition to basic educational reasons for striking this balance, there is another motive. You can more effectively plan the media for teaching if you know what your instructional objective is and where it fits in the taxonomy. For example, if you are operating at the knowledge level in the cognitive domain with an objective that indicates a student should ". . . be able to list the first ten presidents of the United States," you will find that a simple list, either from a text, the chalkboard, or on a duplicated worksheet, will suffice as a teaching technique. On the other hand, comparing the characteristics of the policies on the first ten presidents with respect to world trade requires a higher level of understanding. An objective of this sort might be at the analysis level, requiring more complex media to portray the characteristics of these policies. Finally, asking students to formulate a policy of their own for that period of time in our history would require their skill at the synthesis and evaluation levels, and might involve the production of materials such as a film or slide set by students to demonstrate what they had learned.

One way to incorporate the use of a taxonomy is as a guide when objectives are written. You can construct a matrix of objectives at each level in each domain and then analyze the pattern for balance and completeness. Another place to begin is with the questions used on examinations. You may remember that we pointed out the similarity between objectives and test questions. Whether a teacher admits it or not, an examination is in reality a reflection of objectives. You can always rewrite questions, turning them into statements, and then categorize the statements by using the taxonomy. An exercise like this provides a simple means for analyzing scope and sequence of content in a course of study.

OBJECTIVES: A PRACTICAL NOTE

You may have the idea by now that goals and objectives are perfectly good, never bad. Unfortunately, there are a few nagging side effects that may cause concern if you stray from your objectives. Students begin to sense a double cross, a breach of trust, a change in the rules. If your memory fails briefly, as you construct a test, and you include an item not covered in an objective, students may become upset. Furthermore, as you work your way into the subject matter, you may discover new connections across concepts—bridges not previously perceived when you wrote the objectives. During teaching—and subsequent testing—you may include areas about which students have not been forewarned. To be fair to students, side trips of this sort in your instruction should be pointed out, with the indication that ". . . the next test *may* include some information that is not implied in the objectives."

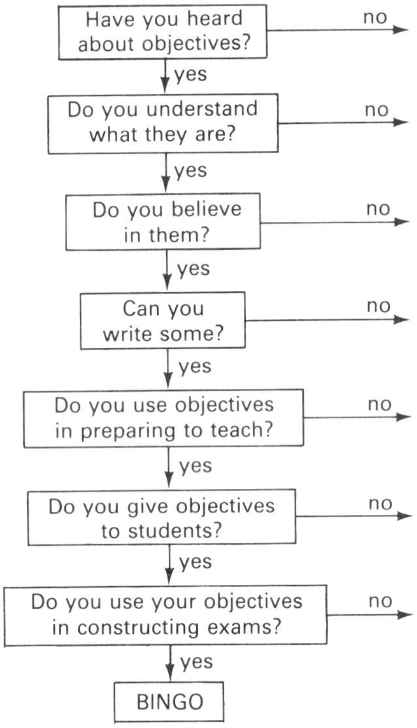

Adoption patterns for objectives

Steps in Instructional Development

Instructional development, the process by which learning events are planned, is characterized by these steps:

- ☐ Narrow the subject/assess the audience.
- ☐ Draft goals and objectives.
- ☐ Match media to objectives.
- ☐ Procure or produce media.
- ☐ Teach.
- ☐ Evaluate.

An additional argument is often brought up in discussions about objectives. Some teachers feel that the educational process is straightjacketed when precise performance objectives are used as guideposts. Students learn only what the objectives specify, and nothing more, they say. There are claims that a sort of three-dimensional luster is missing from the subject matter when objectives are followed too closely. Students get the idea that all content can be distilled to a list of objectives, inducing tunnel vision and narrow-minded understanding.

This obvious pitfall can be avoided if you promote objectives as general guidelines, not air-tight rules. At the same time, you must ensure that your matrix does indeed include all of the areas you consider important. But leave students with the idea that, while objectives and goals are a step in the direction of clarifying both teaching and learning, there must remain some room for reason, a little slippage, and a certain amount of vague, undefined space for acquiring an idea or two that was not on the list.

USING OBJECTIVES IN INSTRUCTIONAL DEVELOPMENT

We have been discussing the advantages of goals and objectives for students who wonder what a subject area is all about. Serving as advance organizers, goals and objectives provide a quick overview for the learner.

Such organizers aid the instructional side of education, too, by keeping the teacher's planning in sharp focus and in the correct sequence. To use performance objectives in the process of instructional development, we will examine some useful connections between types of media and types of objectives.

Research in the analysis of teaching shows that a high percentage of what happens in the classroom is teacher talk, falling into these categories:

- ☐ Explaining
- ☐ Making assignments
- ☐ Answering questions

During much of the remaining time, when students can have the floor, these activities occur:

- ☐ Silent reading, individual work
- ☐ Asking questions
- ☐ Small-group work

The consideration of what teachers and students do can help as instruction is developed, particularly at the stage where media must be matched to objectives. For example, if an objective is in the cognitive domain, at the knowledge or application levels, you might be teaching by "explaining." Question: "Can media be used here?" Answer: "Yes, in several ways." Maybe the knowledge can be gained by students reading a book instead of listening to the teacher. Or perhaps an audiotape and a few worksheets can transmit the information. If you use media to introduce a topic, reinforce an idea, or summarize, you may then have more time to engage in what *you* do best, teacher-student interaction. Also, as learners work or interact with media you may have free time to assist students who need individual help.

The example above illustrates the connections between media, objectives, and instructional development. The chart below shows additional relationships between media and objectives. Approach the chart with caution, however. It may be tempting to assume that there is only one way to teach toward any given objective. Naturally, because of the large number of variables in the educational process, there are several ways to reach an objective. Often each method is equally effective. A certain amount of trial and error will help you settle on your own "best" way, as you accumulate experience.

A FINAL WORD

We have spent these first four chapters providing background information needed for intelligent media production, as well as in emphasizing the entry-level teaching skills that you already have—skills such as speaking and using the chalkboard. Choosing instructional media around the turn of this century was relatively simple because virtually none of the current materials (except for models, realia, and printed material) existed.

Domain	Media That Could Be Used
Cognitive	Print—duplicated handouts, worksheets, short readings, copies of articles, textbooks
	Slide sets—narrated live or combined with audiotape for self-instruction
	Transparencies—used by the teacher to speed up problem solving or to explain complex ideas that require nonverbal illustrations
	Individualized instruction—computer programs, workbook/audiotape sets, filmstrips
	Live instruction—effective, but wasteful (students learn as much from mediated instruction where lower-level cognitive information is concerned)
Affective	Live instruction—quite effective and necessary here. Since attitude change can occur during discussions, it is important that a teacher assist in the orderly and systematic process of conducting a discussion
	Film/videotape—these media may portray contrived events, but the viewer experiences real emotions. Film and videotape are presently our most effective media for representing events in the world
	Audiotape—where only sound is involved in the realism of a situation
	Print—books, magazines, and other print
Psychomotor	Audiotape—instructions can provide a step-by-step recipe for accomplishing a task, freeing students from having to read as they perform
	Film/television—demonstrations of processes that students can then perform are particularly effective if shown in motion
	Live instruction—the familiar lecture/demonstration provides an example for students to follow as they practice the same process on their own
	Print—using stepwise procedures or recipes, students can follow the instructions to perform a task

?

Is the pen still mightier than the sword? Or is television mightier than the sword?

Illustrated lectures have been used for decades. Today the use of visuals in teaching has been greatly simplified because of technological advances.

An important change took place during and immediately after World War I. Educational films, lantern slides, and charts were widely used in the massive training programs for the war effort. These instructional innovations made a far-reaching impact on education. For the first time, instruction on a large scale was carried on without exclusive use of books. Demonstrations, charts, "illustrated lectures," and motion pictures became integral components of the teaching process.

Rapid growth in educational methods in the early 1920s was based on a rather simple "discovery" about the nature of learning: the discovery that learning is quickened and sharpened when students are confronted with the sights and sounds of the subject matter. "Visual education" became a catchword, although it was apparent that it was a pretty accurate description of what was going on.

As recorded sounds and "talking movies" became available—along with improved methods for photographing and projecting still pictures—an appropriate name for this new approach seemed to be *audiovisual education.* The roots of the educational media of today are found in the audiovisual education programs established between 1920 and 1960.

LOOKING AHEAD

The next chapter describes a medium that is quite distinctive when compared to others. The effectively written microcomputer lesson can be interactive. It can respond to a student based on needs or questions expressed to it. Certainly, only humans are truly interactive; however, the computer can approach this ultimate characteristic of good teaching more closely than any other technological device. In the future, a larger portion of classroom events will be coordinated by computers. The next chapter of this text discusses this valuable teaching tool.

REVIEW QUESTIONS

1. What is the difference between a goal and an objective?
2. How are concepts related to goals and objectives?
3. How is a taxonomy of objectives helpful in organizing your ideas for teaching?
4. How do goals and objectives fit into the process of instructional development?
5. List the three elements that should comprise a behavioral objective.
6. Which of the three domains (cognitive, affective, psychomotor) do you believe is the most significant to be reached in your teaching?

ANSWERS TO REVIEW QUESTIONS

1. A goal is a broad, general, nonbehavioral statement that implies some sort of desirable end product or educational outcome. Objectives are subsets of goals; one goal may require a number of objectives in order to be reached. Objectives contain specified behaviors that can be observed, thus making it possible to measure whether the objective has been reached.
2. Concepts are broad, fundamental laws that assist in our understanding of events and phenomena. By reaching certain objectives—which in turn lead to goals—one is able to apply an understanding of concepts to specific situations. The concept that a gas expands when it is heated is related to the goal that a student should gain an understanding of how matter and energy interact. An objective might require that one should be able to devise an experiment to show the relationship between temperature and volume of a gas.
3. A taxonomy implies a hierarchy of relationships; one level leads to another, more complex level. In constructing objectives and exam questions it is important to progress from simple levels to more complex ones, so that a student has the necessary elementary information before attempting more complex understandings.
4. Instructional planning begins with the goals and objectives that are desired by the teacher. It is nearly impossible to devise an instructional strategy unless the outcomes of the instruction are specified beforehand.
5. The instructional experience, the student's performance, and the conditions for assessing the extent to which the objective has been reached.
6. Actually, they have equal significance because it is not possible to reach competence in any one without using the others. But the content area makes some difference—cognitive ability in mathematics may be more important than psychomotor ability, for example. In the arts, the ability to communicate in the affective domain may be the most important.

REFERENCES

Anderson, Ronald H. *Selecting and Developing Media for Instruction.* American Society for Training and Development, Madison, Wis. (New York: Van Nostrand Reinhold 1976).

Biggie, Morris L. *Learning Theories for Teachers* (New York: Harper and Row, 1964).

Bloom, Benjamin S., ed. *Taxonomy of Educational Objectives; Handbook I: Cognitive Domain* (New York: David McKay Co., 1956).

Gagné, Robert M., and Briggs, Leslie J. *Principles of Instructional Design,* 2d ed. (New York: Holt, Rinehart and Winston, 1979).

Krathwohl, David R.; Bloom, Benjamin S.; and Masia, Bertram B. *Taxonomy of Educational Objectives; Handbook II: Affective Domain* (New York: David McKay Co., 1964).

Mager, Robert F. *Preparing Instructional Objectives* (Belmont, Calif.: Fearon Publishers, 1970).

Morris, L. L., and Fitz-Gibbon, Carol T. *How to Deal with Goals and Objectives* (Beverly Hills, Calif.: Sage Publications, 1978).

GOAL

☐ Come to understand the uses of microcomputers in education.

OBJECTIVES

☐ List and explain the uses of computers in education.
☐ Define computer terminology.
☐ Locate, load, and run a microcomputer program.
☐ List and explain the components of a computer system.
☐ List the key dates in the history of computers.
☐ Write an algorithm and flowchart.
☐ Write a simple program in BASIC.

In 1973, an advertisement for a hand-held calculator in a national magazine illustrated a device that would "add, subtract, multiply, and divide" for $79.95. It could be held " . . . in the palm of your hand and carried in a coat pocket."

That price seems absurd today when you can buy a calculator in the supermarket for $6.95, or get one as a gift for opening a new checking account at the local bank. Of the few items actually decreasing in price (but increasing in value) these days, calculators are among the most notable.

While supply and demand may have something to do with calculator price and popularity, the primary reason that calculators are so inexpensive is that they are easy to produce because of innovations in technology. Most of the mathematical functions done by a calculator can be performed in a chip of metal and plastic that would fit on the fingernail of your little finger.

The same sort of miniaturization and price reduction has taken place in full-scale computers. The electronic equipment that would fill an office in 1970 can now be put into a typewriter-sized box that will fit on a desk top. Because of their low price and small size these microcomputers are being widely used in education for a number of purposes.

Many educators feel that every school should have a computer, and some teachers think that every classroom should have one. It is not impossible to imagine every student having a personal computer in the foreseeable future. In 1965, there were 30,000 computers in operation in the United States. In the early 1980s that many computers were being produced each month. Companies such as Apple, Tandy-Radio Shack, and Commodore have taken their place next to the computer giants IBM, Sperry, and Xerox as innovators in the computer field. Computer literacy is becoming as important as any other type of literacy for the teacher.

This chapter approaches computer literacy by first discussing the functions of the computer in the classroom. A section is included on how to use commercially available computer lessons. The basic operation and components of computer equipment will also be discussed. More advanced concepts, such as algorithms and flowcharts, that are necessary for writing original computer instructions, will be described so that simple programming using the BASIC language can be carried out.

5

MICROCOMPUTERS

WHAT ARE COMPUTERS?

A computer is a machine that is capable of accepting information in the form of numbers, letters, or symbols, processing that information at a high rate of speed, and then producing some result. There are three categories of computers—mainframe, mini, and micro. Each is defined according to its size and speed of operation. *Mainframe* computers are those large, complex machines most of us traditionally visualize when we think about computers. A computer like the one at the University Computation Center, and HAL in the film *2001: A Space Odyssey,* are examples. Usually mainframe computers control many smaller devices called input/outputs (I/Os), and are accessed by many users simultaneously on a time-sharing basis. Several operators and programmers are needed to run a mainframe computer. Only the largest school districts or businesses use computers of this size.

A typical mainframe computer, IBM 3081 (courtesy IBM).

A typical minicomputer, IBM 4341 (courtesy IBM).

Minicomputers are like mainframes, except that they are smaller and less powerful. They are designed for many users to share and usually have several input/output locations so that several jobs can be run at the same time. Minicomputers must be operated by a specially trained technician. Many schools and companies purchased minicomputers during the last decade, usually to handle their record-keeping and payroll needs. Some classroom instruction is also done using minicomputers.

Microcomputers, also called personal computers, are the focus of this chapter. Micros have the same capabilities as their larger cousins, but are slower (although still remarkably fast), and cannot manage as large a quantity of information. Micros handle only one job at a time, and do not require a highly trained operator.

An IBM micro home computer with printer, monitor, disk drive, and keyboard (courtesy IBM).

A Commodore PET microcomputer (courtesy Commodore Business Machines).

An Atari 800 microcomputer with disk drives and tape drive (courtesy of Atari).

An Apple II+ microcomputer with dual disk drives, printer, and monitor (courtesy Apple, Inc.).

The TRS-80 microcomputer (TRS-80 is a trademark of the Radio Shack Division of Tandy Corporation; photograph compliments of Radio Shack).

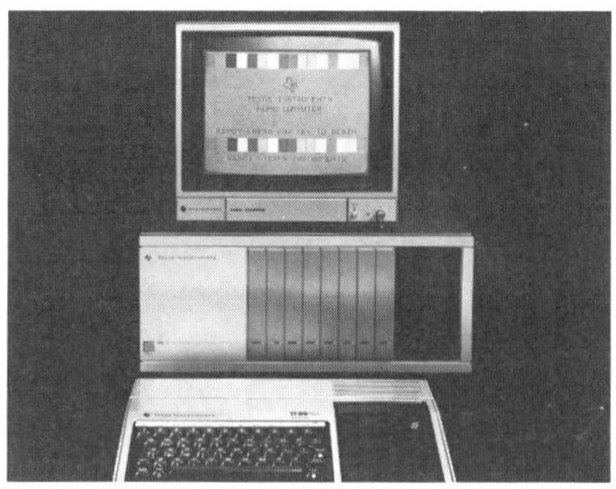

The TI-99/4A microcomputer (courtesy Texas Instruments).

Mainframe, mini-, and microcomputers also have computer relatives in widespread use, such as the pocket calculator, the arcade electronic games such as "Space Invaders" and "PACMAN," and the teaching machines like "Speak 'n Spell." While these devices are controlled by chips like those in larger computer processing units, they do not fit the broader concept of a computer. A computer is a device that accepts number, letter, and symbol information, processes that information, and produces results. Computers can be programmed for different jobs at different times. They can be instructed to carry out some functions on their own. A computer will have inputs (keyboards, tape systems, or disk drives), a central processing unit (CPU), and outputs (TV screen, called a CRT; a printer).

COMPUTER USES IN EDUCATION

Computers may be used in education for three purposes:

1. For organization management (record keeping and word processing).
2. To teach students how computers work.
3. To assist in teaching traditional subjects.

Record Keeping

Record keeping has traditionally been the primary function of the computer. Schools, like other businesses, generate masses of information such as attendance records, grade reports, payroll listings, and supply inventories. Computers can process such data rapidly and can simplify the job of keeping track of all the information generated by a school system about how that organization functions. Word processing is another popular use of the new generation of microcomputers. While record keeping may not be of primary interest to most teachers, this data-processing function can be handled by the same microcomputer that is used in the classroom. The difference is in the software. Some schools purchase their first microcomputer for use in the business office and then find that it is so efficient that it can also be used by the math teacher to show advanced students how computers work.

Classroom teachers also use computers to assist them in the production of teaching materials such as quizzes, spirit masters, and handouts. The computer can be an electronic gradebook to keep track of student progress. It can also be used for diagnosis of students who need remedial work in an area. This function of the computer is called "computer-managed instruction."

Microcomputers are excellent for record keeping (compliments Apple, Inc.).

Selected Dates in the History of Computer Development

ca 3000 B.C.	The Chinese abacus
1630	William Oughtred's slide rule
1642	Blaise Pascal's gear-driven adding machine
1812	Charles Babbage's "Difference Engine"
1854	George Boole's algebraic logic
1868	First patented practical "typewriter"
1887	Herman Hollerith's "Census Machine"
1909	Charles Kittering's accounting machine
1910	James Powers' punched card system
1920	James Smathers' electric typewriter
1937	Howard Aiken's work on Mark I
1944	Mark I operational
1947	Bell Laboratories' transistor
1951	UNIVAC's installation at the Census Bureau
1955	IBM's first large-scale business computer
1959	Transistor circuitry for computers
1963	John Kennedy's & Thomas Kurtz's BASIC
1963	First time-sharing system
1964	Integrated circuits for problem solving with computers
1965	Minicomputers become widespread
1974	Production of microprocessors
1977	Microcomputers become commonplace

In addition to maintaining records, the microcomputer can analyze and display data so that they are easier to understand and use (courtesy Texas Instruments).

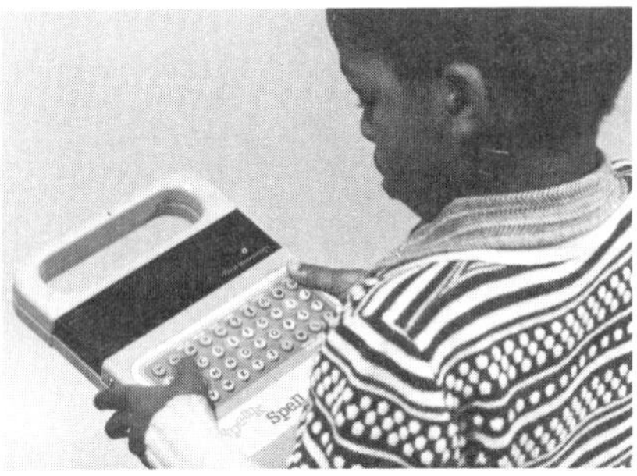

Drill and practice are easily accomplished with a computer-like device, such as a Speak 'n Spell, because it never gets tired and can generate problems for as long as the student is interested.

Tutorials provide each student with an individual approach for learning (compliments of Radio Shack, a Division of Tandy Corp.).

Simulations allow students to participate in life-like activities using the computer.

Computer Literacy

Computer literacy (teaching about computers) is a reason for a school to purchase a computer of its own. Many secondary schools have units in math or science courses on computer programming and operation. The languages used to program computers have now become so simple and easy to understand that even people without formal computer training can write programs of their own. The microcomputer is an excellent machine for teaching students about large, more complex computer systems.

Computer-assisted Instruction

A dynamic and promising use for computers in education is computer-assisted instruction (CAI). In CAI the computer acts as a teacher. Because of its interactive capabilities, the computer can be used to provide the give-and-take type of teaching that other media cannot, and that previously only the teacher was capable of offering. There are five categories of CAI.

Drill and Practice When used this way, the computer provides practice in learning fundamental skills, such as alphabetizing, balancing chemical equations, multiplication table practice, and conversions from one numerical system to another. Drill-and-practice programs usually present quantities of problems for the student to work until a skill has been mastered.

Tutorials Tutorials introduce new topics and provide techniques for mastering them. Usually interactive, a tutorial requires the student to choose a response and then branches to various types of subsequent information as reinforcement, depending on the correctness of the student's response.

Simulations Simulations provide representations of real events, such as the operation of a lemonade stand, management of a nuclear reactor, or flying of a light plane. The student makes decisions and interacts with the computer to learn the consequence of each decision.

Educational Games Games provide skill-building activities, usually in the cognitive or psychomotor domains. Students learn how to manipulate objects, work through puzzles, or compete with others. The computer keeps score and determines the winner.

Problem Solving The computer analyzes data that the student inputs to solve some problem. It performs computations, provides answers, and permits the student to choose alternative solutions to problems. Often the computer is used as a calculator.

How Information Is Stored on a Computer Disk

Microcomputers store data permanently on either a magnetic tape or magnetic disk. The tape system uses standard audio cassette tape and tape recorders. The main advantage of tape systems is that they are inexpensive. However, they are also very slow and prone to malfunction. Many microcomputer users prefer disk storage systems that record information on a 5¼-inch floppy (flexible) disk. Both the cassette and disk storage systems work the same way. However, because the disks spin and store information in concentric tracks, instead of in a linear tracking pattern, the disk-stored information is more accessible.

Disk systems store information on one of a number of tracks. Each track has sectors for storing bytes of data (either 13 or 16 sectors/track for Apple II microcomputers). Each sector can hold a certain amount of magnetically coded information (256 bytes/sector for Apple II).

Computer Terminology

analog computer An analog computer is one that measures continuously. It is used for translating data such as temperature, pressure, and voltage for later study.

BASIC Acronym for the computer language called *B*eginners *A*ll-Purpose *S*ymbolic *I*nstruction *C*ode. BASIC is a widely used language for microcomputers.

binary The actual information handled by a computer is in binary numbers, which consist of zeros and ones.

bug A mistake or problem in a computer program.

byte A unit of information composed of bits, which are simple binary digits, is called a byte. Computer memory is rated according to how many bytes of information it can hold. Microcomputers generally have 8-or 16-bit bytes.

CAI Computer-assisted instruction.

CAL Computer-assisted learning.

CMI Computer-managed instruction.

CPU Central processing unit. The CPU is the place where information is added, subtracted, divided, multiplied, compared, or sorted. It is what most people call the brains of the computer.

CRT Cathode ray tube. A CRT is a vacuum tube that serves as an input/output device. The TV screen is a CRT.

chip A chip is an electronic circuit containing thousands of electronic components photographically produced on a small piece of silicon.

digital Also called a general-purpose computer, the digital computer is designed to input, store, manage, and output units such as numbers or letters.

hardware The equipment generally referred to as the computer. Computer hardware consists of the CPU and the input/output devices.

IC Integrated circuit. ICs are groups of connected electrical circuit elements on a chip of semiconductor material.

I/O devices Input/output devices. Those components of a computer that allow us to communicate with the computer. Cassette tape recorder/players, disk drives, keyboards, CRTs, and printers are examples of input/output devices.

memory Devices in the computer or one of its peripherals where information can be stored.

RAM Random-access memory. Memory used to store information from programs input from the keyboard or other input device. RAM is not permanent. It is erased every time the computer is turned off.

ROM Read-only memory. Permanent memory locations in the computer, such as language compilers and disk operating systems. Language compilers translate programming commands into directions the computer can process. Disk operating systems are directions for storing information onto, or reading information off of, a magnetic disk.

software Coded instructions, written in a computer language such as BASIC or Pascal, that directs the operation of the computer hardware. These instructions are written in the form of *Programs.*

The microcomputer can play numerous educational games. This use of the microcomputer stimulates an interest in learning in many students (compliments of Radio Shack, a Division of Tandy Corp.).

TI LOGO, a child-appropriate computer language, was developed for the TI-99/4A home computer by Texas Instruments, Inc. and Massachusetts Institute of Technology. The language allows students, beginners and advanced, to communicate with the computer in an easy-to-understand language. These activities lead to more complex computer commands, enabling students to develop a wide range of problem-solving and communication skills.

Many computer magazines are now being published that list sources of microcomputer software.

Many publishers are now offering microcomputer programs to accompany their textbook series.

Basically, all magnetic storage systems work in the same way (computer tape or disks, videotape or audiotape). Electromagnets, called heads, place magnetic information on a magnetizable surface in patterns called tracks. This information is retrieved as a disk spins past a *playback head,* which reads out the magnetic impulses that were recorded. These heads convert electrical impulses into magnetic fields, or vice versa.

USING A MICROCOMPUTER TO RUN A PROGRAM

The simplest use of the computer is to run a program written by someone else. This is probably the most important use of the computer for teachers, something like using a textbook written by another teacher. In order to use a prepared program you must locate software and operate the microcomputer.

Locating Software

Locating software can be a problem if a teacher's school or district is not served by a comprehensive media center. Just as with other forms of media, the teacher is limited by what is available. Fortunately, there are a large number of sources for computer software. Computer magazines with countless advertisements presenting new computer software can be purchased at almost any newsstand. Catalogs of programs are also being published (see References at the end of this chapter).

Computer users often form clubs or groups and meet on a regular basis to share programs. The Minnesota Educational Computer Consortium (MECC) is one such group. Members of the Consortium share the programs that they have written. In this way it is possible to collect many programs quickly and inexpensively.

Computer software, as other media, should be evaluated before use. The following form is an example of one that could be used to evaluate computer software.

Worksheet for Evaluating Computer Software

Your Name(s) ______________________________

Program Name ______________________________

1. How appropriate is computer use for this activity?
2. Are the goals of the program clear? (What are they?)
3. Does the activity in the program focus on these goals?
4. Are the instructions clear? (If not, why?)
5. How understandable is the format of the output?
6. How good is the continuity of the output?
7. How vulnerable is the program to input mistakes?
8. How appropriate is reinforcement used?
9. Are the special features of the computer used well? (Where could improvement be made?)

The Origin of "The Bug"

Developed in the 1940s, Mark I was one of the first electromechanical computers. Many of its components were relays that moved to open and close electrical circuits. The story goes that one day the system stopped operating efficiently. The technicians looked around and found a moth caught in a relay. They removed the moth, recorded the incident in their log, and jokingly said that they were "debugging" the system. The tradition continues today; a *bug* usually means an error in human logic.

Source: C. E. Horn and J. L. Poirot, *Computer Literacy* (Austin, Tex.: Sterling Swift, 1981): 17.

Operating the Microcomputer

Operating the microcomputer is always a concern of the new user. The best reference for computer operation is the manual that the manufacturer provides when the computer is purchased. Generally, the set-up and operation of the microcomputer require the connecting of input and output devices to the CPU and turning on all pieces of equipment.

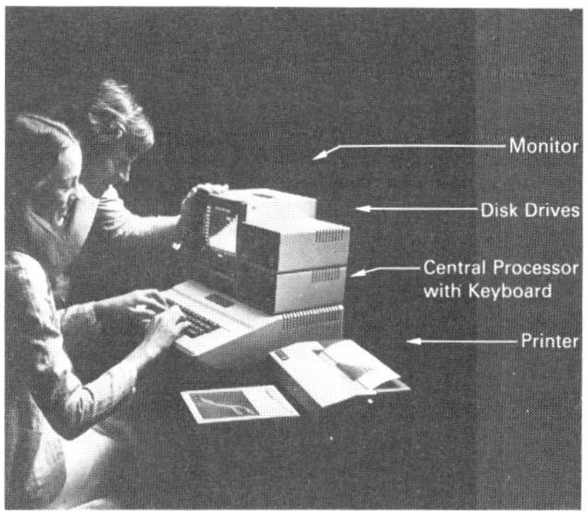

Apple II+ microcomputer, showing monitor, disk drives, central processor with keyboard, and printer (courtesy Apple, Inc.).

OPERATE

Operation of a Microcomputer

1. Collect all pieces of equipment and connector cables.

   Inputs/Outputs
   Cassette tape player
   Disk drive unit
   CRT
   Printer

   Connectors
   Audio cable to connect computer to audio player
   Disk drive cable—furnished with disk drive
   Video cable for connecting CRT to Computer
   RF unit for connecting computer to standard home TV set
   Printer cable to connect computer to printer—furnished by manufacturer with printer

2. Connect inputs and outputs to computer.

3. Turn on all components.

   For some reason, many microcomputers and peripheral devices have their on-off switches on the back panel.

4. Load the program from the input device to the computer.†

   Programs loaded from cassette tapes require setting the volume control correctly. Usually the middle position is correct. If problems occur in loading, try adjusting the volume control.

5. Run the program.

   Programs loaded on disks are generally much easier to load and run. Most commercially prepared programs provide the user with detailed instructions on the computer's output device (CRT), on how the program is to run.

*The specialty multipin cables used to connect disk drives and printers to the computer should be furnished by the manufacturer of that peripheral equipment at the time of purchase. It is important to specify the brand and type of computer with which an input/output device is going to be used when the device is purchased.

†Different computers require different commands for loading and saving programs. See page 78.

Programs can be loaded from or to audio cassette tapes . . .

or magnetic disks. Generally, the disks are faster and easier to use, but the cassette tape systems are less expensive.

OPERATE

Commands for Loading or Saving a Program

Apple System BASIC

Tape	
SAVE	Saves program in memory on cassette tape
LOAD	Loads program stored on cassette into computer's memory
Disk	
SAVE SHORT	Saves program SHORT in memory on disk
LOAD SHORT	Loads program SHORT stored on disk into computer's memory

Radio Shack System BASIC (Level II)

Tape	
CSAVE SHORT	Saves program SHORT on cassette
CLOAD SHORT	Loads program SHORT from cassette into memory
Disk	
SAVE "SHORT"	Saves program SHORT on disk
LOAD "SHORT"	Loads program SHORT from disk into memory

Texas Instruments System BASIC

Tape	
SAVE CS1	Saves program on cassette
OLD CS1	Loads program from cassette into memory
Disk	
SAVE DSK1. SHORT	Saves program SHORT on disk
OLD DSK1. SHORT	Loads program SHORT into memory

Commodore PET System BASIC

Tape	
SAVE	Saves program on cassette
LOAD	Loads program from cassette into memory
Disk	
SAVE "SHORT",8	Saves program SHORT on disk
LOAD "SHORT",8	Loads program SHORT into memory

In the first sections of this chapter the basic function and operation of microcomputers were described. While we realize that currently the greatest use of the computer is in running programs written by others, you will find it challenging to write your own programs. The next section describes the process followed for writing computer programs.

WRITING A PROGRAM FOR A MICROCOMPUTER

Writing a computer program is a process of problem solving. The programmer wants the computer to do something—to solve a problem.

There are three steps in writing a computer program. First, write an *algorithm,* a procedure for organizing a logical series of steps to solve a problem. Next, develop a *flowchart,* a series of symbols and words to describe the solution to a program. It is developed from the algorithm. Last, write the program in a language the computer can understand.

An *algorithm* is a procedure for organizing a logical series of steps with words to solve a problem. Logic refers to the problem-solving pattern for doing things in a specific order. Since computers work logically and sequentially, the programs that are used to guide them must also be organized that way.

Algorithm for Eating Peanuts

☐ Do you want peanuts? If not, finish homework.
☐ If yes, then place hand in the peanut bag.
☐ If there are no peanuts, finish homework.
☐ If there are peanuts, take one from the bag.
☐ Crush the shell and open it.
☐ If the peanut doesn't look good, throw it in the trash and take another.
☐ If the peanuts look good, eat them and throw the hulls in the trash.
☐ If you want another peanut, start over.
☐ If not, finish homework.

PRODUCE

Steps for Writing a Computer Program

1. Write a brief description of what the program is to do—its algorithm.
2. Diagram the strategy needed to solve the problem using graphic symbols, human logic, and human language—its flowchart.
3. Write a program to solve the problem using a computer language that follows the strategy diagrammed in the flowchart and that includes remarks concerning the following:

Name of the author
A summary of what the program is to do
A list of variables used and a functional description of each
Comments to help explain what is happening during the program
"User" instructions to allow someone else to use the program easily; include such things as necessary input and expected output

PRODUCE

Steps for Writing an Algorithm

1. Determine what the specific problem is. If there is more than one problem to be solved, divide them into categories and organize them sequentially.
2. List each step needed to solve the problem or subproblem.
3. Organize each step sequentially.
4. Try the process out to see if the problem can be solved every time by following the sequence.
5. Combine all subcategory solutions sequentially to solve the complex problem, and try it out.

TRY THIS

Write an Algorithm

Write an algorithm for one of the following situations:

- ☐ Washing a car
- ☐ Making a sandwich
- ☐ Opening a beer or pop can
- ☐ Playing a record
- ☐ Introducing yourself to your blind date

A *flowchart* is a way to describe a solution to a problem using words and graphic symbols, and is the second step in writing a computer program. It is a graphic representation of an algorithm.

Flowcharts are usually written so that the direction to the solution is left to right, but complex systems may require more elaborate structuring. There are five primary categories of symbols used in flowcharting—ovals, parallelograms, rectangles, diamonds, and circles. Arrows are used to connect symbols. *Terminal operations* are represented by ovals. Usually, a word such as *start, begin, stop,* or *end* is written in the oval to clarify what is happening.

Begin Start End Stop

Input-output operations are symbolized by using a parallelogram. They indicate that data or information from some outside source are going to be processed, or that data are to be output.

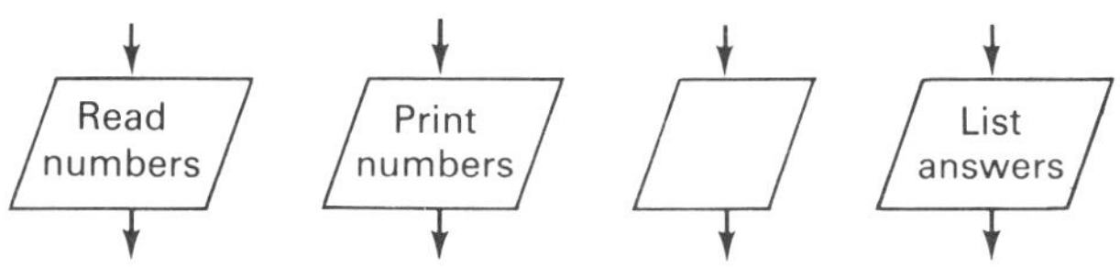

Processing or *definition operations* are visualized by using rectangles.

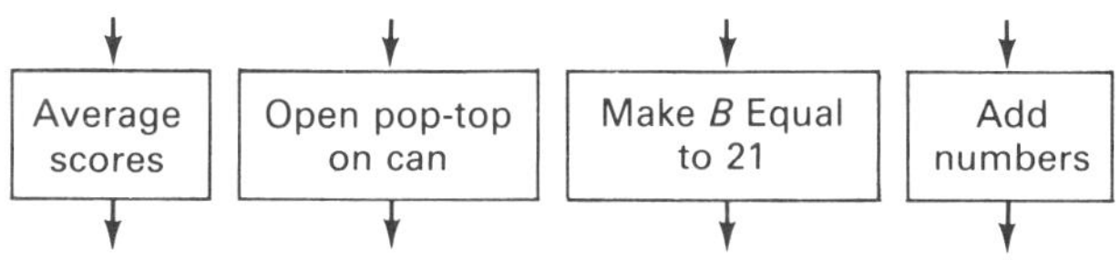

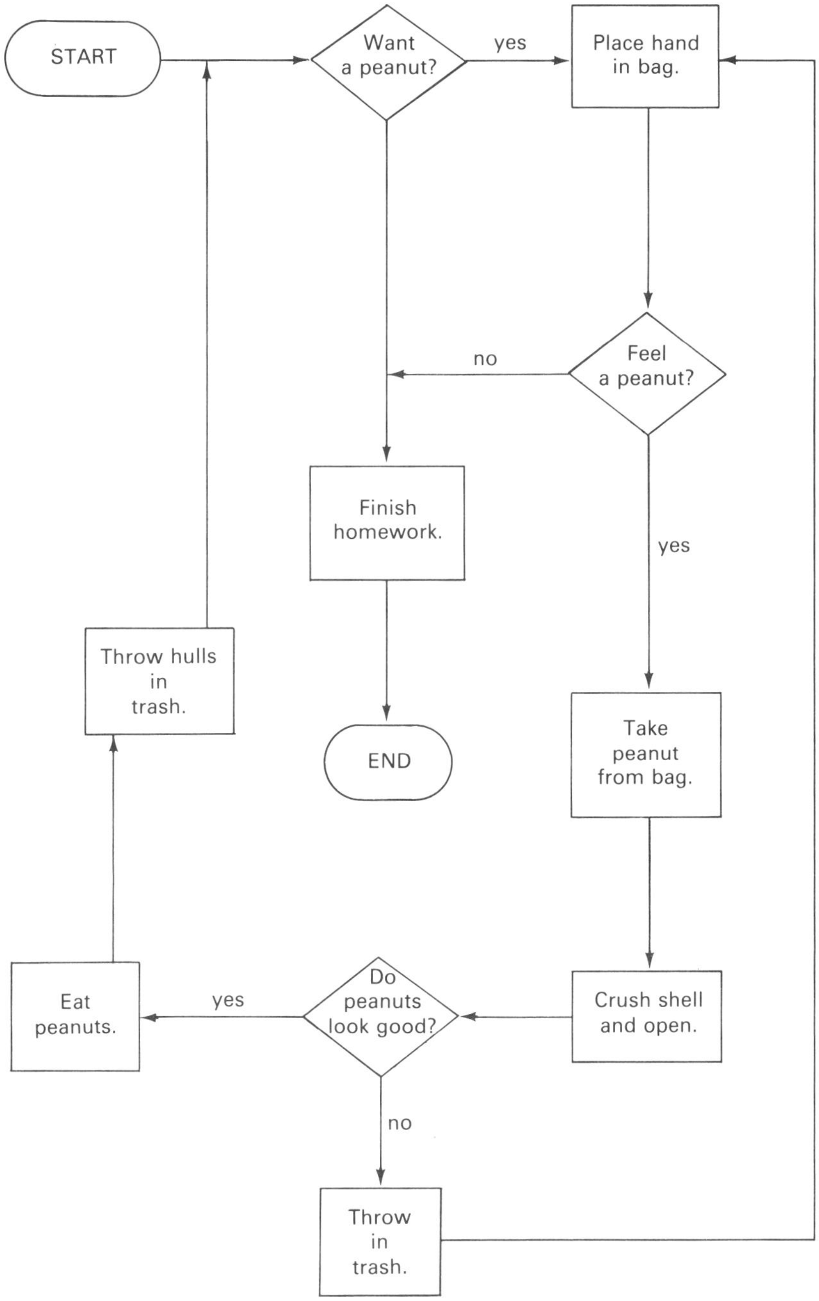

How to eat peanuts.

Decision operations are symbolized by using a diamond. The question to be answered should be written in the diamond. The possible answers to the question should travel from the diamond in two directions, for yes and for no.

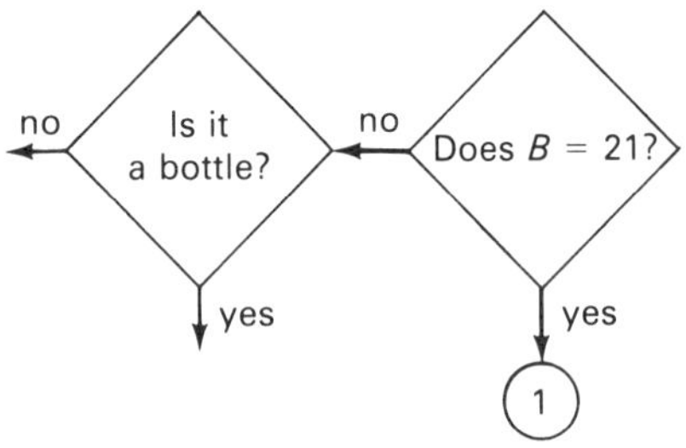

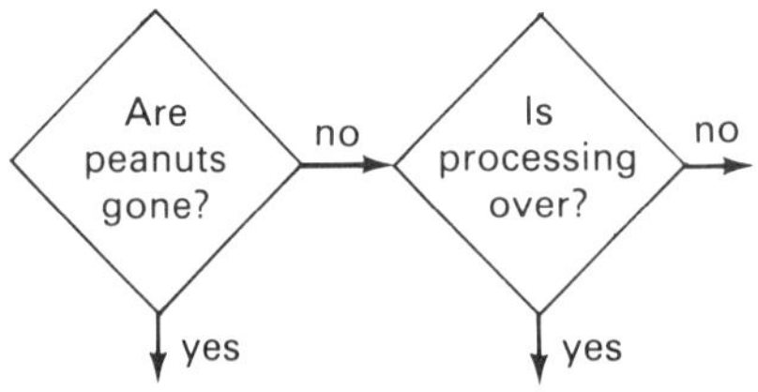

Connectors are used to connect one portion of a flowchart to another without using arrows, and are symbolized by a circle.

Program writing is the next step when developing instructions for the computer to solve problems. Programs are written in computer languages that consist of key words that the computer memory can identify and then translate into machine language. Machine language is the only language that computers really understand. Most computers are made to move and store information in binary code form. Binary code consists of only two symbols, zero (0) and one (1). Any number or letter can be written using binary code.

TRY THIS

Write a Flowchart

Write a flowchart for the algorithm that was written previously.

PRODUCE

Steps in Writing a Flowchart

You need a well-written algorithm:

Procedure

1. Analyze the algorithm and identify the terminal operations, inputs and outputs, processing or definition operations, and decision operations.
2. Develop the flowchart sequentially from left to right and top to bottom on the page.
3. Place each symbol on the page, define the operation with words, and connect each symbol with arrows.
4. Try the process out several times to see if it accounts for all possible options.

6-Bit Binary-coded Decimal Code

1	000001	**A**	110001
2	000010	**B**	110010
3	000011	**C**	110011
4	000100	**D**	110100
5	000101	**E**	110101
6	000110	**F**	110110
7	000111	**G**	110111
8	001000	**H**	111000
9	001001	**I**	111001
0	001010	**J**	100001
		K	111000
		L	100011
		M	100100
		N	100101
		O	100110
		P	100111
		Q	101000
		R	101001
		S	010010
		T	010011
		U	010100
		V	010101
		W	010110
		X	010111
		Y	011000
		Z	011001

Higher-level computer languages were developed so people who do not understand machine languages can still program a computer. There are hundreds of languages. Some of the most commonly used are FORTRAN (*For*mula *Tran*slation), BASIC (*B*eginners *A*ll-Purpose *S*ymbolic *I*nstruction *C*ode), and COBOL (*Co*mmon *B*usiness-*O*riented *L*anguage). Of the newest microcomputers, many use BASIC because it is simple to learn and use, yet is very powerful.

Programming a Microcomputer in BASIC

While each brand of computer has minor differences in how it is controlled by the BASIC language, the similarities are overwhelming. There are five categories of BASIC commands: system commands, operational symbols, variables, program commands, and branching/decision commands.

BASIC Program to Add, Subtract, Multiply, and Divide Two Numbers

```
NEW
 10 PRINT "INPUT TWO NUMBERS"
 20 INPUT X, Y
 30 LET SUM = X + Y
 40 LET DIFF = X - Y
 50 LET PROD = X * Y
 60 LET QUOT = X/Y
 70 PRINT "THE SUM IS";SUM
 80 PRINT "THE DIFFERENCE IS";DIFF
 90 PRINT "THE PRODUCT IS";PROD
100 PRINT "THE QUOTIENT IS";QUOT
110 END
RUN
```

Complex graphics are possible with many of the languages available for microcomputers (courtesy Texas Instruments).

System commands instruct the central processing unit on what the program is to do. The commonly used system commands are NEW, LIST, DELETE, and RUN.

NEW clears all instructions and data from memory to ready it for new information. It is the first command to use before writing a BASIC program.

Example:
```
NEW
10 INPUT A
20 PRINT A
```

LIST is used so a programmer can review what has been written into the computer. It causes the program in memory to appear on the screen or printer. *DELETE* is used to remove lines of code from the computer's memory, and is used to correct programming errors. *RUN* tells the computer to execute the program that it has stored in memory.

Functional or *operational symbols* are punctuation marks or characters used to indicate that something special is to occur. Common operational symbols are:

comma (,) Indicates that spaces are to be skipped.

semicolon (;) Separates variables to be output.

colon (:) Indicates that more than one programming command should be accepted into memory at one time.

parentheses () Indicate that functions within them occur first.

V or ↑ Indicates exponentiation (raising the power).

* Indicates that two variables are to be multiplied by one another.

/ Indicates division.
Indicates addition.
Indicates subtraction.

Variables are quantities that change while the program is being run. They are usually assigned names and values so that they can be stored and recalled from memory. A variable must begin with an alphabetic (A–Z) character, and may be followed by an alphanumeric character.

Examples:
```
A1
N1
TIME
N555
X5
```

Variable values may be set by the programmer, or by the user when the program is run. String variables are words and numbers up to 255 characters long and are followed by a $ (e.g., NAME$).

Program commands are directions for something to happen. All statements in a computer program are executed in order. This order is the order of their numerical sequence. All program commands in BASIC must be numbered. Usually program statements are numbered in steps of 10 or 100, but numbers like 6, 19, or 103 can also be used. The most common program commands are:

END The program is completed and control is returned to the user.

Example:
```
110 PRINT A+B
120 END
```

REM Used for remarks in the body of the program about the program. REM statements are not exekuted when the program is run.

Example:
```
NEW
10 REM This
   program
   averages two
   numbers
20 INPUT A
```

PRINT Used to display data on the printer or CRT.

Example:
```
90 PRINT A*B
```

INPUT Placed in the program when data are expected to be input by the user.

Example:
```
50 PRINT "PLEASE
   TYPE YOUR AGE"
60 INPUT A (The
   computer is
   expecting input)
70 PRINT A "IS A
   NICE AGE TO BE"
```

LET An assignment command that puts data in a specific location in the computer's memory.

Example: `50 LET N=22`

DATA and **READ** Statements that occur in pairs, and are used to assign data to a variable.

Example:
```
10 DATA 5,10,12
20 READ A,B,C
30 PRINT A,B,C
40 END
RUN
```

The output would be 5, 10, 12

Branching and *decision commands* provide the programmer with the ability to jump around the program and to make decisions.

GOTO Used when the programmer wants to branch to another line in the program.

Example: `60 GOTO 40`

IF _______ THEN _______ The most powerful statements in BASIC language. The computer is asked if something has happened, and if it has then something else will occur. Often IF _______ THEN _______ statements are combined with GOTO statements.

Example:
```
 60 IF A = 5 THEN
    GO TO 120
 70
100
110 PRINT A*B*C
120 END
```

Acceptable IF _______ THEN _______ program statements vary. Examples of acceptable uses of this statement are:

Apple System BASIC

IF (expression) THEN (instruction)
IF (expression) THEN (instruction):(additional program statement)
IF (expression) THEN GOTO (line number)
IF (expression) THEN (line number)

AND and OR are acceptable logical arguments.

Radio Shack System BASIC

IF (expression) THEN (instruction)
IF (expression) THEN GO TO (line number)
IF (expression) THEN (line number)

AND, OR, and ELSE are acceptable logical arguments.

Texas Instruments System BASIC

IF (expression) THEN (instruction)
IF (expression) THEN (line number)

ELSE is an acceptable logical argument.

Commodore PET System BASIC

IF (expression) THEN (instruction)
IF (expression) THEN (instruction):(additional program statement)
IF (expression) THEN (line number)

AND and OR are acceptable logical arguments.

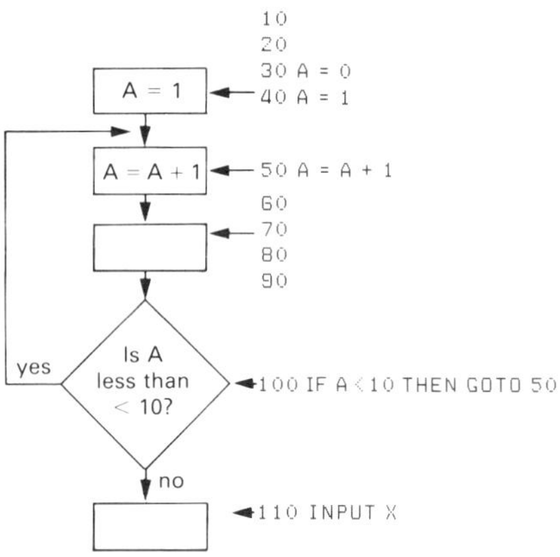

IF _______ THEN _______ statements are used where diamonds were drawn on the flowchart to symbolize decisions, as shown at left.

There are other commands available in BASIC that can be used to write more complex computer programs. Many introductory texts on writing BASIC can be purchased (see References at the end of this chapter). Also, microcomputer companies have produced manuals about BASIC designed specifically for their computers. If you are serious about programming a microcomputer, one of these manuals would be a good investment.

Summary of Common System Commands

NEW Clears all instructions and data from RAM
LIST Causes a listing of program code in memory to appear
DELETE Removes specified line(s) from memory
RUN Causes program in memory to be executed

Summary of Common Program Commands

PRINT Instructs computer to print out or display data
INPUT Computer expects input of variable value(s) by user
LET Assigns value(s) to variable name(s)
REM Indicates comments *not* to be output with solution
DATA Sets values and their order for assignment
READ Sets variable names for the DATA values
GOTO Instructs computer to loop to specified line number
END Indicates the end of a program
IF _______ THEN _______ Permits a decision to be made

TRY THIS

Sample Programs in BASIC

Try copying these programs into your Apple II and see what happens.

A. Colored line across screen * (p. 62, Applesoft manual):

Type this in:

```
NEW
100 GR
110 COLOR = 4
120 HLIN 0, 39 at 20
130 END
RUN
```

Press Return to see it

B. To fill the screen with color (p. 62, Applesoft manual):

```
NEW
200 GR
210 COLOR = 9
220 COLUMN = 0
230 VLIN 0, 39 at COLUMN
240 COLUMN = +1
250 IF COLUMN < 40 THEN GOTO 230
260 END
RUN
```

Press Return

After you've done this, try retyping line 210 this way:

```
210 COLOR = 10
RUN
```

Press Return

Or some other number between 1 and 14 (see p. 15 of Applesoft manual for key to the colors).

C. To see a sequence of numbers (p. 64, Applesoft manual):

```
NEW
100 NUMBER = 0
110 PRINT NUMBER
120 NUMBER = NUMBER + 1
130 IF NUMBER < 300 THEN GOTO 110
140 END
RUN
```

PRODUCE

Steps for Writing a Computer Program in BASIC

You need: a well-written algorithm and flowchart and a microcomputer with input and output devices

Procedure

1. Start writing the program at the top of the page and exit at the bottom.
2. Divide long programs into subprograms so that independent operations can be clearly identified and easily corrected, one module at a time.
3. Use REM statements to clearly label what each module of your program is supposed to do.
4. Select the correct program command, decision command, and variable to match each step in your flowchart.
5. Use recognizable names for variables so you can easily understand their purpose in the program.
6. Number each statement in increments of 10 or 100, so you can go back and add commands later, if necessary, without having to renumber the entire program.
7. Program each module sequentially, then combine modules. Try each module out on the computer to be sure it works before adding it to the master program.
8. Try the entire program out to see if it works.

Hints: Use a limited number of statements so that the computer does not have to work harder than necessary to run your program.

Try not to use GOTO statements more than necessary, because excessive branching makes a program hard to understand.

TRY THIS

Write a Computer Program in BASIC

Write a program in BASIC that will input ten numbers, sum the total, and print the answer.

Hint: You will need to use the IF _______ THEN _______ GOTO _______ statement and a counter.

A FINAL WORD

Computers offer the potential for greatly extending the capabilities of the teacher. Simple computer programming, such as has been described in this chapter, only hints at the potential of the computer in education. Most of us have a difficult time imagining the future because we have such a difficult time trying to understand the present.

Sometimes it is fun to compare the writings of science fiction writers to the realities of the day. Who can forget the almost human computer HAL in the film *2001: A Space Odyssey,* or the robots R2D2 and C3PO from the film *Star Wars?* Possibly the robot is the ultimate computer. Maybe teachers of the future will have a robot (named "Teacher's Pet"?) to assist them.

Science fiction writer Isaac Asimov has already developed the "Three Laws of Robotics":

1. A robot may not injure a human being or, through inaction, allow a human being to come to harm.
2. A robot must obey the orders given it by human beings, except where such orders would conflict with the First Law.
3. A robot must protect its own existence as long as such protection does not conflict with the First or Second Laws.

Science fiction? Certainly, but interesting and fun to think about, too. The computer will most likely be there next to the teacher in the remainder of the 20th and in the 21st century.

This chapter has attempted to provide you with a basic computer literacy by presenting:

- ☐ The uses of microcomputers in education
- ☐ Basic computer terminology
- ☐ How to locate, load, and run a program
- ☐ The components of a computer system
- ☐ The techniques for writing simple computer programs in the BASIC language

LOOKING AHEAD

Another medium used by teachers a great deal is the handout. Puzzles, worksheets, tests, and study problems are examples of some of the many uses of the handout that can be produced for duplication using a microcomputer and printer. As a matter of fact, microcomputer utility programs are rapidly relieving instructors of the need to spend long hours generating homework worksheet originals. Methods for producing and duplicating handouts will be discussed in the next chapter.

REVIEW QUESTIONS

1. The traditional use of computers in education has been for __________.
2. The part of the computer that adds, divides, subtracts, multiplies, compares, and sorts is called the ______________.
3. What are the three categories of components of a microcomputer?
4. What is RAM?
5. What do we call a mistake or problem when we run a computer program?
6. Define *algorithm.*
7. What are diamonds used for in flowcharting?
8. If you want to clear the memory of a microcomputer before writing a program you should type __________.
9. The command in BASIC that allows decisions and branching is the __________ statement.
10. Define *computer literacy.*
11. Write a program in BASIC to have the computer ask the user's name, age, and hometown, and then print the results.
12. BASIC is a machine language. TRUE/FALSE
13. Complete this crossword puzzle dealing with computer terminology. (*Note:* This puzzle was produced by using a special teacher's utility program for the Apple II computer. It took ten minutes, and is an example of how the computer can make the teacher's job easier.)

Computer Literacy Crossword

```
 1XXXXXXXXXX[ ]XXXXXXXXXXXXX[ ]XXXXXXXXX
 2XXXXXXXXXX[ ]XXXXXXXXXXXXX[ ]XXXXXXXXX
 3XX[ I I ]X[ ]XXXXXXX[ ]X[ I I I I I ]X
 4XX[ ]X[ ]X[ ]X[ ]XXX[ ]X[ ]XXXXXXXXXXX
 5XX[ ]X[ ]X[ ]X[ I I I I I ]XXXXXXXXXXX
 6XXXXXX[ ]X[ ]X[ ]XXXXXXXXXXXXXXXXXXXXX
 7XXXXXX[ ]X[ ]X[ I I I ]XXXXXXXXXXXXXXX
 8XXXX[ I I I ]X[ ]XXXXXXXXXXXXXXXXXXXXX
 9XXXX[ ]XXXXXXX[ ]XXXXXXXXXXXXXXXXXXXXX
10XXXX[ I I I I I ]XXXXXXXXXXXXXXXXXXXXX
11[ I I ]XXXXXXX[ ]XXXXXXXXXXXXXXXXXXXXX
12[ ]X[ I I I ]XXXXXXXXXXXXXXXXXXXXXXXXX
13[ ]XXXXX[ ]XXXXXXXXXXXXXXXXXXXXXXXXXXX
14XX[ I I I I I I ]XXXXXXXXXXXXXXXXXXXXX
15XXXXXXXX[ ]XXXXXXXXXXXXXXXXXXXXXXXXXXX
16[ I I I I I I ]XXXXXXXXXXXXXXXXXXXXXXX
17XX[ ]XXX[ ]XXXXXXXXXXXXXXXXXXXXXXXXXXX
   0 0 0 0 0 0 0 0 0 1 1 1 1 1 1 1 1 1
   1 2 3 4 5 6 7 8 9 0 1 2 3 4 5 6 7 8
```

WORDS ACROSS

DOWN, ACROSS	
[3, 2]	PERMANENT MEMORY IN A COMPUTER
[3, 13]	MACHINE LANGUAGE
[5, 8]	A COMPUTER THAT INPUTS DATA CONTINUOUSLY
[7, 8]	A MAGNETIC STORAGE DEVICE
[8, 3]	A UNIT OF INFORMATION
[10, 3]	STORAGE LOCATION ON A DISK TRACK
[11, 1]	COMPUTER-ASSISTED INSTRUCTION
[12, 3]	AN ELECTRONIC CIRCUIT
[14, 2]	CRT OR PRINTER
[16, 1]	A GENERAL-PURPOSE COMPUTER

WORDS DOWN

[1 , 6]	CODED INSTRUCTIONS TO DIRECT COMPUTER
[1 , 14]	COMPUTER-MANAGED INSTRUCTION
[3 , 2]	MEMORY STORED IN COMPUTER BY PROGRAM
[3, 4]	PLACE WHERE DATA CAN BE STORED
[3, 11]	COMPUTER-ASSISTED LEARNING
[3, 13]	A MISTAKE OR PROBLEM
[4, 8]	THE EQUIPMENT
[8, 3]	A SIMPLE-TO-LEARN COMPUTER LANGUAGE
[11, 1]	CENTRAL PROCESSING UNIT
[12, 5]	KEYBOARD OR DISK DRIVE
[16, 2]	GROUP OF CONNECTED ELECTRICAL CIRCUITS

ANSWERS TO REVIEW QUESTIONS

1. Record keeping
2. Central processing unit (CPU)
3. **a.** inputs
   b. central processor
   c. outputs
4. Random-Access Memory—The information entered into the computer memory by the program. It is not permanent.
5. A bug
6. An algorithm is a logical series of verbal steps to describe the solution of a problem.
7. Diamonds indicate that a decision is to be made.
8. NEW
9. IF ______ THEN ______ GOTO ______
10. Computer literacy includes an understanding of the technology used by computers, an understanding of the effects that computers have on society, and an understanding of how computers work.
11. NEW

```
 10 REM Program to ask name, age, and hometown
 20 PRINT "WHAT IS YOUR NAME?"
 30 INPUT A$
 40 PRINT "HI,";A$".";
 50 PRINT "PLEASE GIVE ME YOUR AGE."
 60 INPUT B$
 70 PRINT B$;"IS A NICE AGE TO BE."
 80 PRINT "WHAT IS YOUR HOMETOWN?"
 90 INPUT C$
100 PRINT "GOSH";A$;"IT MUST BE NICE TO
    BE";B$;"AND COME FROM";C$;"."
110 END
```

12. No, machine language is in binary code form. BASIC is a higher-level language used to program microcomputers. The computer will translate BASIC statements into machine language.
13. ACROSS

```
[ 3, 2 ]    ROM
[ 3, 13 ]   BINARY
[ 5, 8 ]    ANALOG
[ 7, 8 ]    DISK
[ 8, 3 ]    BYTE
[ 10, 3 ]   SECTOR
[ 11, 1 ]   CAI
[ 12, 3 ]   CHIP
[ 14, 2 ]   OUTPUTS
[ 16, 1 ]   DIGITAL
```

DOWN

```
[ 1, 6 ]    SOFTWARE
[ 1, 14 ]   CMI
[ 3, 2 ]    RAM
[ 3, 4 ]    MEMORY
[ 3, 11 ]   CAL
[ 3, 13 ]   BUG
[ 4, 8 ]    HARDWARE
[ 8, 3 ]    BASIC
[ 11, 1 ]   CPU
[ 12, 5 ]   INPUTS
[ 16, 2 ]   IC
```

REFERENCES

A. Books

Billings, K., and Moursund, D. *Are You Computer Literate?* (Portland, Ore.: Dilithium Press, 1979).

Coan, J. S. *Basic BASIC* (Rochelle Park, N.J.: Hayden, 1970).

Gately, W. V., and Bitter, G. G. *BASIC for Beginners* (New York: McGraw-Hill, 1970).

Horn, C. E., and Collins, C. L. *Computer Literacy* (Austin, Tex.: Sterling Swift, 1981).

Moody, R. *The First Book of Microcomputers* (Rochelle Park, N.J.: Hayden, 1978).

North Texas State University, Department of Computer Sciences. *Apple II and PET/CBM User's Guide* (Denton, Tex.: Computer Sciences Microlaboratory, 1980).

Pavlovich, J. P., and Tahan, T. E. *Computer Programming in BASIC* (San Francisco: Holden-Day, 1971).

Peckham, H. D. *Programming Basic with the TI Home Computer* (New York: McGraw-Hill, 1979).

Poirot, J. L. *Microcomputer Systems and Apple BASIC* (Austin, Tex.: Sterling Swift, 1980).

Poirot, J. L., and Retzlaff, D. A. *Microcomputer Workbook (Apple II Edition)* (Manchaca, Tex.: Sterling Swift, 1979).

Smith, R. E. *Discovering BASIC: A Problem Solving Approach* (Rochelle Park, N.J.: Hayden, 1970).

B. Periodicals

A.E.D.S. Monitor. Association for Educational Data Systems, 1201 16th St. N.W., Washington, D.C. $12 for nonmembers.

APPLE: The Personal Computer Magazine & Catalog. Quarterly. Apple Computer, Inc., 10260 Bandley Dr., Cupertino, CA 95014. Free to Apple owners.

BYTE: The Small Systems Journal. Monthly. P.O. Box 590, Martinsville, NJ 08836 (McGraw-Hill). $18/yr, $32/2 yrs.

The Computing Teacher. 7 times/yr. International Council for Computers in Education (ICCE). Computing Center, Eastern Oregon State College, LaGrande, OR 97850. $10/yr.

CourseWare: Microcomputer Programs for Learners. 5 times/yr. 4919 N. Millbrook, #222, Fresno, CA 93726. $50/yr.

Creative Computing. Monthly. P.O. Box 789-M, Morristown, NJ 07960. $15/yr.

Educational Computer. Bimonthly. P.O. Box 535, Cupertino, CA 95015. $12/yr.

Hobby Computer Handbook. Annual. Davis Publications, 380 Lexington Ave., New York, NY 10017. $1.95.

INFOWORLD: The Newspaper for the Microcomputing Community. Biweekly. 530 Lytton Ave., Palo Alto, CA 94301. $22/yr.

Interface Age: Computing for Business and Home. Monthly. McPheters, Wolfe & Jones, 16704 Marquardt Ave., Cerritos, CA 90701. $18/yr.

Kiloband: Microcomputing. Monthly. Wayne Green, Inc., 80 Pine St., Peterborough, NH 03458. $25/yr.

Library Computer Equipment Review. Semiannual. Microform Reviews, Inc., Box 405, Saugatuck Station, Westport, CT 06880. $150/yr.

MICRO: The 6502 Journal. Monthly. MICRO INK, Inc., P.O. Box 6502, Chelmsford, MA 01824. $15/yr.

MICROCOMPUTER INDEX. Quarterly (includes abstracts). Microcomputer Information Service, 3070 Adams Way, Santa Clara, CA 95051. $22/yr.

Microsoft News. Northwest Regional Educational Laboratory, 300 S.W. Sixth Ave., Portland, OR 97204.

ON Computing: Guide to Personal Computing. Quarterly. 70 Main St., Peterborough, NH 03458 (McGraw-Hill). $10/yr.

Personal Computing. Monthly. Hayden Publishing Co., 50 Essex St., Rochelle Park, NJ 07662. $18/yr., $33/2 yrs.

Softside. Monthly. 6 South St., Milford, NH 03055. $15/yr.

T.H.E. Journal: Technological Horizons in Education. 6 times/yr. Information Synergy, Inc., P.O. Box 922, Acton, MA 01720. $15/yr.

SOURCES FOR COMPUTER MATERIALS

A. Computer Software

Apple II Personal Computer Software Directory (Vol. 1—Business, Vol. 2—Games). WIDL Video, 5424 West Diversey, Chicago, IL 60639.

The Book of Apple Computer Software (1981). The Book Company, 14013 Old Harbor Lane, Suite 312, Maria Del Rey, CA 90291.

Commodore Software Encyclopedia. Software Dept., Commodore Business Machines, Inc., 300 Valley Forge Square, 681 Moore Road, King of Prussia, PA 19406.

Educational Software Directory: Apple II Edition. Sterling Swift Publishing, 1600 Fortview Road, Austin, TX 78704.

1980 Catalog of Secondary School PLATO Courses. Control Data Corp., HQA02Q—Box O, Minneapolis, MN 55440.

B. Computer Hardware

Apple

Apple Computer Company
10260 Bandley Drive
Cupertino, CA 95014

Atari

Atari
1265 Bonegas Avenue
Sunnyvale, CA 94086

PET

Commodore Business Machines, Inc.
Computer Systems Division
950 Rittenhouse Road
Norristown, PA 19403

TRS-80

Radio Shack
1600 Tandy Center
Fort Worth, TX 76102

TI-99/4

Texas Instruments, Inc.
Box 225012, Station 84
Dallas, TX 75265

GOAL

☐ Understand how to produce handouts.

OBJECTIVES

☐ Operate a spirit duplicator.
☐ Produce a direct master and copies in two colors.
☐ Produce an indirect master and copies.
☐ List and describe types of handout production techniques.

If you had been a teacher in the early days of this century, you would be unfamiliar with nearly all of the media described in this text. They had not been invented yet. There were no slide presentations, no audiotapes, no television, and no convenient method for duplicating worksheets in small quantities. It may be a little awkward to get along without computers and tape recorders, but the need to hand out multiple copies of locally produced printed material can hardly be avoided.

In the early days the production of these "handouts" was a messy business, involving a gelatin-and-ink arrangement. You cooked the gelatin and poured it into a cake pan to harden. A special master copy was prepared with an indelible pencil, then laid on the gelatin, writing side down. When the ink soaked in, the master was removed, leaving ink traces. Individual sheets of paper were placed on the gelatin, one at a time, to pick up the ink.

Later, a process was invented that made handout production easier. The copies were more distinct and you could get more of them. A company—The Ditto Corporation—went into business to produce the materials and machines. While we sometimes call these worksheets "dittos," a more appropriate term is "spirit-duplicated" since methyl alcohol ("spirits") is used. In this chapter we will discuss the design and production of handouts and worksheets using the spirit duplicator, and mention other copy processes such as the offset press and the mimeograph process.

6

HANDOUTS AND WORKSHEETS

PREPARING HANDOUTS FROM COMMERCIAL SPIRIT MASTERS

Because spirit duplication is a widely available production process, there are a number of companies that produce completed spirit masters for teachers to use. Both direct and indirect masters, presenting concepts in all subject areas, can be purchased. *Direct spirit masters* are ones on which you write, type, or draw *directly* on the working surface to transfer aniline dye *directly* to the back of the working surface to create a master. Direct spirit masters are pressure-sensitive. The pressure of a marking instrument transfers the dye to the master sheet.

Commercially prepared masters are popular because many complex concepts are difficult for a teacher to present effectively on a handout. For example, almost every elementary teacher uses outline maps of states and countries in social studies units. Preparing these maps from scratch is fairly difficult. However, it is possible to purchase spirit masters, prepared by professional artists, that can be used to make copies of outline maps. Reproduction is easy—all the teacher has to do is run the masters on the duplicator to produce copies.

Producing Copies from Direct Spirit Masters

Direct spirit masters have three parts—the working surface, the aniline dye-covered sheet, and the tissue paper between the two. Commercially prepared direct spirit masters have the dye transferred from the dye sheet to the back of the working surface when they are purchased. To make copies, merely separate the working surface (the master) from the dye sheet, remove the tissue sheet, and run copies using a spirit duplicator. A commercially prepared direct spirit master can be used to produce up to 300 copies. When your commercial master is used up, do not throw it away; use it as the basis for tracing a new master.

In general, the process of spirit duplication is quite simple. Colored aniline dye is dissolved from the back of the master using copy paper that is moistened by an alcohol-based liquid as it passes into the duplicator. After moistening, the copy paper is squeezed against the dye side of the master and a small amount of the dye is dissolved onto the paper.

Uses of Handouts

- ☐ Worksheets
- ☐ Study guides
- ☐ Announcements
- ☐ Quizzes and exams
- ☐ Note-taking guides
- ☐ Workbooks
- ☐ Learning packages
- ☐ School newspapers
- ☐ Tickets to school events

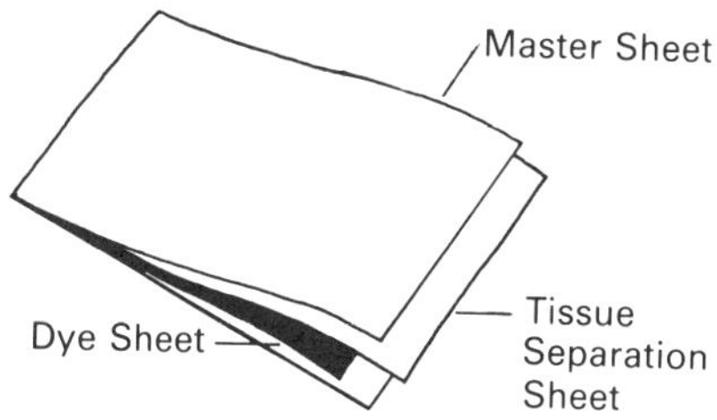

A direct spirit master.

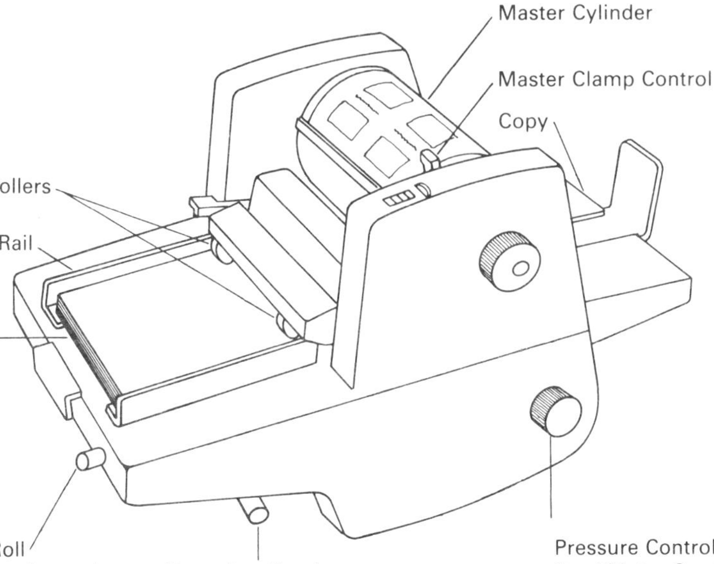

Spirit duplicator.

OPERATE

Operation of a Spirit Duplicator

You need: a prepared spirit master, a duplicator with duplicator fluid, copy paper

Spirit duplicators can be motor-driven . . .

or operated using a hand crank (courtesy of A. B. Dick Co.).

Procedure

1. Prepare the machine for duplicating.
   a. Set wick pressure.
   b. Insert the fluid bottle in its slot in the machine with the wick down.
   c. Set impression-roller pressure to the middle setting.
2. Load copy paper.
   a. Depress paper-loading lever and insert paper until it is against the silver retaining bars in the machine; then release the paper-loading lever.
   b. Position paper guides against the sides of the copy paper.
   c. Set feed rollers against paper, approximately ¾ inch inside the edge of the paper.
   d. Set feed-roller pressure for the type of copy paper being used. Lighter-weight paper requires less pressure, heavier card stock requires more pressure.
3. Attach master to the cylinder—dye side up.
   a. Open master clamp.
   b. Position master, centered on cylinder.
   c. Close master clamp to hold the spirit master in place.
4. Duplicate copies.
   a. Be sure that the impression-roller pressure is on.
   b. Turn on motor, if your machine is motor-driven.
   c. Activate feed lever and produce copies.

Producing Copies from Indirect Spirit Masters

Preprinted originals can be used to make copies with a spirit duplicator and a thermal copier by using a special type of spirit master called a thermal, or indirect, spirit master. *Indirect spirit masters* are made by writing, typing, or drawing to produce a faxable image on a sheet of paper. This is called an original. The original is used with a thermal copier and a thermally sensitive spirit master to transfer dye to the master paper *indirectly.* Indirect spirit masters are heat-sensitive rather than pressure-sensitive.

This process works because the thermal spirit master contains a waxed, temperature-sensitive dye that is transferred to the back of the working surface when the correct amount of heat is absorbed. Direct spirit masters are pressure-sensitive. Indirect spirit masters are heat-sensitive. Thermally prepared masters are run on a spirit duplicator in the same way as direct masters.

The type of preprinted original used to prepare an indirect spirit master determines if the process will work. The printing on the original must be heat-absorbing. In other words, the printing must be *faxable,* or capable of being used to create a facsimile. The original—the item you wish to copy—must contain lines or drawings made with a heat-absorbing substance, such as pencil lead, India ink, silver salts found in photographs, or black lines produced by Xerox or other copy machines. Writing with carbon works best.

Sources of Suitable Originals

- ☐ Newspapers
- ☐ Magazines
- ☐ Commercial originals
- ☐ Pencil sketches
- ☐ India ink drawings
- ☐ Workbook pages
- ☐ Xerox copies
- ☐ Transparency originals

Do Not Use:

- ☐ Spirit-duplication copies
- ☐ Ball point pen
- ☐ Felt pen
- ☐ Fountain pen
- ☐ Grease pencil

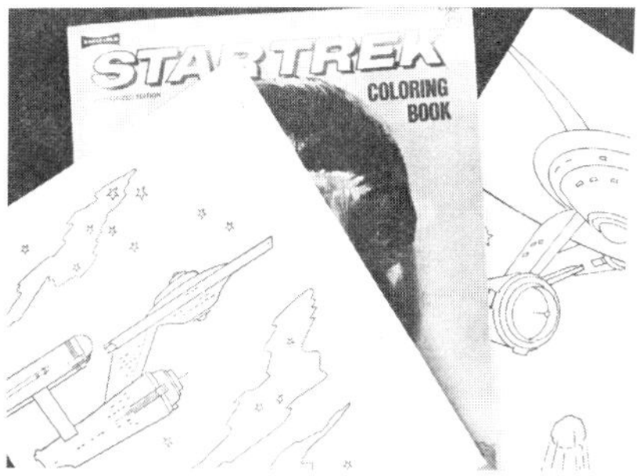

Children's coloring books make excellent originals from which to produce indirect spirit masters.

PRODUCE

Preparing an Indirect Spirit Master

You need: a preprinted original, a thermal spirit master, a plastic "carrier," a thermal copier (see Chapter 8 for copier operating instructions)

Procedure

1. Remove the tissue separation sheet from your indirect spirit master package.
2. Place your original face-up and put the dye side of the indirect spirit master on top of it.
3. Put the original and indirect spirit sheets into the plastic carrier with the thermal master on top of the original.
4. Set the thermal copier to the proper heat setting.
5. Run the carrier and thermal master through the machine.
6. Separate the carbon sheet from the master sheet and run copies on the spirit duplicator.

Note: Remember to save your original so you can reuse it, because you will be able to produce only about seventy-five copies from a single indirect master.

An indirect spirit master.

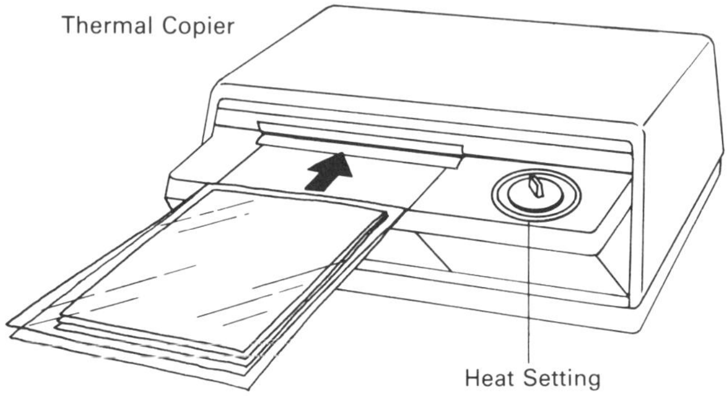

Insert sandwich inside carrier into the thermal copier.

Before running the sandwich of the original and master through the thermal copier, place them in a protective plastic carrier.

Hectographic The correct, generic name for the spirit-duplication process

TRY THIS

Cut an editorial cartoon from a newspaper and use it to create a thermal spirit master. Run ten copies using a spirit duplicator.

PREPARING HANDOUTS FROM MASTERS YOU CREATE YOURSELF

Direct spirit masters can be purchased with dye sheets in several colors including purple, red, green, or black. Purple is the most popular because it seems to produce the clearest and greatest number of copies. Direct spirit masters are similar in appearance to indirect spirit masters, except that they are heavier and more substantial. There is an aniline dye sheet attached to the working surface (master sheet), separated by tissue paper. The master is produced by writing, drawing, or typing on the top of the working surface so that dye is transferred from the dye sheet to the back of the working surface. The color of the dye determines the color of copies. The working surface is separated from the dye sheet and attached to the spirit duplicator to produce copies. A good direct master can produce up to 300 copies.

PRODUCE

Preparing a Direct Spirit Master in Two Colors

You need: an idea for a handout, a marking device (such as a typewriter, medium-point ball point pen, or pencil), two direct spirit masters in different colors, a flat, hard surface to write on

Procedure

1. On a sheet of paper, sketch a rough layout of the desired handout.
2. Decide what portions of your layout you want in each color.
3. Remove the tissue separation sheet and place the master on a flat, hard surface or in a typewriter with the white, working surface face-up.
4. Write, type, or draw on the working surface. Press smoothly and with even pressure so that the dye is removed from the dye sheet to the back of the working surface.
5. When you have finished with the first color, remove the working-surface sheet from the dye sheet and place it over the top of the second colored dye sheet.
6. Complete your master in the second color.
7. Separate the original from the second dye sheet and produce copies in two colors on the spirit duplicator.

Correcting Mistakes

Spirit duplication works because dye comes in contact with moistened copy paper. The alcohol on the paper dissolves small amounts of the dye from the original. To correct mistakes on a master the dye representing the error must be kept from coming in contact with the moistened paper. This can be done by scraping the dye from the back of the master with a razor blade or knife, taping over the mistake with cellophane tape, or erasing small errors with an eraser. A radical technique for correcting mistakes is to cut them off the master with scissors. Remember that if you correct a mistake, and then want to rewrite or redraw over that portion of your master, you should tear off an unused corner of the dye backing and place it face-up under the error. Then when you make your correction there will be unused dye behind it.

Combining Direct and Indirect Masters

Often you may want to add your own comments to a preprinted original, or to include colored labels on a chart or map you have copied using the thermal process. To do so, you must combine masters produced with both direct and indirect production techniques.

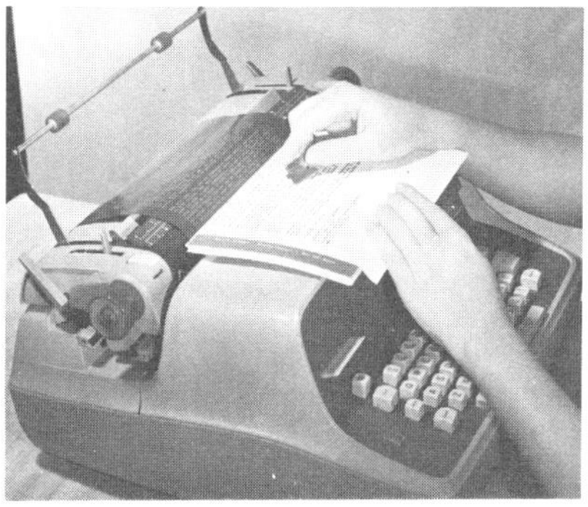

To correct a mistake, scrape the dye from the back of the original using a razor blade or knife. Remember to place a piece of unused dye behind the mistake before correcting it (compliments A. B. Dick Co.).

Clip-Art books are excellent sources for direct and indirect masters. They can be thermally copied or traced.

Tape the cutout of the information produced using an indirect master to the back of the master made using direct production. Remember to tape it dye side up.

PRODUCE

Preparing a Spirit Master Using Direct and Indirect Production

You need: a preprinted original, a thermal spirit master, a direct spirit master, a marking device, a flat, hard surface to write on, a thermal copier, a spirit duplicator, cellophane tape

Procedure

1. Select your preprinted original and make a thermal spirit master using the thermal copier.
2. If your thermal master is small enough (less than 5 × 8 inches), cut it from the unused master and tape it in the desired spot to the back of the working surface of your direct spirit master, dye side out. Be sure to place it where you want it to appear on the copies.
3. Remove the tissue sheet from your direct master and carefully add whatever information you wish to the working surface of your direct spirit master using a pen, pencil, or typewriter. Separate the direct master from its dye sheet and run copies using the spirit duplicator.
4. If your original was fairly large and there was insufficient room to add information around the edges of the direct master, you will need to add your comments directly on the thin thermal master sheet. Separate it from the thermal dye sheet and place it over the direct dye sheet. Carefully mark or draw on top of the thermal master sheet. Fine-tipped ball point pens or sharpened pencils may not work well because they tend to tear the thin thermal master sheet. When you are finished adding your comments, place the spirit master dye side up on the drum of the spirit duplicator and make copies.

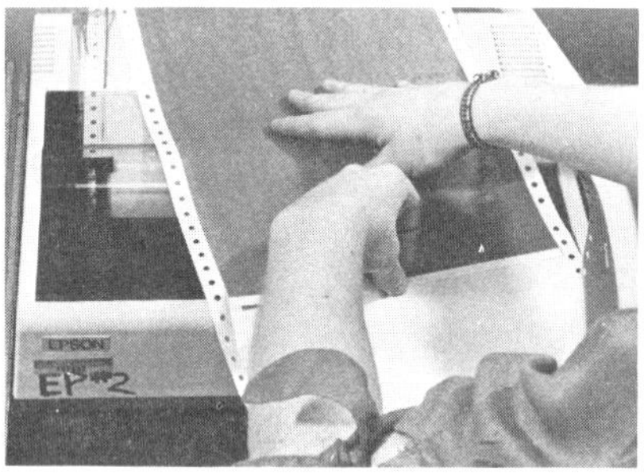

A microcomputer printout can be used to make handouts and worksheets. Just tape a direct master to the printer's copy paper before threading it through the machine.

TRY THIS

Find an editorial cartoon in the newspaper and cut it out. Use it to make a thermal master. Add your own comments using a direct mastering process to explain the purpose of the cartoon. Run copies.

Using a Microcomputer to Create Handouts

Many utility programs and graphics programs are available for microcomputers. These programs allow you to produce crossword puzzles, work jumbles, charts, graphs, and maps with ease. If a printer is attached to the computer, it is possible to make spirit masters. Spirit-duplication master paper can be purchased for many microcomputer printers, and it is easy to use thermal spirit masters to create handouts from computer-generated printouts since most printers type using faxable carbon ribbons. If you have an "impact" printer (one that works like a typewriter) you can tape a direct spirit master to the paper already in your printer and create a direct spirit master. Just position it under the typing element of the printer. (Remember to remove the tissue sheet.)

MIMEOGRAPH AND OFFSET COPYING

One disadvantage of spirit duplication is that it is possible to produce only a few hundred copies from each master. Two processes for producing larger quantities of copies are the rotary stencil process (mimeograph) and the planograph process (offset).

Mimeograph printing presses have been used in schools for decades, but many educational institutions are replacing their rotary stencil machines with offset presses because this fairly recent printing process is more versatile and produces higher-quality copies. Usually, neither printing process is used by the classroom teacher. Rather, they are services provided by building or regional media centers. Teachers are responsible for producing masters, however.

Offset presses are used for large-quantity, high-quality copy needs.

Rotary Stencils

Rotary stencils are wax-impregnated sheets that do not allow ink to penetrate them unless the coating has been broken by a stylus or typewriter type. The penetrated area leaves an opening that ink can pass through. Thermal rotary stencils are also available.

The rotary stencil duplicator or press is somewhat similar in appearance to the spirit duplicator. It consists of a cylinder that has openings for ink to pass through. Inside the cylinder there is a container for ink. An impression roller is located below the cylinder. As the cylinder rotates, ink is forced through the openings in the cylinder onto an ink pad and then through the openings in the stencil onto the copy paper. A special porous copy paper is needed to absorb the ink without smearing.

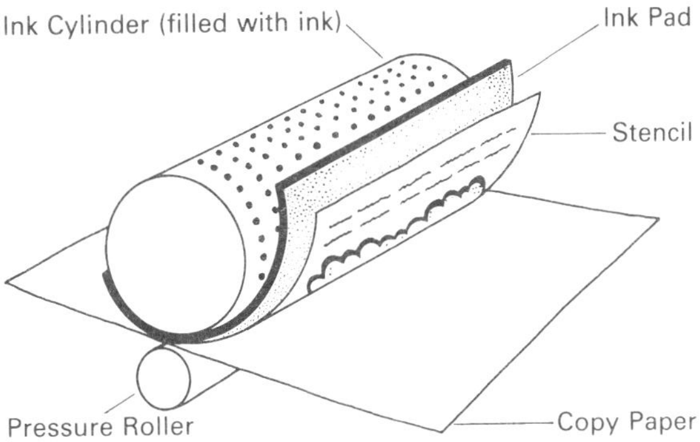

A rotary stencil press.

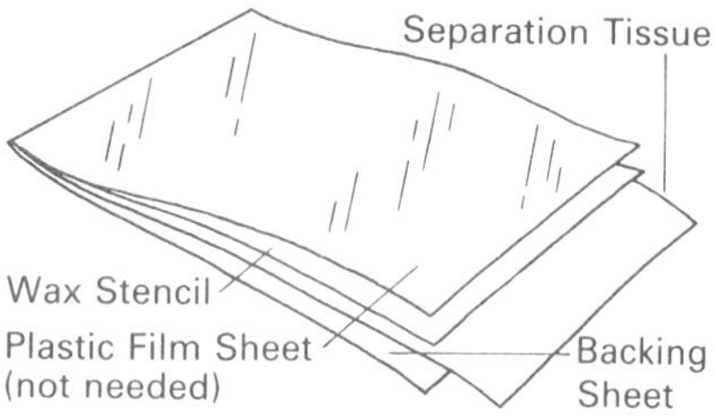

A rotary stencil.

PRODUCE

Preparing a Rotary Stencil Master

You need: a stencil, a typewriter or stylus, correction fluid, a flat surface

Procedure

1. Shift the typewriter to the stencil position so that the ribbon does not move in front of the keys when typing. Clean the type keys if they are clogged.
2. Push the rubber rollers on the typewriter to the outside edge of the stencil so that they do not cut into it.
3. Remove the separation tissue from behind the wax stencil and insert a cushion sheet (or typing plate) between the stencil and the backing paper. Insert the entire sandwich into the typewriter. The glossy side of the cushion sheet should be next to the wax stencil.
4. Type, using a uniform touch. Use an electric typewriter, if possible.
5. Proofread your copy while it is in the typewriter.
6. Make corrections by rubbing gently in a circular motion with the rounded end of a paper clip on the wax stencil. Then apply a thin covering of correction fluid to the error. Use vertical brush strokes. Allow the correction fluid one to three minutes to dry, then retype using a slightly lighter touch than normal.
7. You can write or draw on a stencil by placing it on a flat surface and using a medium-point ball point pen or stylus.

Offset Printing

Planographic printing is one of the most widely used processes for large-quantity copy runs. This technique is not one a classroom instructor would use. Rather, originals would be sent to the central school or office copy center for duplication.

Offset originals should be neatly printed, typed, or drawn as you wish them to appear as copies. Since this process is usually more expensive than spirit duplication, it is usually reserved for printing jobs where high quality is important or when large quantities are needed.

A FINAL WORD

In this chapter we have presented techniques for producing handouts and worksheets. The handout can be used alone as a study guide for the student, or as a supplement to instruction with other forms of media. The simplest and least expensive technique for producing handouts is spirit duplication. Two types of spirit masters are available. Direct masters are so named because you write, type, or draw directly on the master's working surface. Dye is transferred to the back of the working surface from the dye sheet behind it. Direct masters are activated by pressure.

Indirect spirit masters are temperature-sensitive. Dye is transferred from the dye sheet to the back of the master when heat is absorbed from a preprinted original. Both types of masters are used on the spirit duplicator in the same way. Spirit copies are produced because paper moistened by alcohol dissolves the dye from the back of the master when the two are squeezed together as the machine runs. Large-quantity copies are usually produced using mimeograph or offset presses.

LOOKING AHEAD

The display is another traditional and widely used medium. Chapter 7 will present techniques for producing displays such as bulletin boards, hook and loop presentations, flannel displays, and flip charts.

REVIEW QUESTIONS

1. Explain the difference between original, working surface, and master.
2. How many copies can you run with a direct spirit master? With an indirect spirit master?
3. Explain how copies are produced by a spirit duplicator.
4. What are the three parts of an unused spirit master?
5. What is meant by the term *faxable?*
6. What color are the copies produced from an indirect spirit master?
7. What duplication process should be used when several hundreds of copies are needed?

ANSWERS TO REVIEW QUESTIONS

1. The *original* is a preprinted item that a teacher wishes to duplicate. An editorial cartoon from a newspaper is an example of an original. The *working surface* of a spirit master is the top of the white sheet that is attached to the aniline dye sheet. The working surface is the area on which the teacher writes, types, or draws to create a master. The back of the working surface receives the aniline dye removed from the dye sheet as the master is created. The *master* is what is used to create copies on the spirit duplicator. It is attached to the revolving cylinder, dye side up.
2. Direct spirit masters can produce up to 300 copies. Indirect masters are normally good for only 75 copies.
3. Spirit copies are produced as paper moistened by an alcohol-based fluid dissolves small amounts of the aniline dye from the back of the spirit master. The spirit duplicator squeezes the moistened copy paper against the master as it runs.
4. The working surface, the aniline dye sheet, and the tissue paper that separates the two make up a spirit duplication master.
5. Printing that is faxable can be used to make copies by a thermal copier. Faxable printing contains some heat-absorbing component such as carbon, silver salts, or graphite.
6. Indirect spirit masters are purple.
7. Either offset printing or mimeograph printing can be used to produce large quantities of copies.

REFERENCES

A. Information about the Duplication Process

Atkinson, S., and Locatis, C. *Media and Technology for Education and Training* (Columbus, Ohio: Charles E. Merrill, 1984).

Brown, James W.; Lewis, Richard B.; and Harcleroad, Fred F. *AV Instruction: Technology, Media and Methods,* 5th ed. (New York: McGraw-Hill, 1977). (With lab instructor's manuals.)

Bullard, John, and Mether, Calvin. *Audiovisual Fundamentals,* 2d ed. (Dubuque, Iowa: W. C. Brown, 1979).

Bullough, R. V. *Creating Instructional Materials,* 2d ed. (Columbus, Ohio: Charles E. Merrill, 1978).

Kemp, J. E. *Planning and Producing Audiovisual Materials,* 4th ed. (New York: Harper and Row, 1980).

Pett, D. W. *Copying and Duplicating Processes* (Bloomington: A-V Center, Indiana University, 1973).

Volker, R. P., and Simonson, M. R. *Media for Teachers,* 3d ed. (Dubuque, Iowa: Kendall/Hunt, 1981).

Wittich, W. A., and Schuller, C. F. *Instructional Technology: Its Nature and Use,* 5th ed. (New York: Harper and Row, 1973).

B. Sources of Duplication Materials and Equipment

A. B. Dick Company, 5700 West Touhy Avenue, Chicago, IL 60648

3M Company, Visual Products Division, 3M Center, St. Paul, MN 55101

Tersch Products, Inc., Industrial Boulevard, Rogers, MN 55374

GOALS

☐ Consider uses for passive displays in a classroom.
☐ Develop instructional strategies that involve sequential development, using large surfaces.

OBJECTIVES

☐ Design a variety of bulletin boards.
☐ Create 3-dimensional displays.
☐ Use the dry-mount press.
☐ Use the opaque projector.
☐ Outline some instructional sequences that can be delivered with the use of flannelboard or hook and loop board.

Because of the development of technology for teaching, many teachers have become accustomed to using various electrical devices to bring the sights and sounds of a subject into the classroom. Projectors make it possible to show small pictures to large groups. Tape recorders re-create events by allowing the original sounds to be used. Realism, through sights and sounds, can be portrayed in a classroom with film or videotape. These modern media tempt educators to forget about some of the more traditional forms of teaching material and resources, such as chalkboards, flannelboards, and bulletin boards. In this chapter we will examine methods for creating and using displays and display surfaces.

Certain types of display surfaces can be used in combination with certain types of instruction. "Instruction" is used here in the broad sense, to mean "the transmission or delivery of information." Teaching—the process of delivering information in person or through interaction with other people or technology (programmed instruction and computer-assisted instruction are examples)—can be accomplished through the use of display surfaces that a teacher uses in front of a class. Additionally, nonlive, noninteractive instruction can take place through the use of display surfaces.

These two categories of use are explained in this chapter. We have chosen to call the noninteractive use of display surfaces *passive instruction,* and the interactive use of a display surface, when a teacher is involved, is called *active instruction.* Thus, the display surfaces used in these types of instruction are *passive* or *active.* While the use of these two categories is somewhat limiting, they are helpful as ways of organizing your instructional methods.

7

DISPLAYS

PASSIVE SURFACES

Examples of passive instruction, or information delivery, are nearly everywhere, in the form of street signs, billboards, bulletin boards, and signposts. The use of a bulletin board in a classroom is quite similar in principle to the use of a large highway billboard. In both cases some information is being displayed for instructional purposes. Whether the purpose is to convince the reader to purchase something, take action on an issue, or accumulate ideas for a test or quiz, some type of surface is used to display information.

Apparently, the use of large surfaces that convey information is effective, if you consider the amount of money that is spent on outdoor advertising. Principles that make this form of delivery effective apply as well to bulletin boards, exhibits, and displays in a school, museum, or other instructional environment.

The road signs we see every day are really displays. They give us directions and basic information.

Characteristics of Large Display Surfaces

- ☐ Attention-getting
- ☐ Bold graphics
- ☐ Extreme close-ups
- ☐ Low cost per "viewing" or "viewer"
- ☐ Bright colors
- ☐ Familiar settings, people actively involved

Bulletin boards can be used to display notices and announcements, . . .

they can be used to record information during a meeting to facilitate discussion, . . .

or they can be used to support instruction.

Some Uses of Informal Posting of Information

- ☐ Current events
- ☐ Information on holidays
- ☐ Student scores on tests
- ☐ Books I've read
- ☐ Famous quotations
- ☐ Newspaper and magazine pictures

Considering Needs

There are differences between the design and construction of a bulletin board or display that teaches a specific idea, and the use of large, flat surfaces for the mere recording of information. In either case, the surface may be passive, but the amount of planning and the degree of detail might be different. In order to establish a sequence in developing the use of display surfaces, we will begin with the simplest case—where a surface may be used merely to record information. Then we will consider the design of a formal bulletin board, finishing with the design and use of a 3-dimensional display. The elements of this section of the chapter include:

Informal Use

☐ Posting of prepared information, clippings, photos, and notices.

☐ Recording information generated during a meeting or discussion.

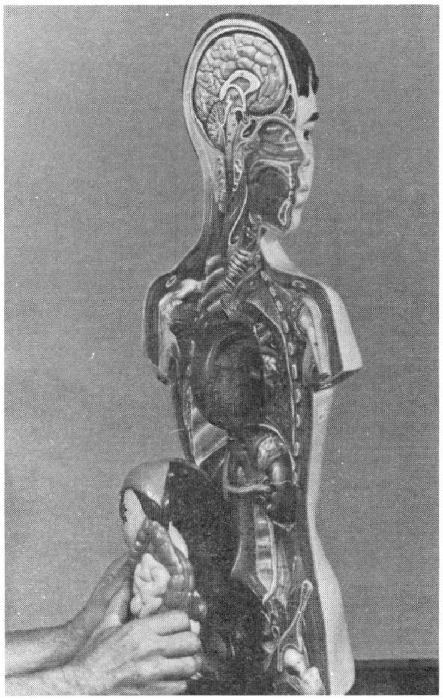

Three-dimensional displays provide an additional element of realism.

Formal Use

☐ Designing and producing a complex 2-dimensional display involving titles, captions, and information to meet specific instructional objectives.

Informal displays can be put up and used without a great deal of prior design. If the purpose of the display is merely to post information for others to see, nearly any type of arrangement is satisfactory. The surface merely acts as an area for holding information, making it available to the intended audience.

These two applications may be considered when informal displays are called for:

Informal Posting of Prepared Information When it is necessary to provide information in the form of clippings, photos, drawings, or notes intended for casual observation, a simple bulletin board may be used. A title or caption is helpful, as well as some indication of the significance of the display and its relationship to other material, content, or activities. There may be no need for a formal display that follows accepted rules of layout and design. The purpose of a display of this type is merely to provide a location and arrangement of information.

TRY THIS

Study the informal displays on this page and comment on their effectiveness, using these criteria:

☐ *Theme* Is the central purpose of the bulletin board clear?
☐ *Arrangement* Are the primary materials in a prominent position, surrounded by supporting material?
☐ *Captions* Are appropriate titles used to direct attention to important elements?
☐ *Unity* Are the clippings all related to each other and to the theme?

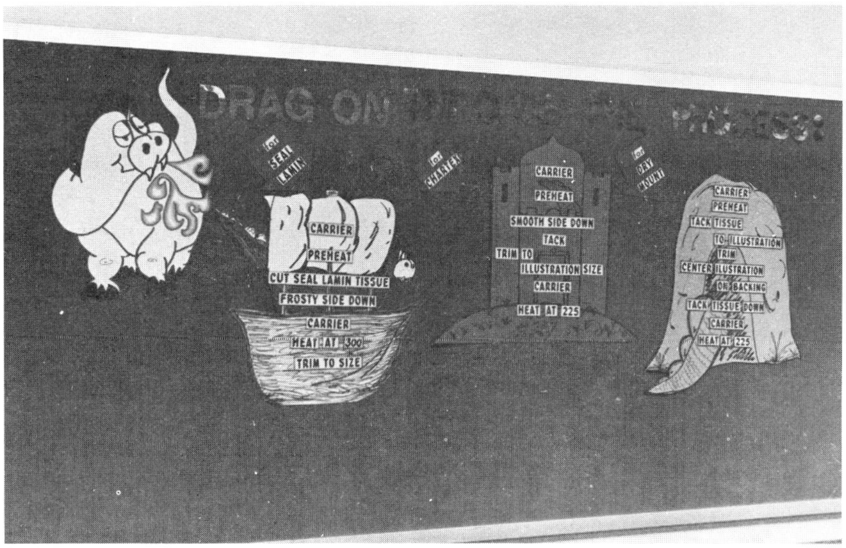

Evaluate these bulletin board displays. Some are more effective than others. Why?

The chalkboard is an easy-to-use, often available, way to list discussion points during a class or meeting.

Typical teaching format, using small-group discussion

- ☐ Assign topics to groups.
- ☐ Ask groups to discuss.
- ☐ Call for summaries.
- ☐ Reward groups with a coffee break but suggest they browse the lists *during* the break.

There are a number of bulletin board idea books available for purchase.

Bulletin board idea books give verbal ideas for displays in various subjects and grade levels.

Informal Summaries of Meetings and Discussions When ideas are being discussed, questions are being asked, and conclusions are being drawn, it is often necessary to summarize and list major elements of the discussion. The simple act of writing an idea on paper tends to help others clarify points related to the idea. Documenting a discussion as points are agreed upon serves to move the meeting along, save time, and give participants a chance to see the points on which they have agreed.

Suppose you are in charge of a small-group discussion that has been included as part of a teaching strategy. The group has discussed a variety of alternatives, and it is now time to bring them to the attention of others in the room. Other groups have done the same. It is necessary for each of the discussion groups to report its suggestions. While it is possible to ask for verbal reports, there are advantages in having each group list its ideas on paper. The process makes it possible for each group to read the others' lists, and it forces each group to provide a succinct and relevant list of the ideas discussed. The lists can be shared by reading from them or they can be posted on the wall in the meeting room so that others can browse and examine all of the information.

TRY THIS

List two or three topics in your subject area that could be discussed and summarized by means of an informal display.

Informal displays fill important needs in circumstances such as those described above. But there may be occasions in which more attractive, attention-getting displays are needed. Two types of such displays are described here:

Formal 2-dimensional Displays There is a difference between *displaying information* and *merchandising ideas.* An effective campaign to sell concepts in your subject area will include the use of labels, captions, appropriate colors and shapes, and the design of the layout (formal or informal balance). All contribute to the effectiveness of the display. If a 2-dimensional display, such as a bulletin board, an announcement, a panel of information, or a visual information center, is to be meaningfully presented, certain techniques should be used during production. They include:

- ☐ Designing the layout
- ☐ Making titles and captions
- ☐ Preparing the materials for display
- ☐ Putting up the display

In the next portion of this chapter we will describe each of these techniques and provide instructions for using equipment to achieve an appealing and effective display.

Designing the Layout

Effective displays don't just happen! They are carefully designed, using the following approaches.

Study Books Containing Bulletin Board Ideas A variety of commercially available idea books will provide suggestions for handling certain subjects and using color, styles of titles, and captions. You can incorporate ideas from several example bulletin boards as you design your own.

Examine Texts in Your Subject Area Charts, diagrams, lists, and other visual materials that may be found in your students' textbooks may suggest ways of creating bulletin boards of special interest. Develop uses that assist students in bringing together a series of related concepts.

Specify Relationships between Behavioral Objectives and the Display
Even though a bulletin board may be a less vital aspect of your instruction than other forms of media, it is important to consider the objectives that relate to the display. Otherwise the bulletin board may be viewed by students as no more than a pretty decoration.

Incorporate Elements of Design
Keep in mind the elements of design that were described and illustrated in Chapter 3. The elements are:

- ☐ Line
- ☐ Shape
- ☐ Space
- ☐ Texture
- ☐ Value
- ☐ Shading

Determine Size and Proportion
Create the layout to fit a surface of a specific size by designing a miniature version of the display, using the same proportions as the full-size board. (If you have a 4 × 8 foot bulletin board, work on a 4 × 8 inch layout.) You can estimate the size and placement of materials more effectively on the smaller layout. Then use an opaque projector to project your layout on the actual bulletin board. This technique not only provides a preview of the finished product, it assists in placing of materials by indicating final spacing of letters and other elements.

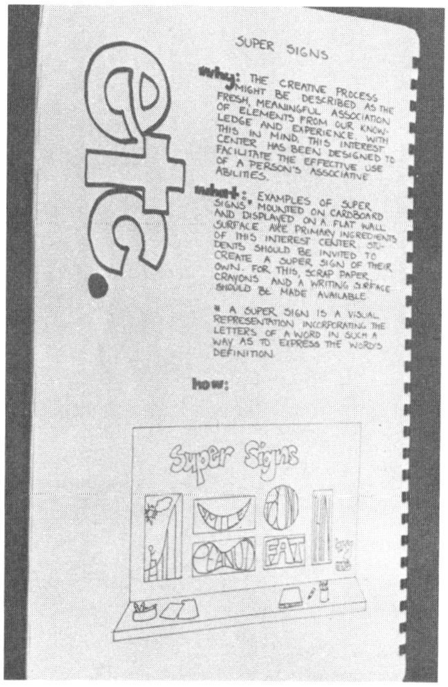

An effective display should be based on an objective, i.e., it should teach the student something.

OPERATE

The Opaque Projector

1. Examine the particular model of projector that is available in order to determine how materials should be inserted; some projectors are loaded from the top, while others require that materials be inserted from the side or rear of the machine.
2. Move projector nearer to the screen to create an image that is smaller and brighter; moving the projector away from the screen will make the image larger and dimmer.

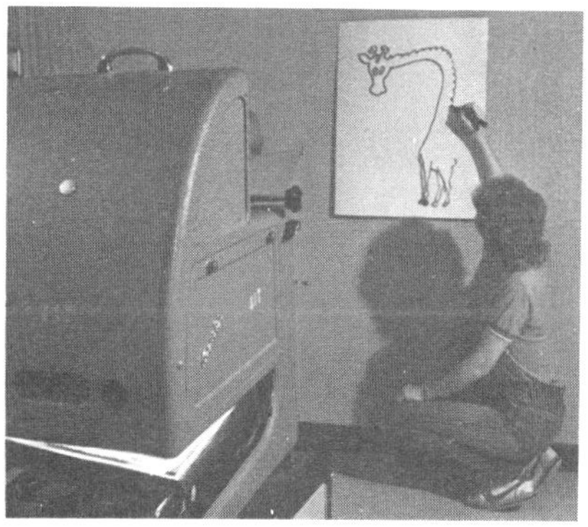

The opaque projector can be used to enlarge small objects so they can be traced onto paper and used as part of a display.

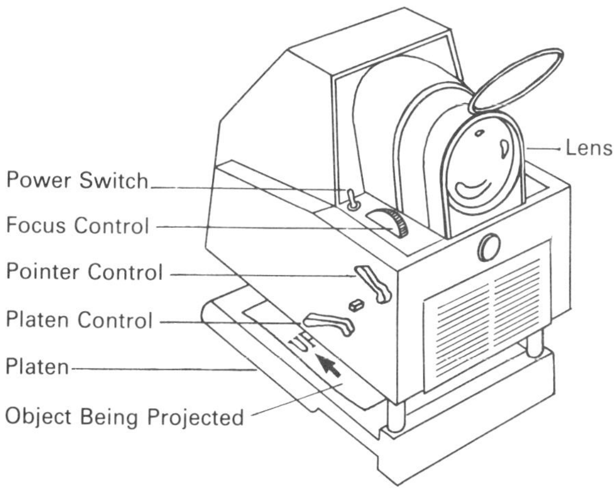

Opaque projector.

An extension on the front of the opaque projector enables the objective lens to slide out farther than usual from the projector, so that smaller images can be projected.

TRY THIS

Design a bulletin board that is either square or rectangular. Create your design on an ordinary sheet of 8½ × 11 paper. Include titles, captions, estimated size and shape of items to go on the board, and the layout of the items. Then enlarge your design to actual size using the opaque projector. Show the projected image on an empty bulletin board if possible, or—if not—on a large empty space of wall.

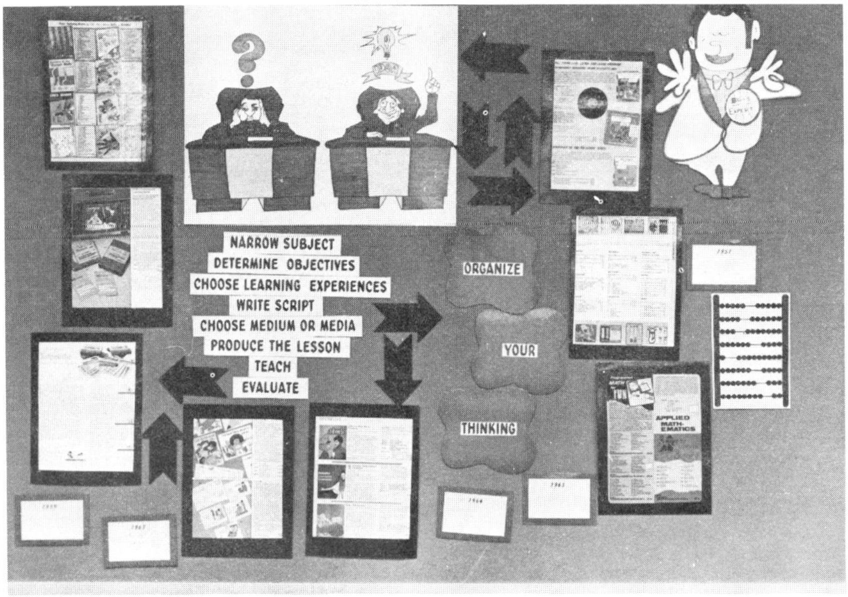

A cluttered, poorly organized bulletin board may not be very effective at presenting a topic.

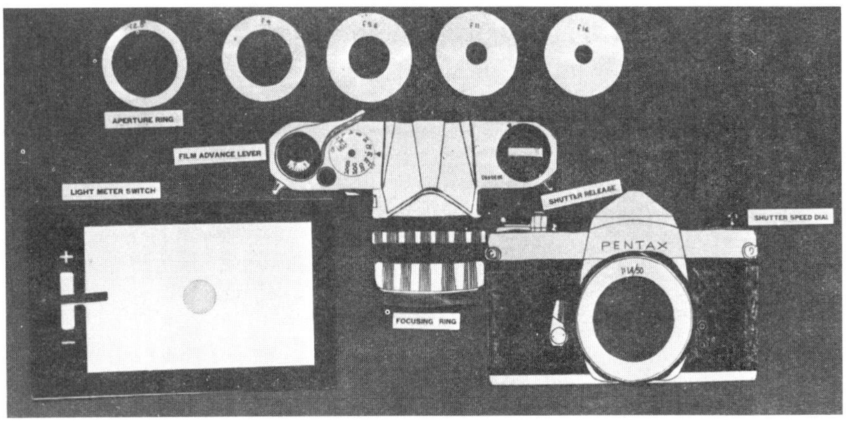

An effective bulletin board display should have large, easy-to-see visuals and captions, and should present information related to what is being taught in the classroom.

Consider Balance and Esthetics Formal (symmetrical) and informal (asymmetrical) balance refer to the placement of materials on the display surface. In addition to balance it is necessary to consider overall appearance and esthetic appeal of the layout. Some suggestions include:

1. *Strive for simplicity* Select the theme with care and carry it out without excessive decoration or embellishment.
2. *Avoid clutter* Leave some breathing space so that important portions can be set off and emphasized. Stay away from a bargain basement look.
3. *Encourage logical, systematic eye movement* Use color, size, line, and texture to direct attention to the major, and then to the minor, elements of the display.

Making Titles and Captions

You want your viewers to know what the items on display mean. After all, the idea is to instruct! There are various ways of making titles and captions that will give the relevant information.

Freehand Lettering This may be satisfactory, if you are relatively skilled. Definite guidelines are not too practical when it comes to titles and captions. Generally, it is important to keep verbal information sparse, uncluttered, and easy to read from a distance. Color, shape, and simplicity should be kept in mind.

Lettering Techniques A variety of machines and devices will make lettering quick and easy. Choose the equipment or technique that is available in your media center. Dry-transfer, rub-on letters may be purchased locally and are used without specialized equipment. Instructions follow for operating lettering machines or producing letters by other means. Some are mentioned in other chapters. You may wish to review the instructions as you complete projects in this chapter:

- ☐ Dry-transfer lettering
- ☐ Cutout letters
- ☐ Prepared letters
- ☐ Leteron machine letters
- ☐ Varityper headliner lettering
- ☐ Varifont or Kroytype lettering
- ☐ LeRoy lettering
- ☐ Wrico lettering

Burnish the dry-transfer letter off its carry sheet and onto the paper with an object such as a pencil point.

PRODUCE

Dry-Transfer Lettering

1. Draw guide lines on work surface, or put lined paper underneath and work on a light table.
2. Commence lettering by applying pressure to letter with a ball point pen. Do not overapply or a waxy build-up will create a halo around letters.
3. Remove sheet from surface.
4. Mistakes can sometimes be lifted from the paper by using a small piece of masking tape to "pick up" the transfer letter.

Carefully remove the carrier sheet so the letter will stay on the surface.

PRODUCE

Cutout Letters

1. Select material for letters: construction paper, typing paper, etc. To increase durability, consider laminating first using a dry-mount press.
2. Insert material and letter cutter; proceed to cut.
3. Remove letter from its background. (You can save the background and use it as a letter too.)

"Cookie-cutter" cutout letter presses produce large letters for display captions.

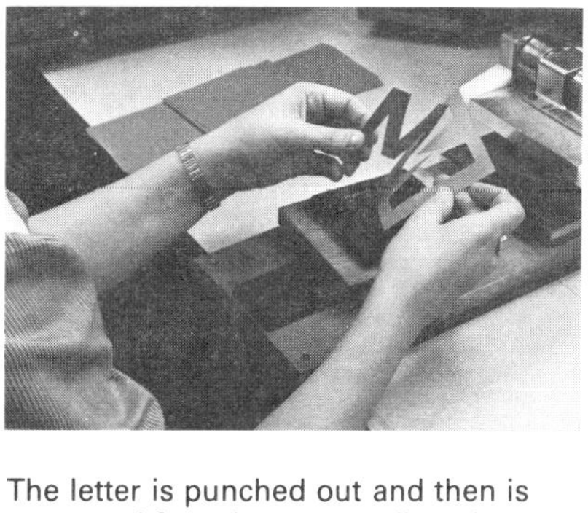

The letter is punched out and then is separated from its surrounding sheet. Either the cutout letter or the letter outline can be used for a caption.

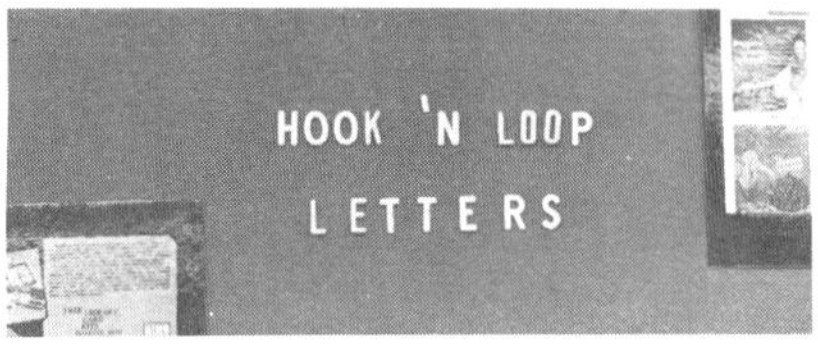

It is possible to purchase magnetic letter sets for metallic chalkboards, or hook and loop board lettering sets for Velcro displays.

TRY THIS

Prepared Letters

Using letters of various sizes and styles it is possible to create titles on magnetic boards. Certain porcelainized writing surfaces are metallic and can accommodate magnetic letters. Hook and loop letters are available in various sizes for creating titles.

Leteron machines produce plastic letters in various colors and sizes.

OPERATE

Leteron Machine

You need: Leteron machine, Letertype, Leterfilm

Procedure

1. Select the color Leterfilm desired and load the machine.
2. Select the Letertype size and style and set machine spacing as specified.
3. Place each Letertype letter in the machine and press the LETTER bar to cut the letter.
4. Stamp out all letters. Use the Space Letertype between words.
5. Remove the tape from the machine. Tear, using the serrated edge.
6. Peel the cutout letters from the plastic backing so the letters themselves remain on the clear plastic.
7. Place the clear plastic strip with the colored cutout letters on the prepared surface. Use the roller or your fingers to press the letters to the surface.
8. Carefully peel the clear plastic off of the letters.

Various font styles and sizes are available.

Preparing the Materials for Display

You are now ready to take the next step—preparing the materials for display. The following methods will help you in this preparation.

Using a Dry-Mount Press Illustrations often require captions, a larger colorful backing, or other special treatment as they are prepared for display. The dry-mount press offers three possibilities for the preservation of bulletin board materials, as illustrated.

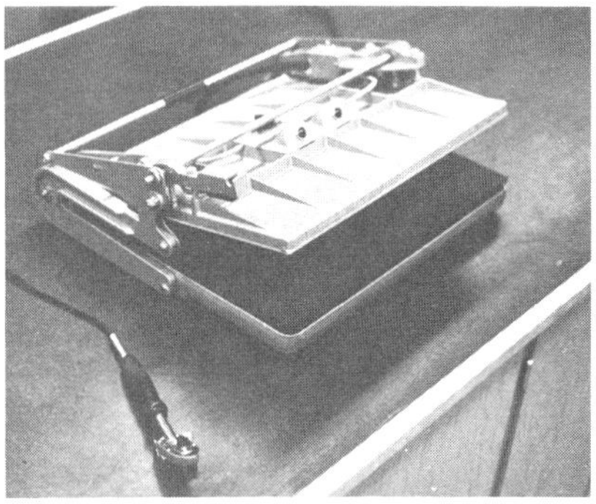

A dry-mount press and tacking iron.

OPERATE

Using the Dry-Mount Press to Laminate

1. Preheat the item to remove moisture; place the item in a newsprint carrier and close, but do not lock, the press and warm the item for 10 to 20 seconds.
2. Place the lamin on the picture with the frosty, or dull, side against the picture. Laminate both sides at the same time if you want both sides sealed.
3. Place the resulting item/lamination film combination in a clean newsprint carrier and seal in the press for about 30 seconds at 275°F.
4. After the item is removed from the press, you may wish to flatten it under a weight during cooling.
5. Trim laminated item, as needed.

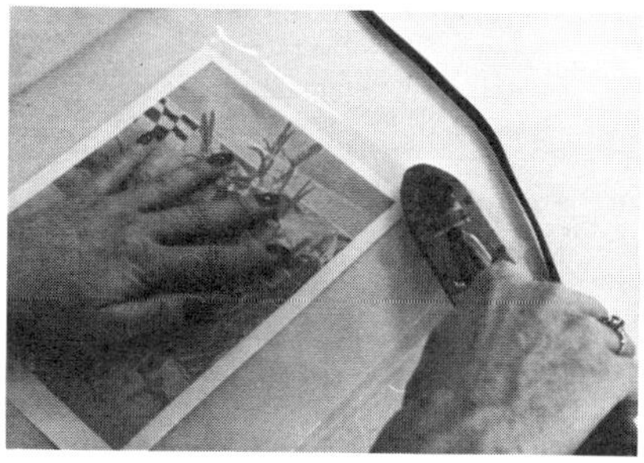

Lamin is a clear plastic covered with a heat-sensitive glue on one side. When it is placed next to a picture and placed in a dry-mount press set at 275°F, is will adhere and provide a protective plastic covering.

OPERATE

Using the Dry-Mount Press to Chartex

1. Preheat the item to remove moisture; place the item in a newsprint carrier and close, but do not lock, the press and warm the item for 10 to 20 seconds.
2. Tack Chartex to back of the item by using a tacking iron. The shiny, smoother side of the Chartex is the adhesive side and should be placed against the item.
3. Place item and Chartex in a newsprint carrier and seal in the press for about 10 seconds at 225°F.
4. After removing item from press, remove it from the newsprint carrier immediately. If you fail to do so, portions of the newsprint may stick to your Chartex because the glue melted through it.
5. Trim Chartex and item, as needed.

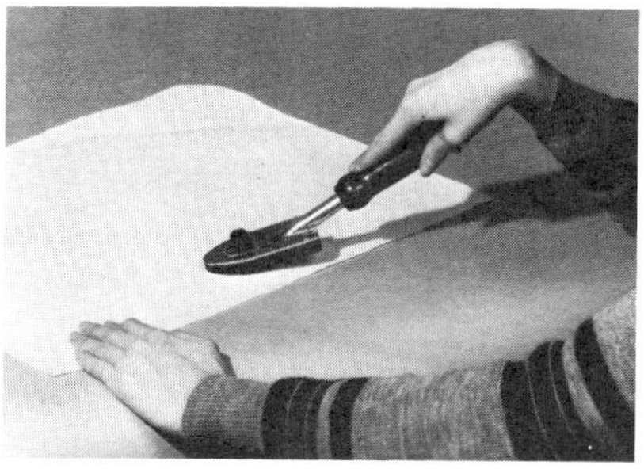

Chartex may be tacked to the back of the chart before placing it in the press, or it can be placed carefully on the back of the object without tacking.

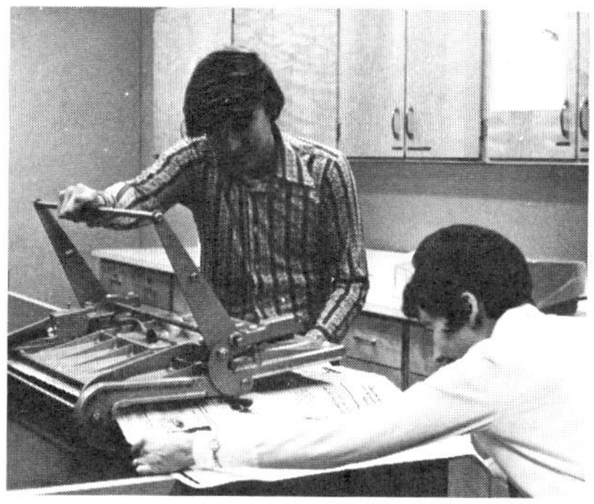

Objects larger than the press can be mounted by doing a portion at a time and overlapping, as shown for this oversized poster that is being Chartexed.

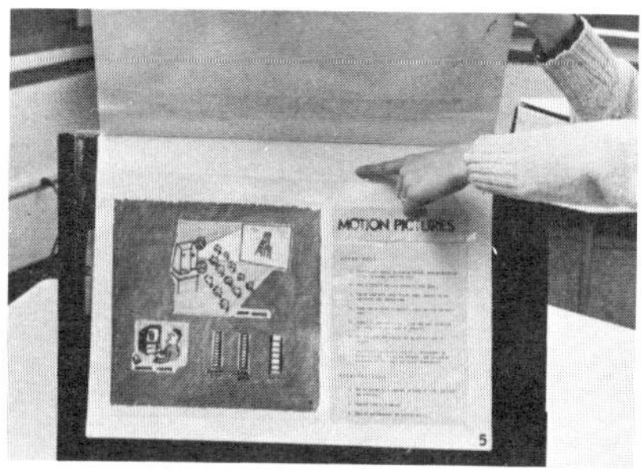

Chartex makes an excellent hinge for a flip chart.

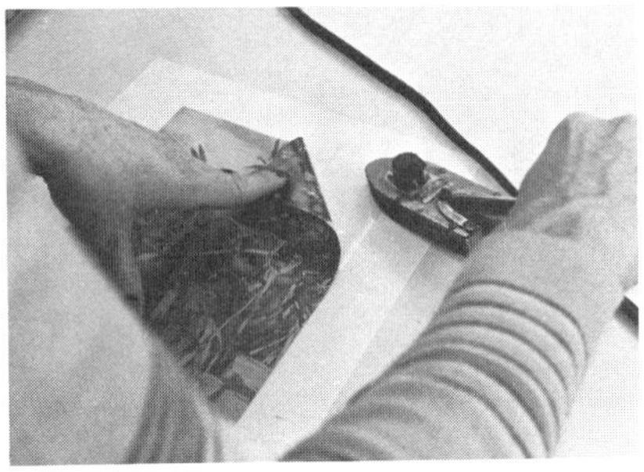

Tack the tissue to the back of the visual and to the backing before placing in a press.

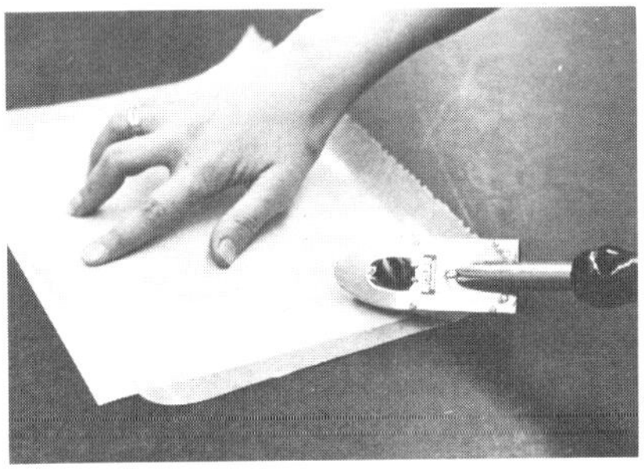

You may find it easier to tack the tissue to the back of the object so it does not slip when you trim it.

Continuous-roll laminators make the task of lamination quick and simple.

OPERATE

Using the Dry-Mount Press to Apply Dry-Mount Tissue

1. Preheat the item to remove moisture; place the item in a newsprint carrier and close, but do not lock, the press to warm the item for 10 to 20 seconds.
2. Cut the tissue slightly larger than the item. Tack tissue to the back of the item in the middle, then retrim so that no excess tissue is visible on the borders of the item.
3. Place item/tissue combination on the selected backing and tack two opposite corners of the tissue to the backing.
4. Place in a newsprint carrier and seal in the press for about 10 seconds.
5. After removal from the press you may wish to flatten the item under a weight during cooling.

OPERATE

Continuous-Roll Laminator

1. Preheat the rollers, following instructions accompanying the machine you are using.
2. Plan the sequence and position of the materials to be laminated, to best use the size of the lamination film. Lay out the materials in order so that they may be fed into the machine without delay.
3. Activate the rollers and insert the materials, being careful that materials do not overlap as they are inserted.
4. When lamination is completed, trim as necessary.

Using an Iron instead of a Dry-Mount Press The three standard materials—tissue, Chartex, and lamination film—adhere to a surface when heated to a temperature of 200° to 300°F. It is more satisfactory to heat a large area so that the entire item achieves an even temperature. A dry-mount press with a large platen is designed to make the process as successful and easy as possible. However, any hot surface of any size can melt the heat-sensitive contact cement that is found in the three materials. A tacking iron, for example, is used to seal small areas.

An ordinary iron and ironing board can be used at home if you want to prepare teaching materials using the dry-mount method. You may need to experiment with a variety of surfaces on which to do the work. The ironing board is suitable if it does not have a thick, padded surface that absorbs heat too rapidly. Alternatives to the ironing board include a formica countertop, a wooden table, or a piece of wood or hardboard (masonite). Some amount of trial-and-error will be needed to derive the appropriate combination of surfaces/time/temperatures. Set the iron on "wool," which is in the 225° to 250°F range.

Alternatives to the Use of Lamination Film If your objective is to preserve the surface of an item, it is possible to cover that surface with a clear plastic film. One technique already mentioned is to use a hot dry-mount press to cause lamination film to adhere to the surface of the item. Two other techniques are available, and neither requires a hot press. They are:

1. *Clear contact film* A plastic film, available at a hardware or variety store, can be attached to the surface of an item. The plastic has a pressure-sensitive adhesive on one side, and will adhere firmly to many types of surfaces.
2. *Wide cellophane tape* A thin plastic type (4 to 6 inches wide) may be used to cover labels on notebooks, instructions posted near apparatus in a laboratory, or in other situations where protection is important. Since the tape is rather thin, it must be applied carefully so that wrinkles do not occur.

Some Additional Techniques for Mounting Items on a Backing Mounting tissue is widely used by professional photographers and others who desire a smooth appearance and a long-lasting product. However, there are cases in which other mounting techniques are more convenient, faster, and equally satisfactory. For example, if the item to be mounted is irregularly shaped, or quite small, it may not be practical to use dry-mount tissue. Two options are offered because of their ease of use:

1. *SprayMent* A commercial adhesive in spray form developed to provide long-lasting, good-quality adhesion.
2. *Double-backed cellophane tape* A clear plastic tape with adhesive on both sides, suitable for holding an item in place. While the item may be attached to a backing of some sort, the final product is not the same as that achieved with dry-mount tissue or SprayMent because the tape cannot cover the entire rear surface of the item, so the bond is only partial.

What to Do about Bubbles and Wrinkles For reasons unknown you may find that the items you are preparing with the dry-mount press develop bubbles, wrinkles, or other surface imperfections during production. Problems like this usually occur because there was not enough heat and/or pressure applied during mounting. First, try placing the materials back into the press. If this does not work, the most satisfactory remedy for these difficulties is to use a tacking iron to apply pressure to the problem area. A hot iron and considerable pressure, applied with a scrubbing motion, will tend to smooth wrinkles. If bubbles or larger unsealed areas remain when you remove the work from the dry-mount press, try using a pin to puncture the bubble. Then apply pressure with a hot tacking iron. While you may not be entirely pleased with the final product, no significant lack of learning can be traced to these imperfections.

Clear contact paper can be used to protect the surface of a visual.

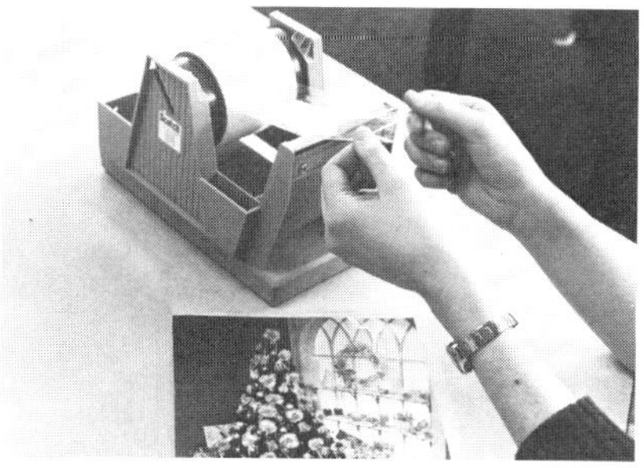

Special wide cellophane tape can be used to protect the front of visuals as an alternative to lamination in a dry-mount press.

Spray glue is an alternative method for mounting a visual to a backing.

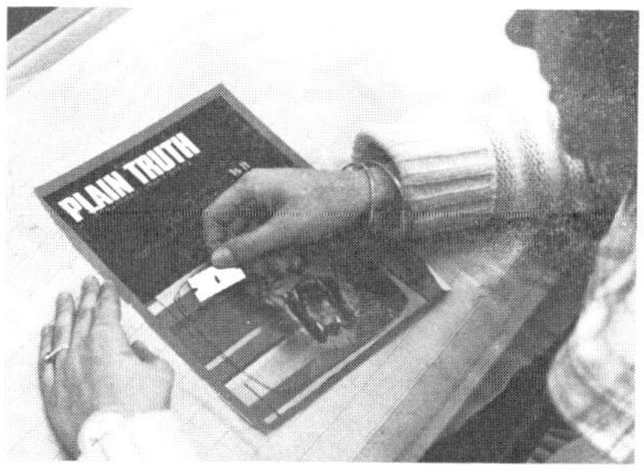

Air bubbles often occur during laminating. They may be removed by poking a hole in the plastic with a pin, or slitting it with a razor blade, then replacing the picture in the dry-mount press.

PRODUCE

Laminated Materials

Consider an item that needs protection from moisture, soiling, or other harmful conditions that may occur during its use. Laminate the surface. If you are planning to laminate materials that are to be put on a bulletin board, remember that the shiny surface of the lamination film may reflect light in such a way that it might be difficult to view the bulletin board. Since bulletin board items are rarely handled by students, there is often no need to laminate items for subsequent posting on a bulletin board. The dry-mount tissue technique for mounting items on poster board is usually sufficient in preparing bulletin board displays.

PRODUCE

Chartexed Materials

Select items that require a substantial and durable backing that can be provided by Chartex. Some suitable items include large maps that must be rolled up for storage, flimsy newspapers, or thin paper goods that might be torn or wrinkled in storage or use. Chartex may be used in conjunction with lamination film to provide an impervious surface with a durable backing.

PRODUCE

Dry-mounted Materials

Select an item of printed material, such as a photo, chart, map, drawing, or clipping, and mount it on a suitable backing. Construction paper, poster board, or other types of backing may be used. Consider the addition of captions or other labels, and arrange the item so that appropriate margins remain.

PRODUCE

Dry-mounted/laminated Materials

Choose an item that should be laminated and then dry-mounted to a suitable backing. Generally, you should mount the item first, then laminate it. Because an item will tend to curl if only one side is laminated, it may be difficult to mount it after lamination film has been applied. (Do not try to laminate and dry-mount in one step.)

PRODUCE

Laminated/Chartexed Materials

If the item you are preparing for use is somewhat thin, if you want the surface protected, and if you want to avoid wrinkles during storage and use, consider laminating the front and using Chartex on the back. Chartex the back of the item, then laminate the front.

TRY THIS

The "Oil Painting" Effect

If you want the surface of your laminated item to take on the appearance of an oil painting, with a surface that is somewhat crinkled and textured, crush the lamination film into a small wad. Put as many wrinkles in it as possible, then flatten and apply to the surface of your photo or drawing, following the usual lamination procedure. Be careful that you do not spread the wrinkled laminating film too much or the fine wrinkles will be removed during mounting.

TRY THIS

Laminating Unusual Objects

A variety of materials such as leaves, cloth, or papers with unique surfaces may be laminated to produce interesting and useful teaching materials. If leaves are to be laminated, they must be flat and dry. Collect the leaves and press them between blotters or soft cloths, then press them flat by putting a book or two on top. In a few days the leaves will be dry enough to laminate. One method for laminating both sides at once is to use a continuous-roll laminating machine.

Including Captions Photographs, line drawings, and other nonverbal materials are more effective when accompanied by some descriptive information. Remember that the person viewing the display may be standing a few feet from it, so the lettering should be larger than that available on an ordinary typewriter (at least 1 inch high to be seen from 30 feet). Locate a typewriter that produces larger lettering or use the rub-on lettering technique.

Putting up the Display

Your materials are now ready. What are the best methods for displaying them?

Determine Placement of Materials Using the sketch you made earlier, during the design stage, you can now project that sketch (using the opaque projector) on the surface that will become your bulletin board.

Attach Materials to the Board Various methods are available. It is more effective if you can avoid using large, colorful thumbtacks because they detract from the visual appearance.

Formal 3-dimensional Displays Well-designed 2-dimensional displays may serve as effective information-transmitting devices, but in some instances it may be necessary to create additional impact. If the display surface is folded, curved, or pleated to take on an aspect of depth—and if 3-dimensional features are added to the display surface—it is possible to create a feeling of realism that carries a good deal of visual impact. The technique of incorporating 3-dimensional items into a display has been used in museums, depictions of historical events, commercial exhibits, sales meetings, retail merchandising, and teaching.

Some of the characteristics of effective 3-dimensional displays include:

1. *Size* Consider how tall and how wide the display should be. Will it be on a tabletop or freestanding on the floor?
2. *Lighting* Unless the display is to be located in a well-lighted area, it will be necessary to use clip-on lights, spotlights, or other means of providing light.
3. *Graphics* Use of bold colors and large block lettering should be considered to make the display attractive and to convey information.

Special, large-type primary typewriters can be used when large typewriter lettering is needed.

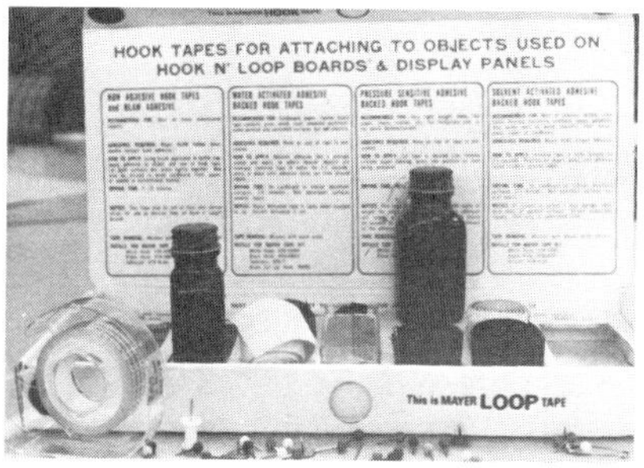

Hook and loop tape, double-stick tape, and pins can be used to attach objects to a display board.

A diorama can be used for a realistic 3-dimensional visual display.

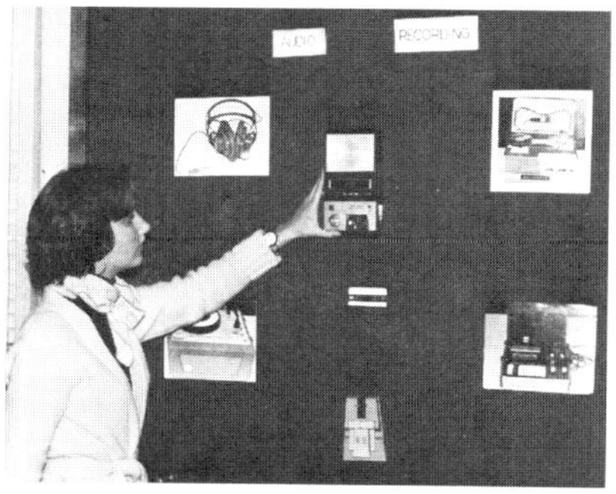

Three-dimensional displays can be made using hook and loop. Real objects, such as an audio cassette recorder, an audio cassette tape, and an audiotape splicer, can be hung as part of the display.

This is the HOOK Tape . . .

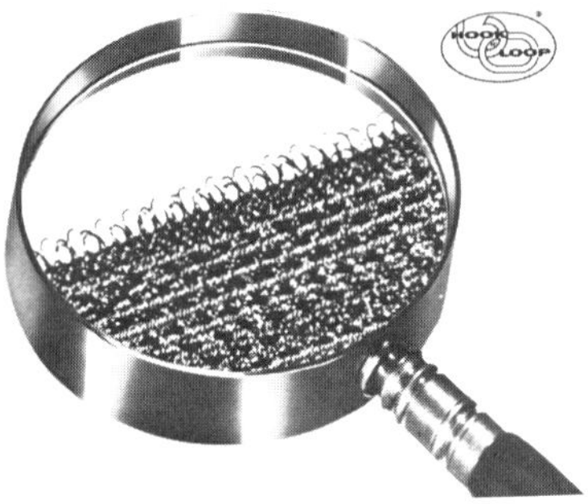

This is the LOOP Fabric . . .

Plastic letters that have hook tape on the back are available for purchase.

4. *Realia* Incorporate as many items of real material as you can so that the display is both meaningful and attractive.

The information up to this point has included techniques and examples for creating displays that treat the viewer as a spectator. Each type of display can be viewed, studied, or assimilated without the need for a live interpreter or teacher. Displays like this fill an important need, and various versions of them are in use in a wide variety of circumstances. Bulletin boards, airline schedules, placards in store windows, and other passive signs are of top importance in many situations today.

However, display surfaces often can be used to even greater advantage if they are designed to implement the presentation of content by a live teacher. Using the large surface as a kind of canvas upon which a scenario can be sketched, one step at a time, instructional sequences can be made to unfold in a systematic series of steps. The need for a teacher to direct and implement this type of instruction, and the nature of the instruction, is discussed in the next section.

ACTIVE SURFACES

Two types of fabric-covered surfaces, suitable for sequential development of instruction, are *hook and loop boards* and *felt-* or *flannelboards.* A metal surface for holding objects containing tiny magnets is known as a *magnetic board.*

Fabric-covered Surfaces

Fabric-covered surfaces offer ease of use and adaptability. The two most commonly used are described below.

Hook and Loop Boards Hook and loop fabric (trade name: Velcro) is a two-part system, with a fabric composed of tiny loops of nylon fiber, and a companion material consisting of tiny nylon hooks. The hooks resemble those found on cocklebur seeds and, upon being pressed into the loops of the fabric, grip tightly. Heavy objects, with a small amount of hook material glued on, are easily held in place on a hook and loop board. Large cardboard cutouts, 3-dimensional objects, posters, or other display materials can be put up without fear that they will fall off.

Flannelboards *Flannel* or *felt* is often used in teaching sequential ideas or concepts, much as hook and loop boards are used. The primary difference is in the size and weight of the objects to be put on the board. Feltboards are adequate to hold lightweight, flat objects but cannot hold heavy 3-dimensional objects.

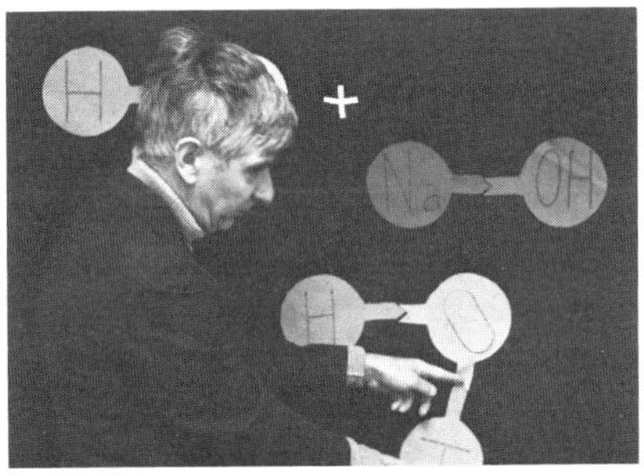

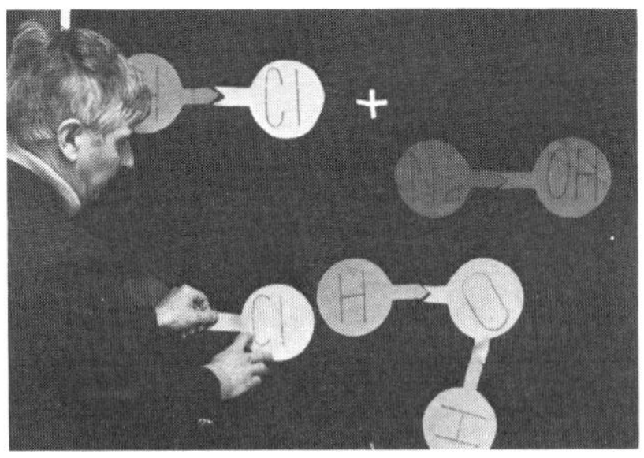

Flannelboards allow the instructor to build a sequence as a presentation develops.

If you must choose between acquiring a feltboard or a hook and loop board, take the latter. You can use feltboard materials on a hook and loop board. Unfortunately, objects backed with the hooks designed for the hook and loop system will not grip sufficiently into a felt-covered surface.

The nature of instruction that can be carried out with either a flannel- or hook and loop board is fundamentally the same, even though the surfaces themselves are different. These large surfaces lend themselves to the development of a sequence, the labeling of parts, the build-up from simple to complex. Sometimes a passage in time is needed to explain a concept. A change in space may be required—or a change in size. This conversion or movement through time, space, size, or complexity is possible on a large surface by operating in an active teaching mode.

Some of the situations in which display surfaces are particularly suitable are illustrated below:

Other Types of Boards

Although fabric-covered boards are popular, they have limitations. Other types of boards should then be considered.

Magnetic Boards Available in a variety of shapes and sizes, these may be designed specifically for use as display surfaces—or may serve the dual purposes of chalkboard and magnetic board. (Some types of chalkboards are constructed from steel backing sheets and can hold objects that are backed with magnets.)

The types of objects generally used on magnetic boards include styrofoam figures and shapes, some commercially prepared materials such as numbers, and other forms of manipulatable objects. It is relatively easy to construct your own magnetic board and the objects that are placed upon it. The magnets may be purchased locally, as can the metal backing sheet.

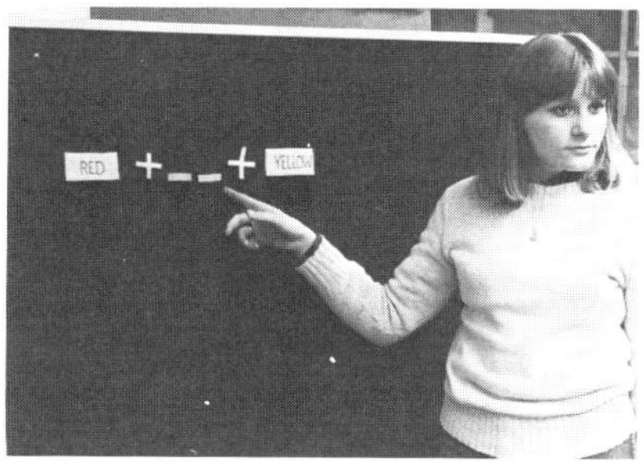

Flannelboards are an effective teaching device for showing a sequence of some activity, from its beginning . . .

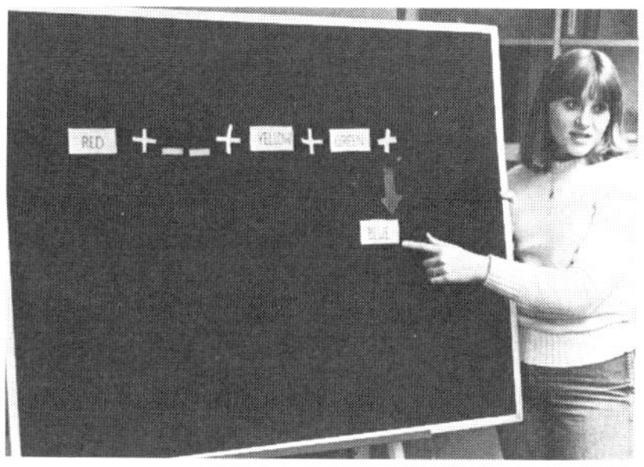

through the middle of the concept presentation . . .

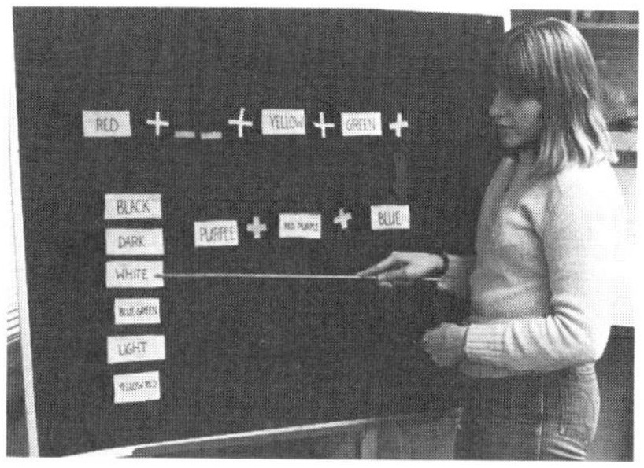

to its conclusion.

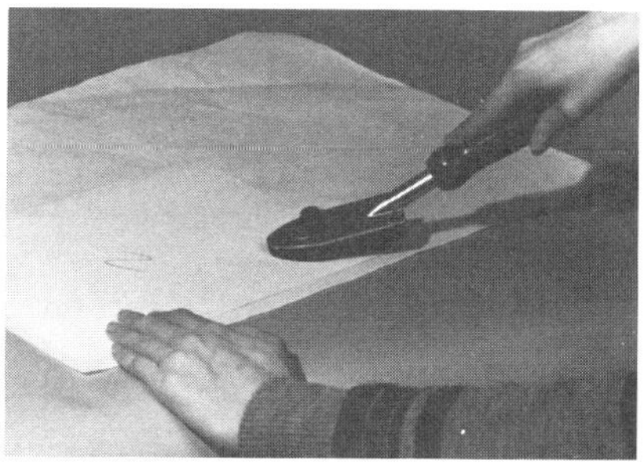

A display can be used to show a sequence of events, beginning with step No. 1—Tack the tissue to the visual.

Step No. 2—Trim excess tissue.

Step No. 3—After tacking to the backing material, place the entire "sandwich" of display items in the dry-mount press.

Cuisenaire rods provide realistic instruction on fractions, and how parts make up the whole.

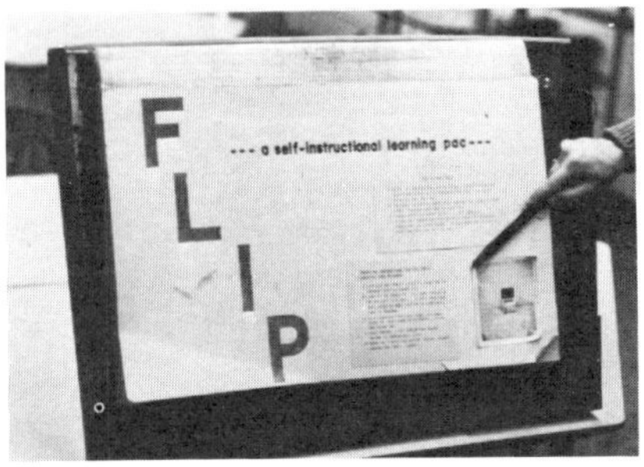

Flip charts can be self-instructional or used for group presentation, if they are large enough.

Large scheduling boards make it easy to plan for some complex activity and to review its progress quickly.

Electric boards provide immediate visual feedback to the students as they progress through self-instructional flip charts. This photograph of the back of an electric board shows how easy they are to build.

Self-Instruction Boards Although large surfaces can be found in use for large-group instruction, adaptations to smaller groups can be equally effective. Several types of self-teaching lessons using electric boards, flip charts, or other forms of manipulative surfaces incorporating ideas of gaming are possible.

Flip Charts Even though locally produced flip charts are not particularly useful for teaching groups of twenty or more (the small charts are too difficult to see if the group is too large), a great deal of small-group teaching or self-teaching can occur with them. The panels of a flip chart are treated like pages of a book. Ideas, processes, and concepts are developed sequentially so that, as the lesson unfolds page after page, a gradual build-up of concepts can take place. In some cases, flip charts are used to hold objects, photos, or other manipulatable items that may be removed for study.

Electric Boards Based on the old idea of multiple choice, in which the correct answer causes a bulb to light, a student can use an electric board for drill and practice and review of ideas. A simple electric circuit is completed when the probe is touched to the metal stub that corresponds to the correct answer.

Scheduling Boards Consisting of a simple surface with pockets to hold small pieces of cardboard, a scheduling board can be used to manage an inventory, determine which students have completed certain units in a block of instruction, or keep track of people and events.

Hinges for flip charts can be made with Chartex.

PRODUCE

A Flip Chart

A flip chart is versatile enough to be used as a small-group teaching device, or for individualized instruction. The three major dry-mount press techniques (use of Chartex, lamination film, and tissue) can be incorporated.

Design the panels of the flip chart so that they may be used conveniently by the group they are intended for; 8½ × 11 is sufficient for individual instruction but a larger panel, up to 18 × 24 inches, for example, might be more easily viewed by a small group. Use lettering techniques for titles and captions, and attach materials to the panels with dry-mount tissue or lamination film.

To hinge the panels, use strips of Chartex. You may wish to make an A frame to hold the flip chart in an upright position. A variety of methods can be used to hinge the panels if you decide not to use Chartex hinges.

A FINAL WORD

This chapter has provided a number of ideas leading to the production of a variety of bulletin boards and displays. While these teaching methods are not as glamorous as some of those involving high-technology devices, they nevertheless fill an important role in learning and instruction. As you accumulate techniques for creating instructional strategies, do not overlook the unique place that displays may occupy in the instructional process, because they represent one of the few techniques for offering information to students on a continuing basis. The dependability and availability of displays make them an important consideration when instruction is being formulated.

LOOKING AHEAD

The teaching strategies in this chapter have been concerned with the use of large surfaces—areas that occupy many square feet. The use of color, bold graphics, accepted principles of layout and design, and sequential development of ideas have been pointed out as ideal methods for taking advantage of these media in teaching.

This chapter on display surfaces serves as an introduction for the next chapter in this section of the text, in which we explore the use of projected visuals. In Chapter 8 the nature and use of transparencies and the overhead projector will be explained, and you will have the opportunity to create a wide variety of teaching materials for projection on a large screen.

REVIEW QUESTIONS

1. List some advantages of asking discussion groups to summarize their main points on large sheets of white paper.
2. Describe the steps necessary to design the layout of a display.
3. List some lettering techniques that do not involve complex processes or expensive machines.
4. Describe an objective or a teaching need for which a flannel- or hook and loop board can be used.
5. Describe two types of display surfaces that can be used effectively for self-instruction.

ANSWERS TO REVIEW QUESTIONS

1. Saves time in reporting results, keeps group activities focused on the issues, permits everyone else to see and discuss results.
2. Consult texts in the content area, examine source books containing bulletin board ideas, determine objectives, incorporate elements of design, determine size and proportion, consider balance and esthetics.
3. Dry-transfer letters, magnetic or hook and loop letters, LeRoy lettering, Wrico.
4. Answer should include the characteristics of developing ideas sequentially, progressing from small to large, or showing historical change.
5. Electric boards and flip charts.

REFERENCES

Bullough, Robert. *Creating Instructional Materials* (Columbus, Ohio: Charles E. Merrill, 1978).

Dwyer, Francis. "Exploratory Studies in the Effectiveness of Visual Illustrations," *AV Communication Review* (Fall 1970): 235–249.

Kemp, Jerrold. *Planning and Producing Audiovisual Materials,* 4th ed. (New York: Harper and Row, 1980).

Minor, Ed. *Handbook for Preparing Visual Media* (New York: McGraw-Hill, 1978).

Minor, Ed, and Frye, Harvey. *Techniques for Producing Visual Instructional Media,* 2d ed. (New York: McGraw-Hill, 1977).

SOURCES FOR SUPPLIES

A. B. Dick Co. (SprayMent)
5700 West Toughy Ave.
Chicago, IL 60648

Kroy Industries, Inc.
1687-T Woodlane Dr.
St. Paul, MN 55164

LeRoy Lettering Equipment
Keuffel & Esser Co.
29 Whippany Rd.
Morristown, NJ 07960

Seal Inc. (Chartex)
550-T Spring St.
Naugatuck, CT 06770

Varigraph Inc.
P.O. Box 690
Madison, WI 53701

VELCRO USA, Inc. (Velcro)
521 Fifth Ave.
New York, NY 10175

Wrico Lettering Guides
Warren-Knight Instrument Co.
2000 Bennett Rd.
Philadelphia, PA 19116

GOAL

□ Gain an understanding of transparency production and use.

OBJECTIVES

□ Produce a layout that presents a concept visually.
□ Make a direct transparency using clear plastic and felt markers.
□ Operate an overhead projector.
□ Prepare an original and produce an indirect, thermal transparency.
□ Incorporate selected special techniques, such as overlays, color lifts, diazo production, and polarized film into transparencies.
□ Use opaque and 3-dimensional objects in overhead projection.

You may recall the example in Chapter 1 in which you were teaching a class in social studies, using your voice and the chalkboard. You found it necessary to illustrate some of the ideas associated with the causes of the Civil War, and stepped to the board to make some hurried notes for the class.

This chapter on transparency production is an appropriate place to go back to that earlier teaching moment and consider some techniques that might have improved your teaching—and student learning. Perhaps, during your planning of that lesson, you included notes and sketches that you intended to put on the board. (It is natural to think about chalkboard work as you make lesson plans because you know the board is always in your classroom and there is no need to plug it in or turn it on.) Preplanned use of a chalkboard may not occur a large part of the time, however. More frequently you will find that a thought coming to mind, a question from a student, or a last-minute idea occurring while you are teaching accounts for your major use of the chalkboard. It is often used as a work surface for your ideas, and those of students.

In the case of our Civil War example, suppose you worked out an outline to follow during your teaching. Rather than put it on the board, consider making a transparency as you prepare the lesson. While you may spend a few minutes doing so, that time will be saved during classroom teaching. Other benefits of preparing a transparency include:

□ Avoiding mistakes you might make in doing board work during classroom teaching.
□ Creating several transparencies, thus conveying more information than might fit on the board.
□ Reviewing information from previous transparencies, since they are not erased as is a chalkboard.
□ Including sketches and drawings that might be too time-consuming, complex, or detailed to draw on the board.

8 OVERHEAD TRANSPARENCIES

In this chapter we will be explaining how to make quick and simple transparencies without machinery. We also include production of more complex transparencies such as thermal and diazo visuals, as well as a number of special techniques. These additional ideas may be particularly helpful if you wish to add motion, make projected backdrops for a school play, or demonstrate some principles used in the teaching of science.

DIRECT PRODUCTION

The easiest way to make and use transparencies involves a process called *direct production.* You write or draw directly on a plastic sheet.

A Word about Layout

The key to effective production and use of transparencies is not necessarily in the making—the lettering, the adding of color, or the use of overlays. Layout is the most critical step. Layout design is the process of organizing your teaching information logically and efficiently.

While we have not summarized guidelines for cartoons, sketches, diagrams, and other graphic elements, both verbal and nonverbal information should be *big, bold, simple,* and *clear.* If you have a great deal of detail or a large number of words, the most appropriate medium is NOT a transparency. Instead, put the information on paper; make a spirit-duplicated handout.

Traditionally, teachers have used the chalkboard to illustrate ideas during instruction.

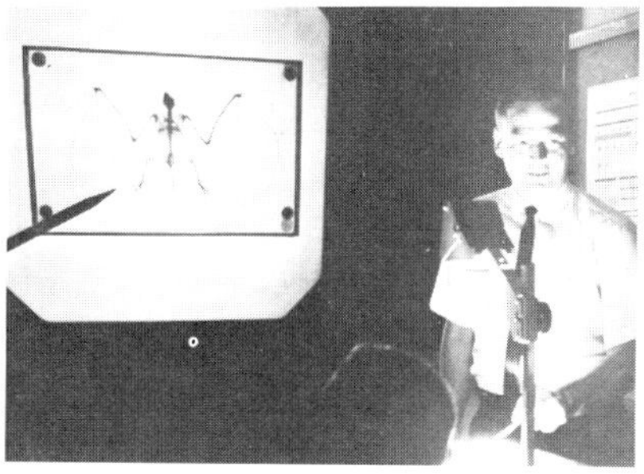

The overhead projector can be used to show real, 3-dimensional objects, such as this bat skeleton.

Most overhead projectors can be equipped with a roll of plastic that can be cranked across the stage. The instructor can write on the plastic with a felt marker during a presentation.

PRODUCE

Producing a Layout

You need: a written description of some concept or idea, paper/pencil or chalk/chalkboard

Procedure

1. Identify the key concept or idea you wish to present.
2. Select a two- to five-word title that describes the concept.
3. Identify the basic elements of the concept, and think how they might be shown using a picture, chart, graph, line drawing, or some other kind of visual.
4. Design the layout. The title should be at the top in bold lettering. Arrange the visual elements (refer to Chapter 3's discussion of the elements and principles of design, and of informal and formal balance). Arrows may be used to direct viewer attention.
5. Transfer the completed layout to the medium you plan to use for instruction: chalkboard, transparency, duplicated handout.

In the PRODUCE box we have suggested how to go about developing a layout, and below we have presented an idea for you to experiment with to develop your layout skills. Remember—there is usually more than one way to do something correctly. Do not be concerned about "right" and "wrong" at this point. Follow the guidelines in the procedure for layout and see what you can come up with.

TRY THIS

At the chalkboard, or on a sheet of paper, produce a layout of the information below. The layout you design should have a title, some graphic or visual information, and labels or descriptions of parts of the visual.

To make sure that no branch of the federal government can become too powerful, a system of checks and balances is included in the Constitution. Each branch of government (legislative, executive, and judicial) has powers that check, or limit, the powers of the other branches. In this way, the powers of government are balanced, or equally divided three ways.

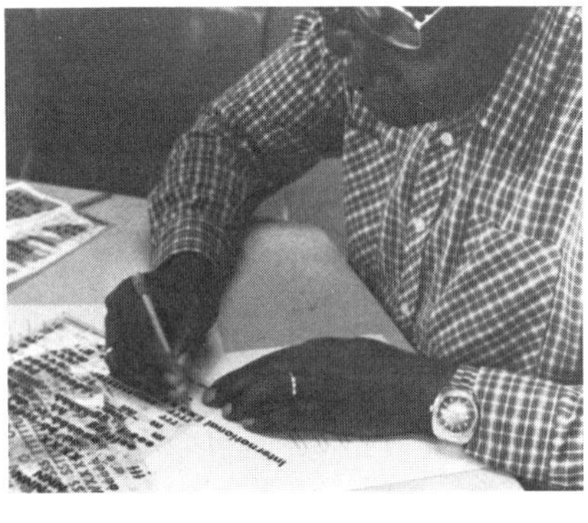

Dry-transfer letters are added directly to the transparency plastic.

Making a Direct Transparency

With your layout completed, you can transfer the design to plastic and make a direct transparency. A wide variety of writing surfaces and writing materials is available. Choose a combination that suits your needs and experiment with several versions if you wish.

Writing Surfaces

- ☐ Clear plastic sheets
- ☐ Plastic rolls
- ☐ Plastic notebook dividers
- ☐ Lunch-wrap plastic
- ☐ Plastic photo holders

Writing Materials

- ☐ Lettering: Felt pens
Grease pencils
Dry-transfer letters
- ☐ Drawing: Felt pens
Grease pencils
- ☐ Color: Felt pens
Color adhesive film
Color dye

PRODUCE

Producing a Direct Transparency

You need: an 8 × 10-inch sheet of clear plastic, and one or more of the following lettering devices: water-soluble felt marker, permanent felt marker, grease pencil, dry-transfer letters

Procedure

1. Produce a layout following previous instructions.
2. Place the plastic sheet on a flat surface and letter the title with dry-transfer letters or felt marker. If you use dry-transfer letters follow these procedures:
   - ☐ Place the letter you wish on top of your clear plastic.
   - ☐ "Burnish" the letter off the carrier sheet onto your clear plastic using a pencil. Rub on the top of the carrier sheet. Be careful not to rub too hard or the letter might crack.
3. Letter or draw the remainder of your transparency as neatly as possible with your colored felt markers.
4. Place your transparency on the overhead projector to see how it projects.

TRY THIS

Using the layout you designed earlier, produce a direct transparency. You will need a clear sheet of plastic and colored markers or dry-transfer letters.

Teaching with an Overhead Projector

You have just created some software and now you must consider teaching with it. In the case of transparencies, there are important suggestions to follow to gain maximum impact. They include the following.

Revelation The process of revealing one line at a time as you teach from a transparency. The principle here is to prevent students from reading ahead, which distracts them from your teaching.

Turn the Projector off while Changing Transparencies A lighted screen attracts attention. You want to capitalize on that attraction, but *only* when the screen shows relevant information. Your teaching will become more rhythmic and smooth if you turn off the projector between transparencies.

Look at the Class, Not at the Screen Use the advantage that eye contact with your students gives you and keep yourself as the focal point of instruction. Yielding to the screen, and talking to it instead of to students, weakens your assertiveness and effectiveness as a teacher.

Ways of Adding Color

Color adhesive Colored plastics that can be cut to the proper size and placed on the transparency

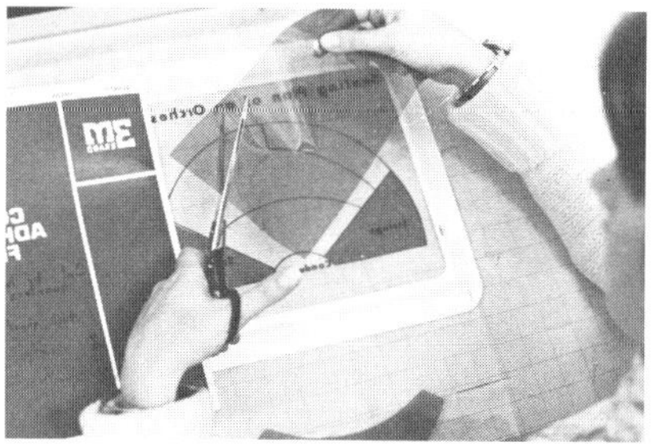

Color adhesive is an excellent way to color large portions of a transparency. The adhesive plastic is cut to the size desired and placed in contact with the transparency plastic.

Color dyes Colored liquids that can be "painted" onto the plastic to add color to small portions of the transparency

Color dyes are an excellent way to color small portions of a transparency. The dye is applied with a brush directly onto the transparency plastic.

Color stripe tapes 1/16 to 1/8-inch wide plastic tape in various patterns that can be placed on the transparency plastic for graphs, outlines, and other colored lines

Dry-transfer letters Translucent colored letters of various styles and sizes that can be rubbed onto the plastic

Felt markers Permanent and water-soluble markers that can be used to color directly on the plastic

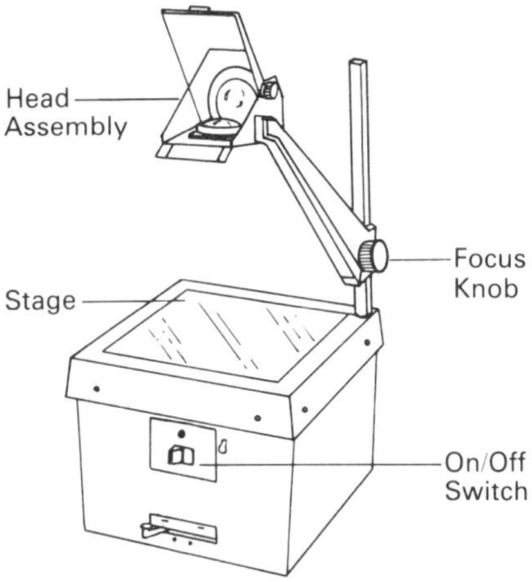

Main components of an overhead projector.

OPERATE

Operating an Overhead Projector

1. Plug in projector, turn on, and fill the screen with a square of light.
2. Tilt the *head assembly* to raise or lower image on screen.
3. Focus by rotating the focus knob. Lay a pencil, comb, or torn edge of paper on the stage and focus the image projected on the screen.
4. Put transparency on stage.
5. Turn off projector between use of each transparency.

Helpful Hints for Operating an Overhead Projector

☐ Tie power cord to a table leg to prevent the projector from being pulled off the table if someone trips over the cord.

☐ If the lamp does not light, check the following:

- **a.** Is the power cord plugged in correctly?
- **b.** Is the lamp compartment door on the projector securely closed?
- **c.** Is the lamp OK? If it is burned out, replace it with a similar type.

☐ The transparency should be placed so the teacher can read it while facing the audience. During use, glance from transparency on stage to students in class. *Avoid looking at screen*—though you may have formed this habit by working with a chalkboard.

☐ Use a pencil as a pointer, and lay it on transparency to draw attention to areas needing emphasis.

☐ Focus is adjusted by varying distance from the head assembly lenses to the stage. This is usually accomplished by turning a knob on the arm that supports the head assembly. Sometimes the focus knob is on the front of the projector below the stage.

☐ Most newer projectors have a thermostatically controlled cooling fan that will shut off when the projector has cooled. Older models with a three-position switch (on-fan-off) should be left in the fan position until the air coming from the fan is cool.

Projection Screens

There are three reflective surfaces commonly used on projection screens: matte, glass-beaded, and lenticular. The matte screen surface is most suited to a square room. The glass-beaded works best in a long, narrow room, and the lenticular screen combines some of the advantages of both, but is more expensive.

When setting up a portable screen, try to position the bottom of the screen at about eye level. Optimally, no one should sit closer than two screen widths from the screen, or farther away than six screen widths. This is often called the "2W-6W rule."

Placement of the projection screen in the front corner of the classroom nearest the windows is generally considered the optimum location because:

1. The least number of students is obstructed by the projector or teacher when the overhead projector is in use.
2. Light from the windows will not fall on the screen and wash out the projected image.
3. The chalkboard and the screen can be used together.
4. Students can see around one another more easily.

The *keystone effect* is sometimes a problem when operating an overhead projector. This effect is described as a distortion of the projected image that is caused because some portions of the image travel farther to reach the screen. As light travels it spreads out and a distorted, keystoned image appears. (A keystone is the main, central stone in an arch.) To correct for keystoning, the screen and/or projector should be adjusted so that the screen is parallel to the front lens of the projector. Move the top of the screen forward, or the bottom of the screen backward, to correct for vertical keystoning.

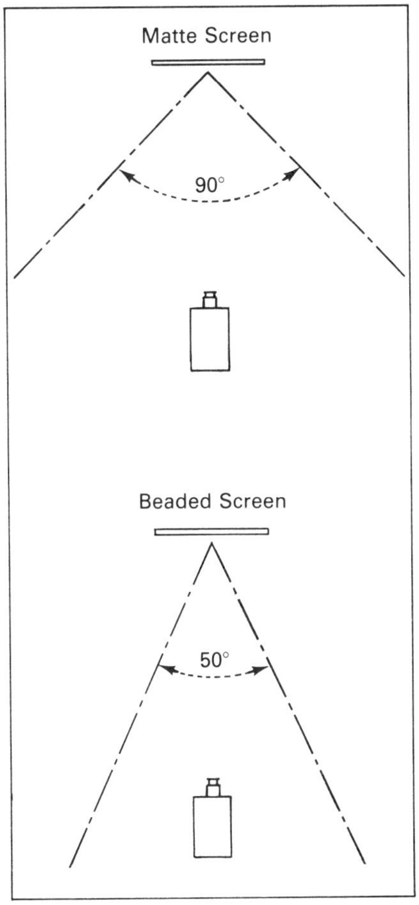

Optimum screen-viewing angles.

For best viewing, the screen should be placed in the front corner of the classroom.

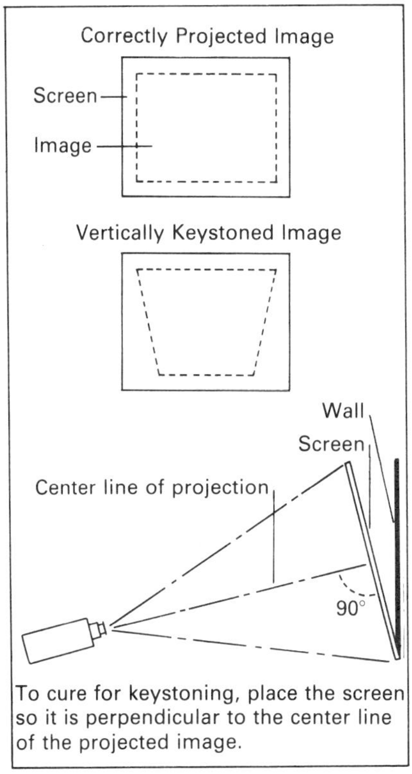

To cure for keystoning, place the screen so it is perpendicular to the center line of the projected image.

Curing keystoning.

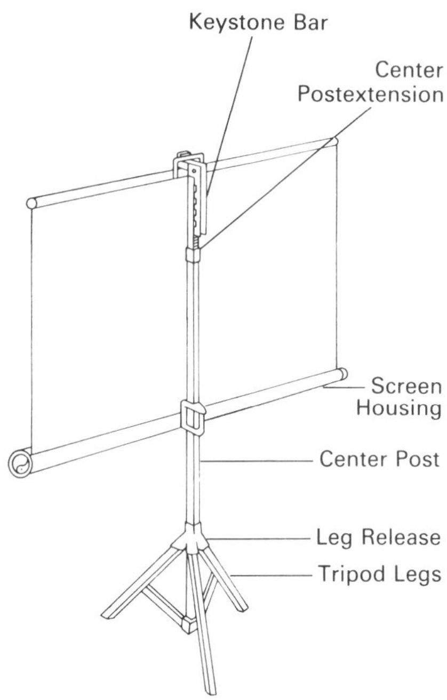

Components of a tripod screen.

Helpful Hints for Use of Projection Screen

☐ It is easier to grip the carry/transport handle and lift the screen while pushing the leg release control. The leg release control will usually be found on the center post where the tripod legs are collapsed, or at the bottom of the center post. When released the legs will spring out to the extended position.
☐ Normally, the screen housing will swing in only one direction without striking a tripod leg.
☐ To raise the screen, depress or pull the housing release button on the center post with one hand and raise the screen and center post with the other.

OPERATE

Setting up a Portable Screen

1. Extend tripod legs.
2. Place screen upright, on legs.
3. Release the screen housing from the center post and swing it to a horizontal position.
4. Extend the keystone bar forward from the top of the center post, if such a bar is provided.
5. Pull the screen out of the screen housing and attach it to the keystone bar, or top of the center post.
6. Raise the center post extension and screen along with it to the desired height.
7. Adjust the bottom of the screen to the desired height.

TRY THIS

Using the direct transparency you produced earlier, set up your classroom for projection. You will need an overhead projector and a tripod screen. Refer to information elsewhere in this chapter to:

☐ Set up screen.
☐ Arrange classroom for use of projector.
☐ Use proper techniques for operating the projector and showing the transparency.

By now you may have discovered some of the limitations of direct transparency production and use. You are forced to hand-letter, or use time-consuming (and somewhat expensive) rub-on letters. Thin lines and fine detail are impossible to reproduce. Prepared, printed drawings can be used only if you trace them, and you have only one finished product; if something happens to the transparency you made, you have to make a new one.

INDIRECT PRODUCTION

The preceding problems can be overcome by using the *indirect* method of transparency production. You create the layout on ordinary white paper, in pencil, and then merely copy the image with a special plastic sheet and a thermal copy machine.

Consider what you gain—fine lines and exact copies without tracing, permanent images that will not rub off or smear, and the chance to erase mistakes on the paper original *before* you make the transparency.

Do not forget what you learned about making transparencies by the direct method, because all of the techniques can also apply here—once you have made the thermal copy on plastic. The use of color, stripe tapes for borders or lines, and writing with felt pens all can be done on the plastic as before. The difference is in the preparation of the *paper original* and the use of *thermal plastic* to pick up the image by sending both through a thermal copier.

Before you think about designing your own original, look around for suitable material already in print. Several publishers offer printed originals in a wide variety of subject areas. While they are designed specifically for use as transparencies (bold letters, clear and simple visuals), you may find some teaching material close at hand. A political cartoon from the morning paper or a line drawing from a laboratory manual makes an excellent original for thermal transparency production.

OPERATE

Operating a Thermal Copier to Produce an Indirect Transparency

1. Prepare an original from preprinted material.
2. Calibrate the thermal copier using test strips:
   a. Select a piece of thermal plastic.
   b. Cut it into five strips. Be sure to note the location of the notch so that the other strips can be cut to indicate the correct orientation when being placed in the thermal copier.
   c. Place a test strip on the original and run it through the thermal copier with the heat setting in the center position. If the test strip copy is too light, move the setting half-way to the darkest setting and run another strip. If the first copied test strip is too dark, move the copier setting half-way to the lightest setting and run another test strip.
   d. Continue this bracketing process until the correct heat setting for the copier and your original has been found.
3. Place a full sheet of thermal plastic on top of your original with the notch in the upper right-hand corner and run through the thermal copier at the predetermined setting.
4. Frame the transparency.

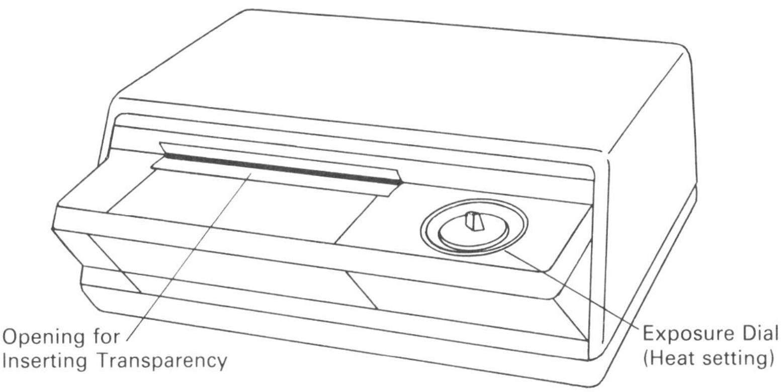

Thermofax transparency maker.

Types of Thermal Plastic

- ☐ Black lines on clear background
- ☐ Colored lines on clear background
- ☐ Black lines on tinted (red, green, yellow, blue) background
- ☐ Colored lines on opaque (black) background

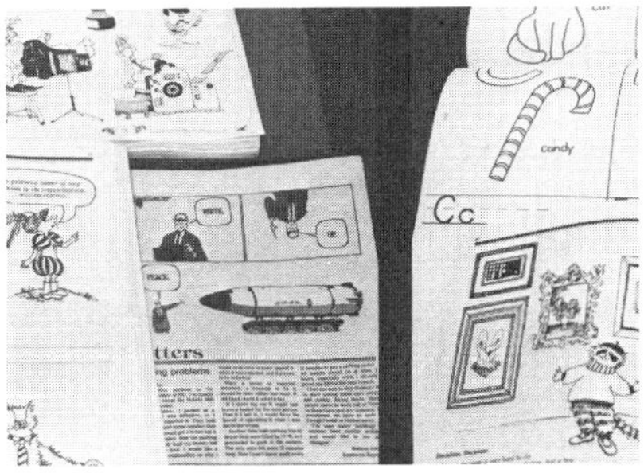

Cartoons from books or magazines can be adapted to show some concept on a transparency.

Helpful Hints for Thermal Copying

☐ The thermal process will reproduce only lines or letters that absorb heat. Writing devices that produce marks that contain carbon, like pencil lead, and are dark, work the best.

☐ If the print on the original is faded, is in colors other than black, or has been made with felt markers, you will need to photocopy the original and use the copy as an original.

☐ If the original is on thin paper it is often helpful to back it with a heavier sheet of paper. This is especially important if the original has dark printing on the back, such as with newsprint. "Printing through" of this information to your copied thermal transparency can be reduced by placing a dark sheet of paper behind your original and carefully adjusting the copier's heat control before running thermal plastic and original through the copier.

Transparency Frames

The stage of the standard overhead projector is 10 × 10 inches. Remember to design your transparency original so that the information on it fits a 7½ × 9½-inch space.

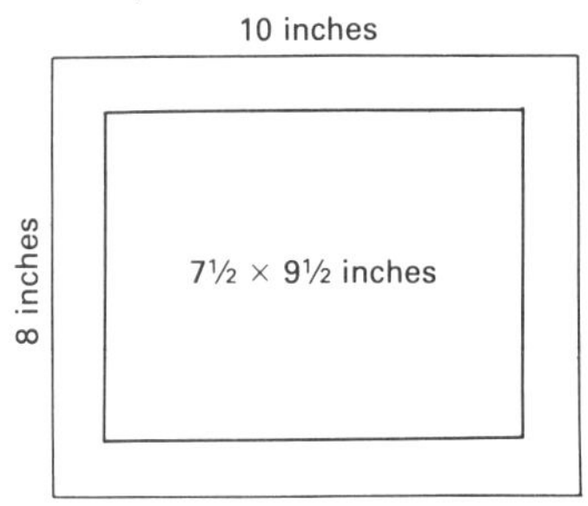

A simple, yet effective, method of lettering is to use a stencil and a No. 2 pencil.

TRY THIS

Look through newspapers or magazines to locate a cartoon, drawing, or other graphic design suitable for making a transparency; attach the item to white paper and make a thermal transparency. Mount in a frame and project.

If you are like many teachers you view prepared teaching materials with a critical eye. Possibly the labels on drawings are more complex than you would really like. Perhaps the captions are wrong, or maybe information is out of order for your teaching plan.

You can correct these problems by making your own original while incorporating work from other sources. At this point it is necessary to introduce a wide variety of equipment and techniques that were not needed before. Many are lettering devices that add a professional touch to your finished product. Some of these devices are:

- ☐ Pencil stencils
- ☐ Large-type typewriters
- ☐ Wrico and LeRoy professional lettering sets
- ☐ Varifont and KroyType mechanical lettering machines
- ☐ VariTyper Headliner photographic lettering machine

Instructions for the use of some of these devices are found below. You will find that more details accompany the equipment. It is always smart to check instruction manuals.

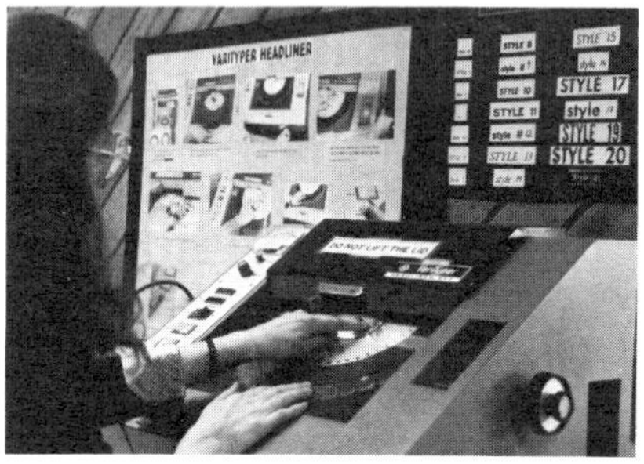

The VariTyper Headliner can be used to produce a variety of lettering styles for the original used to create an indirect transparency.

OPERATE

Lettering with a LeRoy Template

1. Check template label to select correct size of lettering.
2. Set the template against a T-square.
3. Set the pen firmly in the holder in the scriber arm.
4. Fill the well of the pen with two or three drops of ink.
5. Set the back/rear end of the scriber in the long black groove under the letters of the template.
6. Place the scriber arm with the sharp metal point into the groove of the desired letter.
7. Lower the pen to the surface of the paper and guide the scriber through the outline of the letter.
8. Lift the pen from the paper when finished and move the template into position to complete the next letter. "Eyeballing" letter and word spacing works best, once you have perfected the skill.
9. Wash your pen and plunger after you are finished using them.

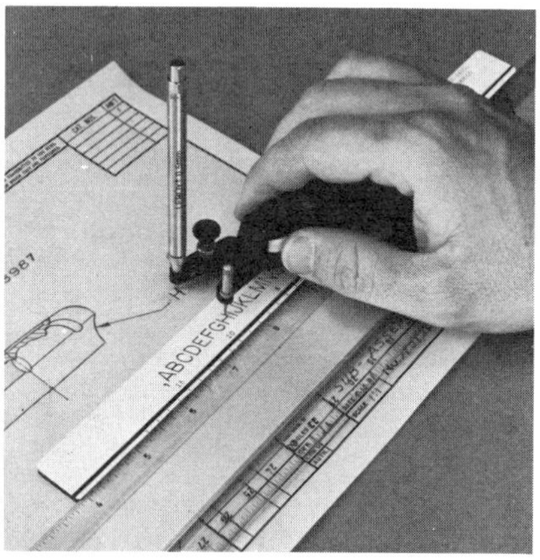

LeRoy lettering system (compliments Keuffel & Esser Co.).

Helpful Hints for Using LeRoy Lettering

☐ It is often desirable to tape the T-square to the table to keep it from slipping during lettering.
☐ To start ink flowing, stroke the pen on scrap paper. Sometimes it is necessary to raise the plunger part of the pen so that ink will reach the tip.
☐ The vertical screw just behind the scriber pen point should be adjusted so the pen is barely touching the surface of the paper.
☐ Begin lettering near the center of a letter or line so that ink does not glob at the ends of strikes.
☐ It is best to judge letter space by eye according to what looks good.
☐ Keep the pen moving. If it sits in one spot, ink may collect to form a large spot.
☐ A reservoir pen, such as a rapidograph pen, can be used in place of the standard LeRoy pen. This eliminates the need to constantly refill the pen reservoir.
☐ Clean pens by soaking them in a detergent solution, household ammonia, or a commercial pen cleaner.

Sample Template Label

Leroy 3240-350CL KEUFFER & ESSER CO. (Pen 4)

CL = capital and lowercase letters
350 = letter size (1000 = 1 inch; 350 = ⅓ inch)
Pen 4 = pen and plunger are numbered. The smaller the number the narrower the line.

OPERATE

Lettering with a Wrico Stencil Signmaker

1. Select the stencil size desired. (See information on stencil label.) Each stencil is lettered so the appropriate pen and stencil can be matched to one another.
2. Select the appropriate cork-backed stencil holder.
3. Align the stencil holder on the page and set the stencil on it.
4. Hold the pen in one hand and depress the plunger on the top. The grooves of the pen brush will protrude from the front of the pen.
5. Release the plunger and rotate the metal collar at the back of the pen shaft until the pen brushes are flush with or slightly recessed from the outer tip of the pen.
6. Fill the pen by depressing the plunger and immersing the exposed brush of the pen so that it goes below the surface of the ink. Ink will be held in the grooves of the brush. After a few seconds, release the plunger. The pen is now ready to use.
7. Position the Wrico stencil over your paper, and then hold the pen in a vertical position so that the tip is flat on the surface.
8. Begin a stroke near the middle of a line and work toward the ends. Keep moving so that puddles of ink do not form on the paper.
9. Set space between letters and words according to what looks good.
10. Wash the pen point and brushes under running water when lettering is finished.

Wrico Signmaker.

Sample Stencil Label

WRICO
Guide NO AVC 150
Use with brush pen C or
felt pen CF

AV = letter style (others are A, D, BF, T, MS)
C = capital letters on stencil
L = lowercase letters on stencil
N = numbers in stencil
150 = letter size (100 = 1 inch; 150 = 1½ inch)
C = pen size. Pens range in size from A to E with A making the thinnest line. Felt pens are coded AF to EF.

Helpful Hints for the Wrico Procedure

☐ Wrico pens are stamped with letter size near their points.
☐ The back of the stencil holder is stamped to indicate which elevation or side to use for each stencil.
☐ The brush is not an ordinary brush with lots of bristles, but since it holds the ink it is called a brush.
☐ Be careful not to let the outer tip of the pen touch the ink while loading.
☐ Retrace over a letter to make all lines crisp and even.
☐ Never leave ink in the pen long enough for it to dry. Always cap a pen when it is not in use.

PRODUCE

Producing an Indirect (Thermal) Transparency

You need: paper, marking devices (such as a pencil), stencils/guides or lettering machines, clip art pictures, thermal plastic, a frame, and masking tape

Procedure

1. Produce a layout following previous instructions.
2. Work within the size of the opening of your transparency frame. The horizontal format is preferable to the vertical.

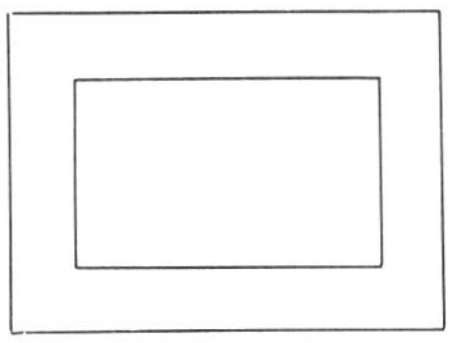

Horizontal

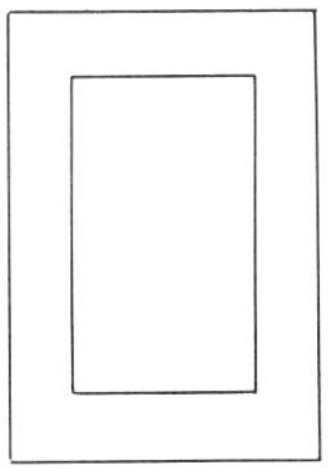

Vertical

Graphics for Transparencies

Use
India ink
Soft lead pencil
VariType
Varifont
Photocopies
Newsprint
Mimeograph copies
Offset printed copies
Boldface primary typewriter

Do Not Use
Spirit-duplicated copies
Ball point pen
Felt pen
Fountain pen
Grease pencil
Regular typewriter

Cutouts of visuals can be taped with double-stick tape or rubber-cemented onto the original.

3. Letter, draw, or paste-up to prepare your original. (*Note:* It is often easier to prepare drawings, sketches, and other information on separate paper and then paste-up this information on your transparency original. Just decide where you want a design and use double-stick tape or some other adhesive to secure it to your original.)
4. Select the appropriate type of thermal plastic.
5. Place the thermal plastic on top of your original with the notch on the plastic in the upper right-hand corner.
6. After the original and thermal plastic have been run through the thermal copier to make the transparency, use direct production techniques to add color or dry-transfer letters, if needed.
7. Mount your transparency on the back of your frame using masking tape. Leave the top of the frame free for notes or to attach overlays. Save your original.

When overlays are used it is possible to build a complex transparency from a simple one as the presentation progresses.

Special Techniques

If you stop here you will have a good grasp of transparency production and use. This next section of the chapter mentions several techniques, however, that you may want to add to your teaching repertoire. Special effects are not universally useful but we offer them because your imagination may lead you to techniques that have not been considered previously. Some of these techniques include:

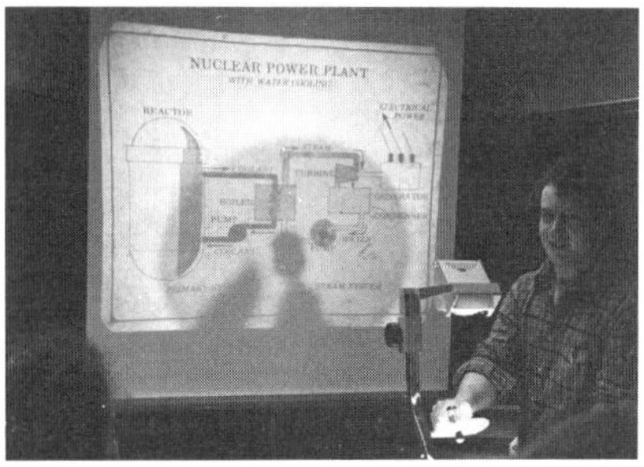

By using polarized film and a polarizing spinner it is possible to portray motion using an overhead projector.

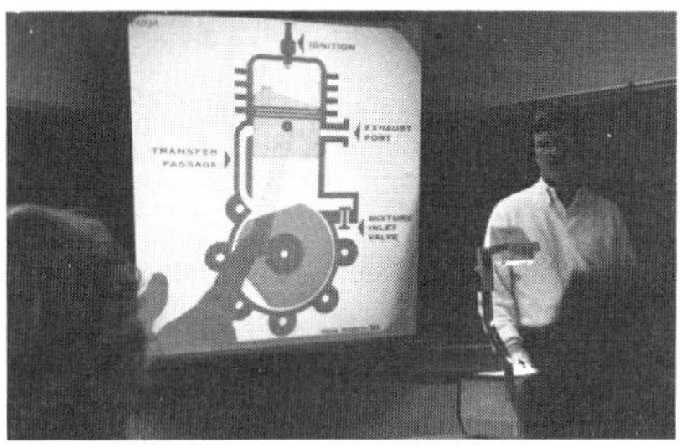

3-dimensional transparencies with moving parts can be used to show complex topics, such as this one on the operation of a two-cycle engine.

Overlays Used for any area in which it is necessary to develop a concept sequentially.

Color Lifts Used for any subject area in which you need a transparency of a colored magazine picture. They can be projected, put on an exterior window exposed to sunlight, or incorporated into a hanging light box.

Diazo This is an alternative technique for indirect production, with some minor advantages over thermal transparencies. The original can contain large areas of black, the lines are more distinct, and a wide variety of colors is available.

Polarized Transparencies Used for any subject in which movement would clarify a concept. The flow of blood through a heart and clockwise/counterclockwise rotation of gears are examples.

Three-dimensional Objects These objects could include a ripple tank, plastic-embedded objects, tools, or hand-held calculators.

Rear-Screen Projection A large rear screen, white sheet, or scrim may serve as a surface on which to project. Placed behind the surface (you will need about 1 foot throwing distance for every 1 foot of projected image width), your transparencies become backdrops for plays, modern dance, and other productions on a stage.

TRY THIS

Using your prepared original, produce a thermal transparency. Since this transparency will be framed, you will have a working area that is less than 8 × 10, so you should plan your layout size accordingly. On the base cell of this transparency include the title and the graphic/visual you designed. This information should be drawn and lettered on a sheet of paper. Use stencil, LeRoy, Wrico, or machine lettering. On the second cell (also called an overlay) place the labels for the visual. Use a different size and/or style of lettering than you used for the title.

Follow the directions for producing a thermal transparency and for operation of a thermal copier to produce your plastic cells, then mount them in a frame. The base cell should be taped to the back of the frame, and the overlay should be taped to one edge on the top of the frame. Add color using direct transparency production techniques. Place the transparency with overlay on the stage of an overhead projector and practice using it.

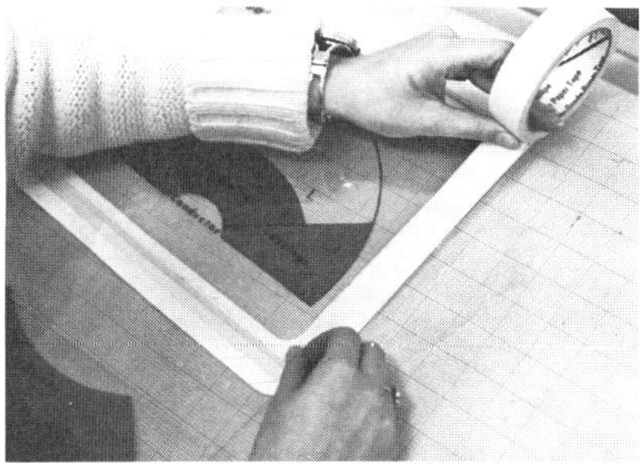

Use masking tape to attach overlays to the top of the frame. The base cell is taped to the underside of the frame.

PRODUCE

Producing an Ink (Color) Lift Transparency

You need: a colored magazine picture that is printed on a clay base, clear contact paper or lamination film, a pan of water, a transparency frame, and masking tape

Procedure

1. Select a magazine picture. Those in bold, bright colors are most suitable. Only pictures printed on a clay base will lift. To determine if a magazine picture is printed on a clay base conduct the following simple test:
   a. Moisten one of your fingers.
   b. Rub a white portion of a picture in the magazine with your finger.
   c. Check to see if a chalky white residue appears on your finger. If you find this residue (chalk), the inks in this picture were printed on a clay base and can be easily lifted to make a transparency. If no residue appears, you cannot use pictures from this magazine for ink lift. (Magazines that are printed using a clay base are *Time, Newsweek, Field and Stream, Ladies Home Journal, Sports Illustrated, Better Homes & Gardens, U.S. News & World Report,* and many, many more.)
2. Remove the picture from the magazine.
3. Laminate the front of the picture. If lamination film and a dry-mount press are not available you can use clear contact paper. Rub the contact paper plastic on the front of the picture.
   a. Laminating Film—Use normal dry-mounting techniques to laminate the front of the picture.
   b. Clear Contact Film—Cover the picture face with clear contact paper, being careful not to let wrinkles or bubbles develop. Rub the entire surface of the contact paper with the flat side of a comb or side of a ball point pen to seal the plastic tightly to the picture. Failure to do so may result in a poor lift that is grainy with blank spots.
4. Soak the picture in water. A minute or two is usually long enough. Sometimes, a drop of detergent in the water may be needed to allow the water to loosen the paper from the plastic.
5. Peel the plastic from the paper. The ink should be lifted by the laminating film when the paper is peeled away.
6. Carefully wash off excess chalk. Hang the lift to dry.
7. Protect the back of the lift by spraying it with a clear plastic spray, or by laminating it again. It may be necessary to laminate the lift again so that it will fit in a frame. Spraying with clear plastic also dramatically makes the lift more transparent.
8. Mount the ink lift in a frame for protection.

Helpful Hints for Ink Lift Transparency Production

☐ Bright pictures with sharp color contrasts work the best for lifts.
☐ The surface of the picture must be kept as clean as possible. Dust, dirt, or fingerprints may prevent inks from adhering to the plastic.
☐ Do not trim the picture before lifting.
☐ A roller, such as a rolling pin, should be used to press the contact paper plastic to the picture.
☐ If the completed color lift is too small to fit a frame, make your own frame from a piece of cardboard. Or laminate the color lift again and mount the finished product in a standard frame.

Soak the laminated magazine picture in water with a few drops of soap dissolved in it.

TRY THIS

Select a brightly colored full-page magazine picture for a transparency. Test the magazine to be sure it is printed on a clay base. Laminate or press contact paper to one side of the picture and produce a color lift by following the directions above. Mount the clean, dry transparency in a frame. Try it out on an overhead projector.

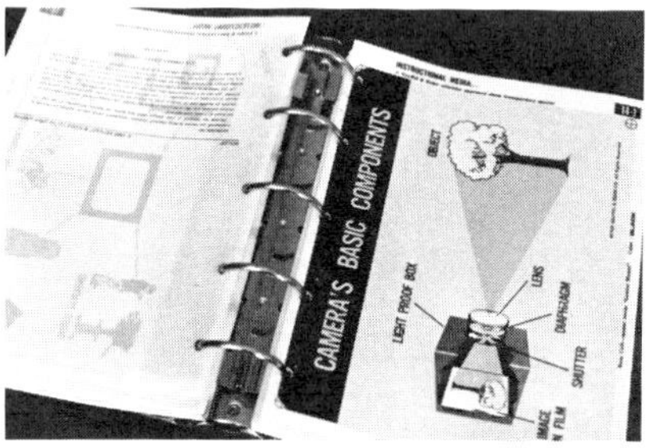

Diazo originals can be produced locally or purchased commercially, such as this collection of originals on media.

Hints on Diazo Transparency Production

☐ Exposure time in the ultraviolet light is critical, but development time is not.
☐ Multiple transparencies can be made from the same original.
☐ Class handouts can be made from the same original the transparency was produced from.
☐ A thermal transparency may be used as the original for a diazo transparency.
☐ Diazo films do not have a long shelf life; they should be kept in a dark, cool place such as a refrigerator.
☐ Diazo originals cannot have imperfections such as smudges. Mistakes may be cut out of the original with a razor blade.
☐ One brand of diazo film (Copyflex Multicolor) gives more than one color on a single sheet of plastic. Color couplers are applied after exposure to form any one of five colors.

PRODUCE

Producing a Diazo Transparency

You need: translucent tracing paper, diazo film of the appropriate color, black drawing ink and pens, guides or stencils for lettering, a diazo machine (ultraviolet light printer and developing chamber), a frame, and masking tape

Procedure

1. From your layout sketch prepare a master on translucent paper using black inks or other materials that make opaque lines. High-grade tracing paper used in engineering or drafting works best. Note: Many commercially prepared diazo originals are available for purchase. If a multicolor transparency is planned, prepare the masters for the other colors of diazo film. These multiple masters will be sandwiched together on the frame.
2. Select the appropriately colored diazo film(s).
3. Expose the diazo film to ultraviolet light. Generally the transparency plastic (notch in the upper right-hand corner) goes under the paper original. The original should be between the diazo film and ultraviolet light.
4. Transfer the exposed diazo film to the ammonia developer. In a short time an image will appear. Keep the film in the ammonia until full color appears. It is not possible to overdevelop. Note: Many types of diazo machines are available. Be sure to read the directions for the one you have.
5. If the transparency image is too *light*, or faint, *reduce* the exposure time for the next sheet of film. If the transparency shows unwanted color, *increase* the exposure time to the ultraviolet light for the next sheet of diazo film. A diazo transparency with unwanted color in spots meant to be clear has been underexposed.
6. Mount the transparency in a frame for projection.

PRODUCE

Producing a High-Contrast Transparency

You need: a line drawing suitable for reproduction (usually one in need of enlargement or reduction), Kodalith Ortho Type-3 film (or equivalent, from another manufacturer) in 35mm and sheet form, Kodalith developer, stop bath, fix, a 35mm SLR camera, a copy stand with lights, a darkroom, a transparency frame, masking tape.

Procedure

1. Load camera with 35mm high-contrast film.
2. Photograph your subject on the copy stand (see Chapter 11). Kodalith Ortho Type 3 film has an ASA of 8.
3. Process the exposed film using recommended high-contrast developer. High-contrast film can be processed in open tanks under a red safelight.
4. Opaque out any light spots or unwanted information on the negative with India ink or some other opaque marker. Note: At this point the high-contrast negative could be used as a master for a diazo slide, or mounted into a 2 × 2-inch cardboard slide frame and used as a high-contrast slide.
5. Place the high-contrast negative in the enlarger negative holder and insert it in the darkroom enlarger (see Chapter 11). Expose an 8 × 10-inch sheet of high-contrast sheet film. Develop, stop, and fix as in step No. 2.
6. Dry and mount the transparency for use. (*Note:* If a colored instead of black and white high-contrast transparency is desired, the 8 × 10-inch Kodalith film can be used as a master to produce a diazo transparency.)

PRODUCE

Producing a Half-Tone (Continuous-Tone) Transparency

You need: a half-tone or continuous-tone original you want to enlarge or reduce to make a transparency, regular continuous-tone black and white film (Kodak Tri-X, for example), high-contrast Kodalith sheet film, paper developer (Dektol), stop bath, fix, a 35mm SLR camera, a copy stand with lights, a darkroom, a frame, masking tape

Procedure

1. Load your SLR camera with continuous-tone 35mm film, such as Tri-X or Plus-X.
2. Photograph your continuous-tone subject.
3. Process the exposed film into a negative (see Chapter 11).
4. Enlarge the negative using the darkroom enlarger and expose an 8 × 10 inch sheet of Kodalith high-contrast film. Note: Make test strips to calibrate the correct exposure time, because this process usually requires shorter exposure times than other kinds of enlargement work.
5. Develop, stop, fix, and wash the exposed sheet of high-contrast film using a paper developer such as Dektol. A darkroom safelight may be used. Image quality can be properly assessed only under room lights, however. Hint: Try a 1:12 dilution of Dektol to water for the developer concentration.
6. When the continuous-tone transparency is dry, mount it in a frame.

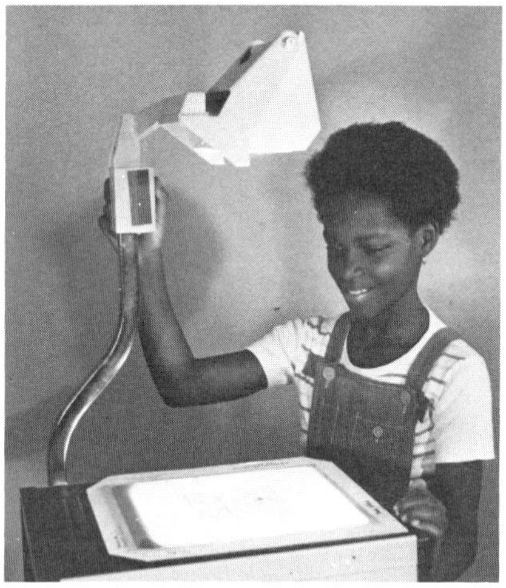

Encourage students to use transparencies when they make presentations to the class.

A FINAL WORD

We consider transparencies and the overhead projector to be among the most important and useful media for teaching. Little artistic ability is necessary, the machinery is not particularly complex or expensive, production steps are easy to follow, and a great deal of teaching mileage is to be gained.

One more thing—do not be stingy with your ability to make transparencies. It is not the private property of teachers. Suppose, for example, you assign a small group of students in your class the task of preparing for a panel discussion. The traditional panel discussion would be entirely verbal, of course. Each member, in turn, would offer views and a moderator would maintain order and flow. You would hope, somehow, that students would say something significant, on target, in a reasonable time.

The entire matter could be handled more effectively with transparencies. Allow the panel members to work out their arguments beforehand on a piece of plastic, using a felt pen. Review the transparencies prior to the discussion, checking for content and pace. Make suggestions for revision if needed, and put in a plug for the use of good communication techniques—eye contact, voice modulation, and so forth. Above all, look for evidence that students have adequately prepared themselves for their roles on the panel. You may be surprised at the interest they show—and their creativity—when transparencies are introduced as an alternative to verbal communication.

LOOKING AHEAD

With these first eight chapters we are ready to move to the next major section of the book—Producing Scripted Media. Up to this point your design and use of teaching material required some planning, but not necessarily a master plan. Each of the media permitted you to extend your senses by showing, telling, and explaining more clearly in less time, but each required your presence and your control.

We do not believe the teacher will ever be replaced. Perhaps re-place is a better description. Your role may change. Your function may shift from providing direct classroom instruction to providing the media for learning. A tape/slide set is a good example. Properly scripted and executed it is possible to offer this combination of media with the same intentions that you offer a book. Students can learn from the media. Your role is that of programmer, of organizer, and of guide. You draw on the necessary elements, orchestrating them into an instructional pattern that can stand alone.

We mentioned the need for scripting. This systematic planning procedure is somewhat like writing a lesson plan. But writing and script writing are not the same—and that is another story for another chapter.

REVIEW QUESTIONS

1. When planning the layout for an overhead transparency, should you design it horizontally or vertically? Why?
2. What should be the minimum letter size used on a transparency?
3. How long should the title of your transparency be?
4. Is it possible to remove the markings made on plastic with a permanent felt marker?

5. List some techniques for adding color to a transparency.
6. Define *keystoning.*
7. Where should the screen be placed in a standard classroom setting? Why?
8. If your classroom is long and narrow, what screen surface would work the best at providing a bright image?
9. What determines what kinds of lettering will work for the original used to make a thermal transparency?
10. Should dry-transfer letters be added before or after the original is run through the thermal copier?
11. If your transparency is light and underexposed what should you do to the setting of the thermal copier before making another copy?
12. When framing a transparency, to which side of the frame should the base cell be taped?
13. When producing a diazo transparency, is exposure time critical?
14. How can you determine if a picture in a magazine can be lifted to make a transparency?
15. Is it possible to make a transparency of a person's face?
16. How are overlays mounted onto a transparency frame?
17. Why might a high-contrast transparency be an important technique for producing a transparency?
18. What is the difference between stencil, guide, and machine lettering systems?

ANSWERS TO REVIEW QUESTIONS

1. Horizontal, because in a low-ceilinged room, where the screen cannot be placed high enough, the bottom portion of the transparency will not always be visible.
2. 1/4 inch
3. Three to five words
4. Yes, with a solvent such as lighter fluid.
5. Colored felt markers, colored adhesives, colored tapes, dry-transfer letters, and color paints.
6. Keystoning is the distortion of a projected image caused when the screen surface is not parallel with the projector (because the screen is not perpendicular to the axis of projection).
7. In the front corner nearest the windows. In this location the light from the windows will not fall on the screen, the least number of students will be blocked out by the teacher and projector, the blackboard will not be obstructed by the screen, and students can see around one another more easily.
8. Glass-beaded, because this screen surface has a fairly narrow viewing angle, but gives off a very bright image, so students sitting farther away than the equivalent of six screen widths will still be able to see what is projected.
9. The lettering must absorb heat. Usually this means it contains carbon.
10. After; dry-transfer letters are a direct production technique.
11. Move the setting to a darker position. Usually this means the original and thermal plastic will move through the thermal copier more slowly.
12. The back, so the front is free for notes or to attach overlays.

13. Yes; development time in ammonia is not.
14. By using the clay test.
15. Yes, by following the procedures described for making a continuous-tone transparency.
16. They should be taped to the front of the frame on one side only.
17. In order to reduce or enlarge the size of some visual information so that it will fit on an overhead projector stage.
18. Stencils are designs or letters cut through a lettering aide. To letter, a pencil or pen is placed through the cutout stencil. (e.g., Wrico Signmaker). Guide lettering requires the use of a special mechanical device, such as a scriber, that allows you to follow the outlines etched into, but not completely through, the surface of the guide itself (e.g., LeRoy guides). Machine lettering is done entirely with a mechanical device. To letter, a key or button is pressed and letters are produced either photographically (e.g., VariTyper) or by the striking of a raised letter on a paper or plastic surface (e.g., typewriter or Varifont machines).

REFERENCES

Adam, Sarah, et al. "Readable Letter Size and Visibility for Overhead Projection Transparencies," *AV Communication Review* (Winter 1965): 412–417.

Brown, James W.; Lewis, Richard B.; and Harcleroad, Fred F. *AV Instruction: Technology, Media and Methods,* 5th ed. (New York: McGraw-Hill, 1977). (With lab and instructor's manual.)

Bullard, John, and Mether, Calvin. *Audiovisual Fundamentals,* 2d ed. (Dubuque, Iowa: W. C. Brown, 1979).

Bullough, Robert V. *Creating Instructional Materials,* 2d ed. (Columbus, Ohio: Charles E. Merrill, 1978).

Kemp, Jerrold E. *Planning and Producing Audiovisual Materials,* 4th ed. (New York: Harper and Row, 1980).

Making Black-and-White and Color Transparencies for Overhead Projection, Pamphlet S–7 (Rochester, N.Y.: Eastman Kodak Company, 1975).

Minor, Ed, and Frye, Harvey R. *Techniques of Producing Visual Instructional Media,* 2d ed. (New York: McGraw-Hill, 1977).

Projecting Ideas on the Overhead Projector. Series of three 16mm motion pictures, sound, color. (Iowa City: University of Iowa, 1964).

Satterthwaite, Les. *Graphics: Skills, Media and Materials,* 3d ed. (Dubuque, Iowa: Kendall/Hunt, 1977).

Snowberg, Richard C. "Bases for the Selection of Background Colors for Transparencies," *AV Communication Review* (Summer 1973): 191–207.

Teachers Guide to Overhead Projection (Holyoke, Mass.: Technifax Corp., 1969).

Volker, Roger P., and Simonson, Michael R. *Media for Teachers,* 2d ed. (Dubuque, Iowa: Kendall/Hunt, 1977).

Wittich, Walter A., and Schuller, Charles F. *Instructional Technology: Its Nature and Use,* 5th ed. (New York: Harper and Row, 1973).

Yeamans, George T. *Projectionists' Programmed Primer,* 2d ed. (Muncie, Ind.: Ball State Bookstore, 1973).

SOURCES FOR TRANSPARENCY-MAKING EQUIPMENT AND MATERIALS

A. Transparency Supplies
Advanced Audio Visual
231 Penn St.
Brooklyn, NY 11211

ED-Tech Services Co.
Box 407
Chatham, NJ 07928

Labelon Corp.
10 Chapin St.
Canandaigua, NY 14424

Scott Graphics, Inc.
28 Gaylord St.
South Hadley, MA 01075

Tersch Products, Inc.
P.O. Box 206
Rogers, MN 55374

B. Diazo Equipment and Materials
Arkwright Inc.
100 Main St.
Fiskeville, RI 02823

Bruning Division, Addressograph-Multigraph Corp. (Multicolor film)
1834 Walden Square
Schaumburg, IL 60196

General Aniline and Film Co.
140 West 51st St.
New York, NY 10020

Keuffel & Esser Co.
29 Whippany Rd.
Morristown, NJ 07960

Scott Graphics, Inc.
28 Gaylord St.
South Hadley, MA 01075

VariTyper Corp.
720 Frelinghauysen St.
Newark, NJ 07114

C. Heat-Process Equipment and Materials
A. B. Dick Co.
5700 West Touhy Ave.
Chicago, IL 60648

Agfa-Gaveart Inc.
275 North St.
Teterboro, NJ 07608

AM International (Copyflex)
1800 Bruning Dr., West
Itasca, IL 60143

Arkwright Inc.
100 Main St.
Fiskeville, RI 02823

Bell and Howell Co. (Ditto Automaster)
6800 McCormick Rd.
Chicago, IL 60645

Labelon Corp.
10 Chapin St.
Canandaigua, NY 14424

3M Company, Visual Products Division
3M Center
St. Paul, MN 55101

Tersch Products, Inc.
P.O. Box 206
Rogers, MN 55374

D. Polarizing Materials
American Polarizers
1500 Spring Garden St.
Philadelphia, PA 19101

E. Xerography Film
Arkwright Inc.
100 Main St.
Fiskeville, RI 02823

Eastman Kodak Inc. (Kodalith, Dektol)
343 State St.
Rochester, NY 14650

Folex-Biotest-Schleussner, Inc.
60 Commercial St.
Moonachie, NJ 07074

Keuffel & Esser Co.
29 Whippany Rd.
Morristown, NJ 07960

Labelon Corp.
10 Chapin St.
Canandaigua, NY 14424

Plastic Suppliers (Zelar)
2400 Marilyn Lane
Columbus, OH 43229

3M Company, Visual Products Division
3M Center
St. Paul, MN 55101

PRODUCING MEDIA FROM SCRIPTS: AN INTRODUCTION

C

Media discussed in Part B, such as the handout, the display, and the transparency, have at least one common characteristic. Often they are used in combination with other media, or are presented by the teacher in a lecture/discussion. Handouts can be used to summarize information presented in films or slide presentations so students will not have to take notes when the lights are off. Displays act to reinforce classroom instruction, and transparencies are often used to extend the ability of the teacher to present visual and verbal information during a lecture.

In Part C, media that are often used as the primary information delivery method are presented. Slide presentations, films, and television programs are often designed to stand alone and to present concepts that do not require direct teacher involvement. These media are based on a script, and a more elaborate production process is followed than for media presented in Part B.

The script is the basis for producing and using the media presented in this section of the text. Audio recordings, slide presentations, film, and video programs begin with a plan. That plan is used as a guide in making the media. Without this plan, called a script, the media product is not likely to be effective at reaching instructional goals and objectives.

Chapter 9 explains how the script is written, and how it is used to guide the production of other media. Chapter 10 presents techniques for making audio recordings. While audio is often used to accompany other media, such as in slide presentations or films, it can also stand alone as an information delivery technique.

All the media discussed in Part C are consumer items. While many households may not have a video recorder, most will have at least one television receiver. Super 8mm cameras are found in many homes, and many of us go to the movies. Most of us have access to still cameras and audio recording equipment. These media are certainly not new to us, and even though some production techniques may be a little complex, the media are familiar. This section will build on the knowledge we already have about audio recorders, still and motion cameras, and television equipment, so these media can be effectively and easily used in teaching and training.

GOAL

☐ Understand the process of planning a media presentation.

OBJECTIVES

☐ Develop a story line.
☐ Write a treatment description.
☐ Write a script.
☐ Produce a storyboard sequence.
☐ Describe record-keeping techniques used during production.
☐ Describe different ways to view action.
☐ List ways to connect visuals.
☐ Describe ways to show continuity when producing a series of visuals.

Actors take their places. The director scans the script. Lights dim. Camera motors whirl. The show begins.

Suddenly, a shot rings out! A body drops. A scream pierces the darkness as music pulses relentlessly.

An educational film? Probably not. How about something like this, then?

Guitar music begins as the screen shows a brightly colored image—"HOW TO SELECT A PATTERN." After five seconds, the teacher refers to a script and changes the image to a close-up of a Butterick #8013 pattern envelope as a pleasant-sounding narrator begins talking. "Obviously you want to learn to sew your own clothes, so let's take the first important step toward that goal, the selection of your pattern . . ."

Do these two productions share anything? One is probably the beginning of a professionally directed and acted movie, maybe a Clint Eastwood detective drama. The other is a lesson for a tenth-grade home economics clothing class, possibly one the teacher made as a student in college. One similarity is that they were both produced from a script.

When media are produced at any level, professional or amateur, it is critical that a strict planning sequence be followed. There is a direct relationship between planning and success. The process of organizing any visual presentation, no matter how technical, or how well financed, is virtually the same. The complexity in producing the medium from the script is what differentiates the amateur from the professional presentation.

This chapter will explain the organizational techniques used when videotapes, slide sets, classroom films, and instructional videotapes are produced. The story line, the treatment description, the script, and the storyboard will be described. Connectives, scene sequencing, and production record keeping are also included.

9

PLANNING AND SCRIPTING PRESENTATIONS

A MODEL FOR PLANNING INSTRUCTION

The term *script* means different things to different people, and has different meanings at different times. Script means one thing to the television producer, another to someone writing a radio program, and still another to the teacher preparing a single-concept film loop. Usually script refers either to the general planning process followed for producing media or to a specific format or model for sequencing visuals and narration that will be produced later. Both uses will be described in this chapter.

There are several plans that can be used for the production of media. The one presented here incorporates ideas from most of them. Steps 1 and 2 in this model are discussed in Chapter 4. The third step is explained in Chapter 3 as part of the discussion of Dale's *Cone of Experience.* Steps 5 through 8 will be the foundation for Chapters 10 to 15. This chapter will concentrate on step 4—the development of the production plan. This is often referred to as the *scripting process.*

There are five substeps to follow when media are planned and produced. A *story line* is developed first. Similar to a lesson plan, a story line sequences media presentations and teacher-student interaction events into a unit. If commercially produced media are not available to fill in the story line, media may need to be produced. The remaining planning steps pertain to the media production process.

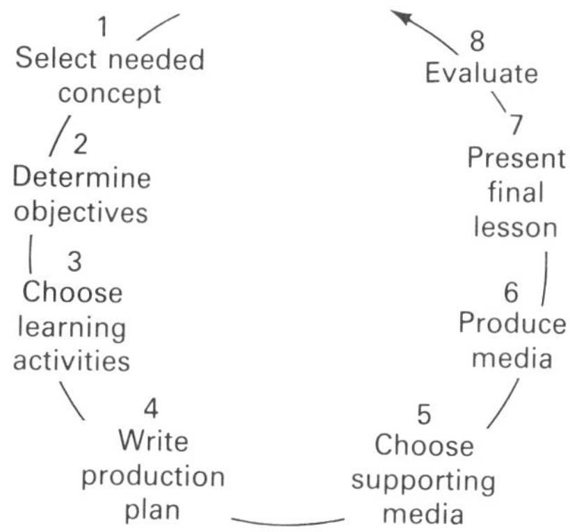

Sequence for planning a media presentation.

Production Planning Process

1. Story line
2. Treatment description
3. Production script
4. Storyboard
5. Teaching script

A *treatment description,* a narrative explaining what the entire presentation is about, is written before any other production is begun. Next, a *production* or *planning script* is written. This script sequences and correlates visuals to narration. Fourth, a *storyboard* is produced. The storyboard is the tool used during production to direct all filming, photography, and recording activities. Once media are produced, the *teaching script* is written by revising the production script.

THE STORY LINE

Sample Story Line Sequence

1. Text study
2. Discussion
3. Slide set with audiotape
4. Discussion
5. Worksheet
6. Film
7. Discussion
8. Quiz

The *story line* is the combination of several instructional events into a lesson. For example, a teacher may plan to begin instruction on a topic, such as pattern selection, by having students read about patterns in their textbooks. Next, a discussion on what patterns are supposed to do for the seamstress or tailor would follow. This discussion would lead the class into the presentation of a series of slides on pattern selection. Again there would be a discussion. The next day the teacher might schedule a film on patterns. During this film the students would complete a worksheet. At selected times the teacher would stop the film and ask questions. Last, the lesson on patterns would conclude with a quiz. The organization of these instructional activities is a story line.

The story line is a general outline of how instructional events fit together to reach a major course goal. For each medium subunit within the story line the teacher would have a teaching script. This script would be a detailed description of what will be shown and what will be said. For a classroom discussion, the teaching script might be a list of questions with possible answers. Previously produced media should have teaching scripts that can be incorporated into newly developed story lines. New presentations must have teaching scripts produced for them.

PRODUCE

Preparing a Story Line

1. Identify an important subgoal in your subject area (e.g., nationalism, supply and demand, pesticide application, company health care programs).
2. List specific goal-related objectives you feel students should be able to accomplish.
3. Determine the instructional technique most suited to reaching each objective.
4. Locate instructional materials designed to reach your objectives.
5. Produce materials when you cannot find media for your objectives.
6. Sequence objectives and instructional activities into a story line using Bloom's Taxonomy.

TRY THIS

Produce a story line for a topic in your subject area.

TREATMENTS, SCRIPTS, AND STORYBOARDS

When instructional materials are to be developed, a *treatment description* is written first. A treatment description is a general outline of what the media presentation is to be about. It is written as a narrative, is fairly short, and tells the story of the program. Some teachers compare the treatment description to an abstract of a document, or to a proposal for a research study.

Proposed Film Treatment

Open with an "art-type" shot of a symbol for an electric company such as a low-angle view of huge transformers in a switchyard.

Narrator speaks about a public utility, such as a light and power company, introducing a situation in which a customer phones to complain about an overcharge on a current bill.

We see and hear the customer telephoning; then we go to a split screen—customer on one side, clerk on the other. Clerk explains that it will be necessary to look up the records and call the customer back.

Narrator introduces a retail store, clerk again responding to an irate customer's complaint—repeat the split-screen technique.

Diagonally split screen is used to see and hear another irate customer.

Continue with a couple of more typical customer complaints treated in a similar manner. Here we switch either to the narrator in person—possibly a "name"—or we cook up a trick montage, with one complaining customer splitting into two, then four, eight, sixteen, thirty-two.

Narrator refers to these situations, saying that undoubtedly the viewer is familiar with this problem, and that we wouldn't have set it up unless we had a solution. The solution is microfilm *plus* rapid retrieval.

When the narrator first utters the word "microfilm," we cut to a microfilm screen with the images racing across. On the word "retrieval," one image stops.

Narrator suggests that we go back a moment.

We again see the last of our original complaining customers talking with store employee. Clerk asks customer for the number on the sales slip the customer is questioning.

In close-up shots, we see:
  the clerk's hand selecting a microfilm magazine,
  inserting magazine into a reader slot,
  pushing a button,
  lines zip across the screen and stop on a sales slip.

Clerk answers customer's question on phone, and customer expresses satisfaction.

Source: *Basic Production Techniques for Motion Pictures,* Kodak Pamphlet P-18 (Rochester, N.Y.: Eastman Kodak Co.): 6.

PRODUCE

Preparing a Treatment Description

1. Select a topic and narrow it.
2. Write behavioral objectives for the topic, and sequence them according to Bloom's Taxonomy.
3. List ideas or topics you want to include in the presentation. Use your objectives as guides.
4. Sequence these topics and connect them with a narrative that explains what you want to present in your program. Write just as you would if you were explaining your presentation to a friend.

TRY THIS

Write a treatment description for a presentation you might make on a topic in your subject area.

The next step is planning the production of a slide show, film, or videotape is to develop a shooting or *production script.* A script is a document that matches visual to verbal information. It lists what is shown and what is said. The basic component of a production script is the *script block.* A script block has three parts: (1) the visual, (2) the medium used, and (3) the narration. There may be two to eight script blocks for each script page.

Sample Script Block

1. The *visual* is shown here

2. The *medium* is listed here

3. The *narration* is written here

. . . .

. . . .

. . . .

Things to Be Considered in Your Script

1. "Readiness" (pretest information if needed)
2. Introduction
   a. Orientation
   b. Motivation
3. Individual differences
4. Variety of presentation (attention span)
5. Student involvement and immediate reinforcement
6. Key questions aimed at the desired thinking level.
7. Examples of use of the concept
8. Exercises which are practical preparation for the behavioral objective stated
9. Activities appropriate to the stated age or grade level
10. Closure
    a. Summary and review
    b. Directions to ensuing student activities
11. Student resources for study and/or reinforcement

An adequate script is one which could be given to a fellow student who could use it as a guide to preparing all needed media and present the lesson in your absence, much as a substitute teacher must do in practical teaching situations.

The production script becomes the map for the medium to be produced. It provides directions for picture-taking, filming, and recording activities. In the box on the left-hand side of the page, the visual is described or indicated by a sketch. This box should portray what the slide should look like, or what the film or video scene will show. A short written description of the visual can be used instead of a sketch. On the line under the box the medium is identified. The narration, in either verbatim or outline form, is included on the right-hand side of the script page. The more complete and specific the production script, the easier subsequent production will be.

Ways of Scripting Various Media

You may wonder how many frames on a script sheet should be used when scripting slide sets, tapes, and films. Here are some guidelines.

Slides Each slide is a frame. Make slides move rapidly—about 6 to 10 seconds per slide. Sketch or describe close-ups to show exactly what the narration refers to.

Prints Each print is a frame. Sketch or describe the scene, showing points of interest as closely as possible.

Film or Videotape Each scene is a frame. A scene is a single shot with a certain camera angle or position and a certain degree of magnification—Long Shot (LS), Medium Shot (MS), Close-Up (CU), Extreme Close-Up (ECU). Scenes should be 5 to 10 seconds long. For super-8 film, each roll lasts 3 minutes 20 seconds, so all scenes must fit that time—unless you make more than one roll of film.

Transparencies Each cell is a frame. Cells are single plastic sheets. The transparency in the frame is a cell. Each overlay is a cell. If you are revealing a single cell using cardboard "windows" or a sheet of paper, each revelation is a cell.

Hook and Loop and Flannelboard Each addition is a frame. As you add items, use a new frame for each—much like a transparency is scripted.

Flip Chart Each panel (page, sheet, etc.) is a frame.

Audiotape Forget about frames, just script what you want to say. You can write it out word-for-word, or make an outline and talk extemporaneously from it. Extemporaneous tapes are a little more natural and students seem to like them better, but they often contain a lot of "oh's," "and's," and "but's."

When more than one person is going to be involved in the production process, a *storyboard* of the presentation is developed next. Often teachers bypass this step in the production process because they script and produce the media themselves.

The Teacher as a Blueprint

☐ If students see the benefit of quality effort, they will come to value quality.
☐ If students see that planning will enhance the execution of skills, they are more likely to commit themselves to preparation.
☐ If students see integrity and sincerity in the efforts of the teacher, they will find a basis for mutual respect.

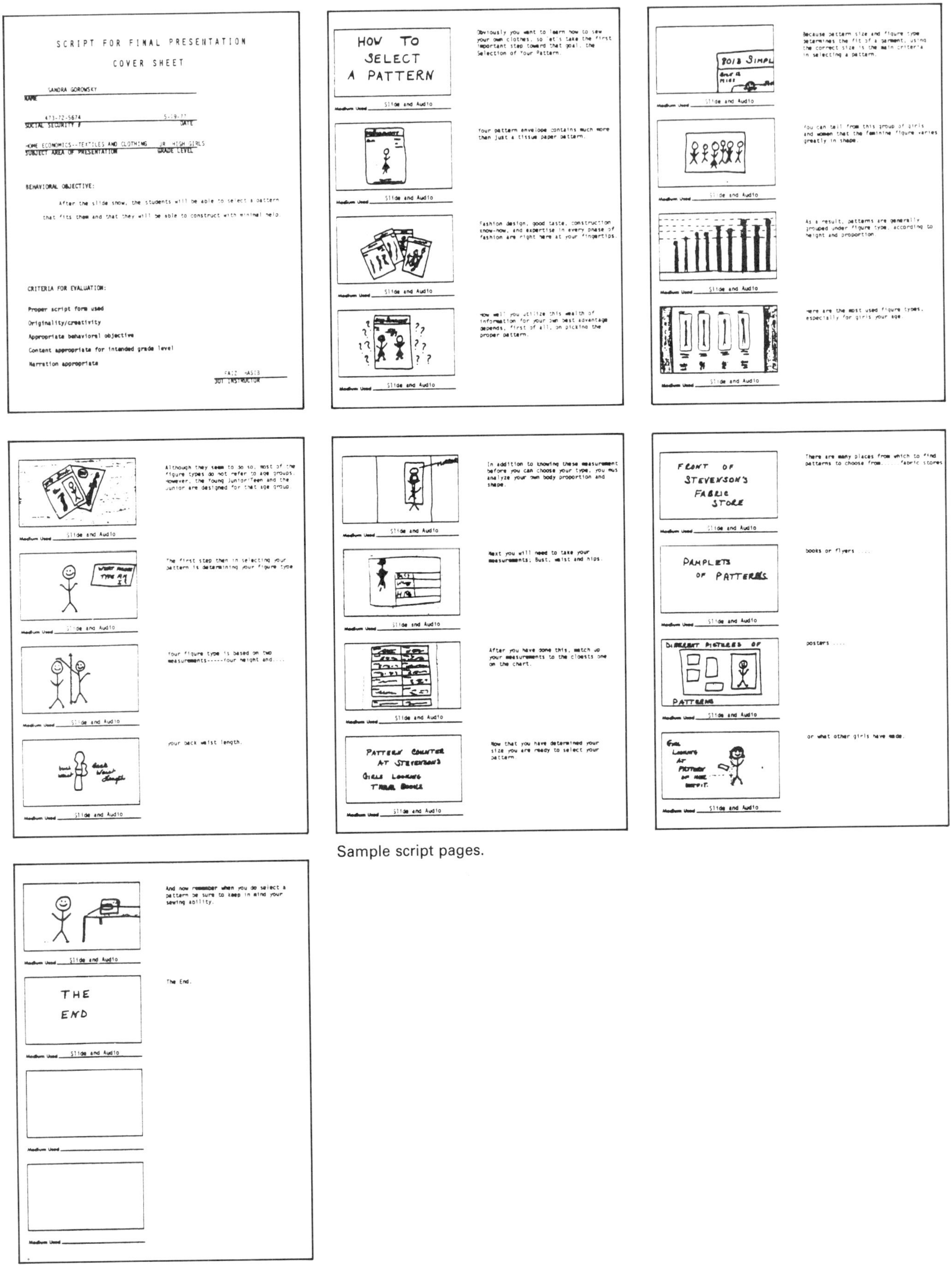

Sample script pages.

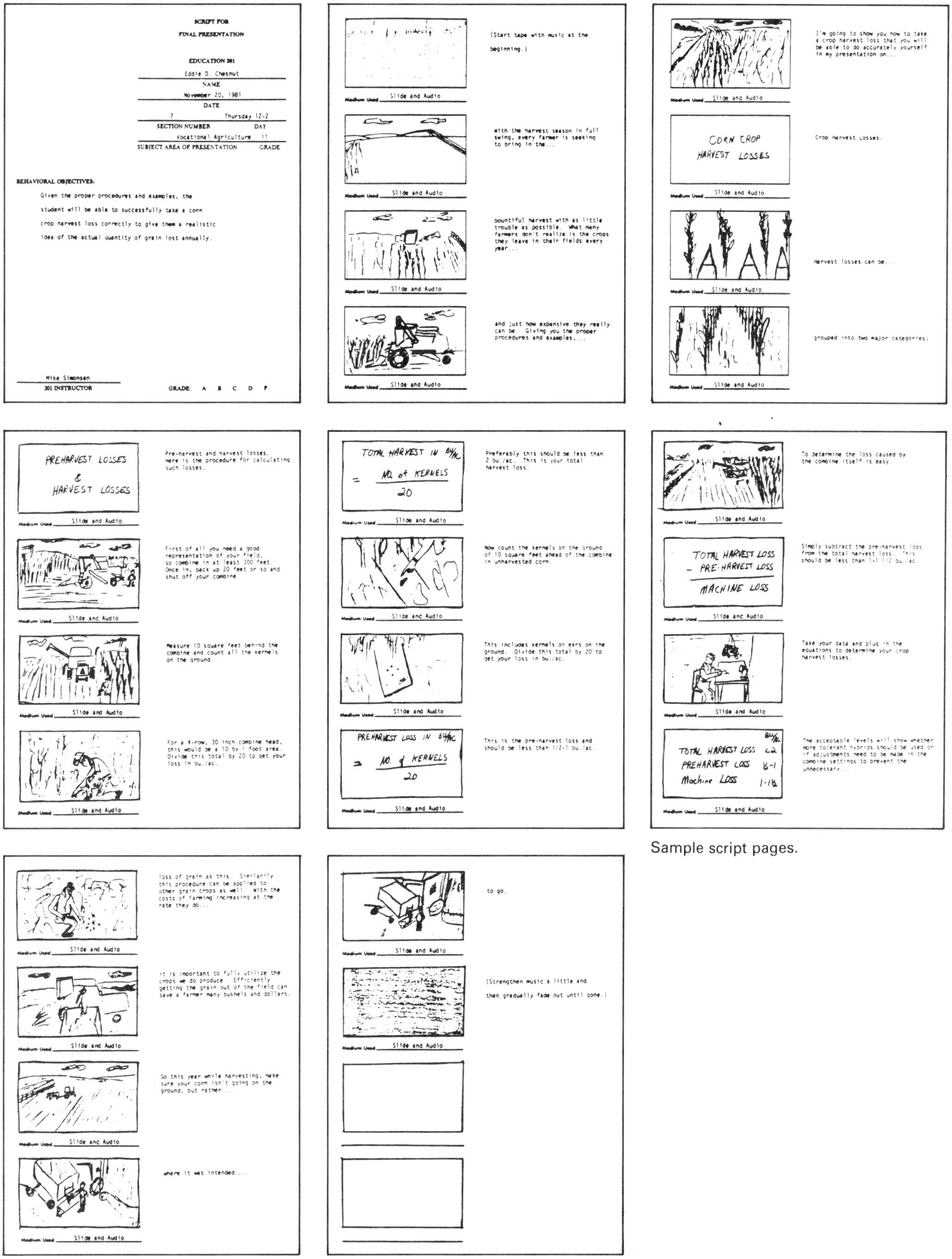

Sample script pages.

PRODUCE

Preparing a Production Script

You need: a treatment description, blank script sheets

Procedure

1. Follow your treatment description and produce an outline of how you want to organize your presentation.
2. Use the outline to write a narration that explains the concept you want to teach. Be sure to include a complete introduction and summary. Use the right side of blank script sheets. Write the narration as if you were writing a speech.
3. Match visuals to narration. Use title scenes with words when appropriate. While titles may not seem to be a very creative way to deliver information, they are very effective at indicating what information is important.
4. Select the ideal image to explain or clarify your narrative.
5. Revise the narration and the visuals. Be sure to consider rules listed in this chapter.

TRY THIS

Write a script for the treatment description you produced earlier. Keep the narrative short, less than five minutes. You should have twenty to thirty visuals.

A storyboard is created by transferring each script block onto a 3 × 5 or 5 × 8 card. By doing so, it is possible to rearrange the scenes of the production so they can be shot out of order more easily, or can be given to specialists to produce. (This out-of-sequence document is often called a shooting script.) For example, title scenes may be given to a graphic artist to design. Copystand scenes might be sent to animators, and live-action sequences might be given to photographers. All technicians can then produce their portions of the presentation at the same time. When production is completed, the script writer can reassemble the numbered storyboard cards and organize the final presentation from them.

A storyboard card contains the same information found on a script block, with two additions: The cards are numbered sequentially, and directions for production are included. Precise production notes

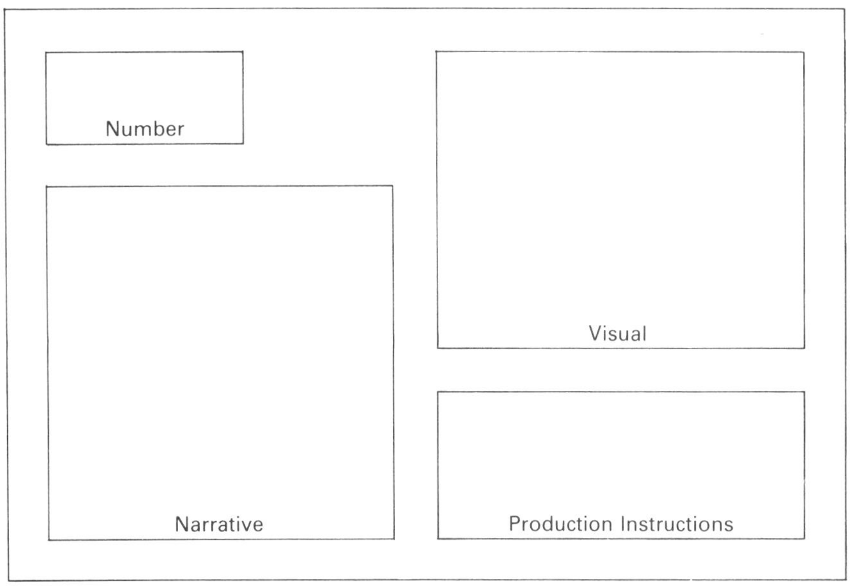

Sample storyboard card.

are important because the photographers, graphic artists, and film makers producing the medium may not be in direct contact with the script writer. As a matter of fact, in many professional production companies photographers and artists may be working on several presentations simultaneously.

After production has been completed, the production script is revised, if necessary, into a *teaching script.* The teaching script should be an exact written reproduction of the medium. This script is used by the teacher as the medium is presented in the classroom, just as a dinner speaker might use note cards. If the production script was carefully prepared, and few revisions were necessary during production, the two scripts may be identical.

RECORD KEEPING AND SCHEDULING

Production can be a pleasant and rewarding experience. Getting out the camera, setting up the lights, and training actors are exciting to do. Unfortunately, many amateurs begin with production and bypass much of the careful planning. The scripting and storyboarding processes are the foundations for what happens during production. Besides the obvious need to know what the show is to cover, the script is the basis for the development of the production schedule, the log sheet, and the release form.

The Production Schedule

Before beginning production, prepare a list of all the scenes that are to be shot at the same time or in the same location. This list is also called a shooting script. Grouping scenes using the same camera angle at one location is also a timesaver. This schedule should then be followed during production.

Sample Production Schedule

Production Schedule: Pesticide on the Farm Page ____

Date	Location	Scene Number	Description
5/5	Copy stand	1	Title
		7	Chart
		9	Chart
		29	Credits
		30	THE END
5/6	Studio	2	2 actors talking
		15	2 actors talking
		12	CU of Don
		16	CU of Frank
5/7	4-H Farm	3	LS of farmyard
		4	MS of farmer at shed

Sample Log Sheet

Presentation title: Date: 5/7

Software: Time: 1:00 p.m.

Equipment:

Scene	Take	Camera Setting	Length	Conditions	Remarks
3	1	f/8,1/250	—	Cloudy & windy	Dog ran thru as scene was shot
3	2	f/8,1/250	—	same	Good
3	3	f/2,1/1000	—	same	Narrow depth of field to reduce clutter in background

Page 1

Sample Release Form

Date ________

I give permission to (producer's name) to use pictures taken of me in an educational presentation titled ______________________.

Name ______________________

Signature ______________________

Address ______________________

Log sheets include:

- ☐ Script scene numbers
- ☐ Take number
- ☐ Conditions of production such as lighting
- ☐ Camera settings
- ☐ Scene length
- ☐ Remarks
- ☐ Date and time of day

Fair use The legal copying of copyrighted materials by teachers for classroom use.

The Log Sheet

During production it is a good idea to keep a record, called a log sheet, of all shots or scenes photographed. The log sheet shows what pictures were taken and in what order. It should also list scene length, the conditions present during production, and an evaluation of how things went. When slides or motion scenes are viewed after production, the log can be used to assist in the evaluation of scene quality and content. Logs are especially important during the editing of films and videotapes.

Release Forms

Taking someone's picture is really doing just that. You are taking something that belongs to someone else. When we shoot a picture of a person we are using something that is his or her property. It is professional and polite to ask permission of anyone who will appear in your productions, and it is prudent to obtain this permission in writing. It is also legal protection. Have the person sign a simple release form. Parents should sign for minor children.

Also, copyrighted materials should not be copied without permission. Pictures from books, magazines, and pamphlets make excellent sources for visuals, but most of these sources are copyrighted. While most lawyers agree that it is appropriate for a teacher to make one copy of copyrighted materials for his/her own use in the classroom, it is best to request permission from the copyright holder before copying. Usually a note on school stationery describing the use you will make of the visual, and requesting permission to copy it, is all that is needed. Requesting permission is the professional thing to do. It is also a legal requirement for any copying except the most inconsequential.

PLANNING TO SHOW ACTION AND TIME

Several terms must be explained before we can discuss how action is scripted. The first is a *shot.* A shot generally refers to one picture taken with a still camera. A *scene* is similar to a shot except scenes refer to production with motion cameras or video recorders. A scene is what is recorded when the camera is operating. Often scene and shot are used interchangeably.

A *take* is the recording of a single scene or shot. Often you may want several takes of a scene, so you can select the best one. A sequence is a series of related shots or scenes that show one idea or one complete activity. Shots and scenes are the basic elements of production. They make up sequences which are used to make up presentations.

Basic Scenes or Shots

The fundamental series of scenes or shots is the long shot (LS), medium shot (MS), close-up (CU) sequence. The long shot establishes the setting where the action is taking place. The medium shot eliminates unwanted detail and orients the viewer to the central point of action. The close-up and the extreme close-up (ECU) bring the viewer tight in on the action so it can be observed easily. The LS, MS, CU sequence is often used at the beginning and end of presentations.

It is often effective to mix up the LS, MS, CU sequence. Once the viewer has been oriented to the action, it may not be necessary to use the LS often. An MS may be all that is needed to reorient the viewer when sequences change. In many cases an LS, MS, CU, MS, CU, MS, LS sequence may be satisfactory.

There is no set rule for use of this basic sequence of scenes or shots. Try to think of what you would want to see as a viewer of the action. If you feel a need to get close to the action, a CU or ECU is probably needed. If you feel a need to see where and in what setting the action is occurring, an MS or LS may be needed. Experience gained from the careful examination of finished presentations will make this type of decision easier.

When any sequence of scenes is used, an important rule must be considered. Cinematographers call it the *General Rule.* The rule states that "when shooting a new scene, change the size of the image, or change the camera angle, or both." If this is not done, a phenomenon occurs which gives the appearance of jumping toward or away from an object. Changing the size of the image being photographed or changing the camera angle between scenes reduces this problem. If this is not done the result will be quite distracting to the viewer.

Camera Position and Angle

When scenes are filmed or shots photographed, camera position and camera angle should also be considered. *Camera position* refers to whether a scene is recorded from a *subjective* or *objective* viewpoint (see page 158). If the camera is positioned so that action is viewed as the actor would see it, the shot is called a subjective one. This is also called an over-the-shoulder shot. Most scenes are recorded from the objective position. For objective filming, the camera is positioned where the audience would be. The camera records action as an observer.

Camera angle can be used to add variety, emphasis, or understanding to a shot. The normal camera angle is at about eye level. Since the intent of most photographers is to show events with a camera as they would be observed if the audience were watching them directly, this camera angle is

A long shot (also called an establishing shot).

A medium shot (MS brings viewer in on the action).

A close-up shows the critical action on which the viewer is to concentrate.

Subjective shots are also called over-the-shoulder shots because the viewer sees the action as the actor would. This position is good for showing how-to-do procedures.

Objective shots show the action as a viewer in an audience might observe it.

A high-angle shot.

A low-angle shot.

A person moving in one direction in a scene . . .

should *not* be shown moving in the opposite direction in the following scene, or the audience will become confused. This is called *maintaining screen direction.*

most often used. Events are seen with the eye so cameras are usually located where the viewer's eyes would be. Sometimes, however, high-angle or low-angle shots are needed. For a *high-angle* shot the camera should be positioned to look down at the subject or down on the action. High-angle shots give the illusion of placing the viewer in a superior or dominant position, and the subject in an inferior position. When the camera is placed lower than normal, for a *low-angle* shot, these perceived roles are reversed. The actor seems more dominant and powerful, and the viewer tends to feel threatened or inferior.

Sequence of Scenes to Show Action

Two important techniques for organizing action shots are *directed* and *chronological* sequences. These techniques are discussed in Chapter 3. The LS, MS, CU sequence is a good example of a directed sequence. The viewer is directed to some action or person by the use of progressively closer shots. Chronological sequences show action as it occurs over time. A commonly used chronological sequence is one that begins with a long shot that shows all the participants in some activity. This is followed by a series of shots or scenes where these actors or objects interact to complete some sequence over time. The chronological sequence is concluded by another long shot of the actors and objects. Directed and chronological sequences can and often are mixed during presentations.

Matching Action between Scenes

Scenes that are next to one another should show action that is related. This is called *matching action,* and it is important so that the audience perceives the film, videotape, or slide series as a smooth-flowing sequence.

Several techniques can be used to match action. One is called *screen direction.* If a person or object in a scene is moving from left to right then the action should move in the same screen direction in the next scene. Similarly, when an actor moves out of view on one side of the film frame he or she should enter the next scene on the opposite side of the frame. Maintaining screen direction is important. If the action in your presentation moves in different directions from scene to scene your audience will become confused, and the delivery of your message will be impaired. If it is necessary to reverse or change the direction of the action in a sequence, be sure to include one or two orientation scenes between the two with action moving in different directions. This orientation scene should be shot so that your audience can see how the action became reversed.

The *imaginary line* (also called the stage line) is a term related to screen direction. When filming, imagine that there is a straight line, like the edge of a stage, across the front of your scene. Your camera should not cross this line. Shots can be taken from any point in a 180° arc behind the imaginary line, and action will be matched. If the imaginary line must be crossed, there must be scenes showing the action occurring during this crossing, so that the audience will not lose its orientation to the sequence.

Action between scenes can be matched in other ways, too. The speed of action in adjacent scenes should be similar. It would be inappropriate in most sequences for one motion medium scene to show rapid movement and the next to have actors standing still.

The placement of objects or persons between scenes is important when matching action correctly. The way people move, or the way action is shown, can be used to provide continuity between scenes. Great variations in lighting between scenes can cause audience distraction. One scene should not be shot at noon and the next at dusk.

If the screen direction for main action of a scene reverses . . .

during a sequence, it is important . . .

to show the viewer how the central point of the scene reversed itself . . .

so that the viewer does not become confused.

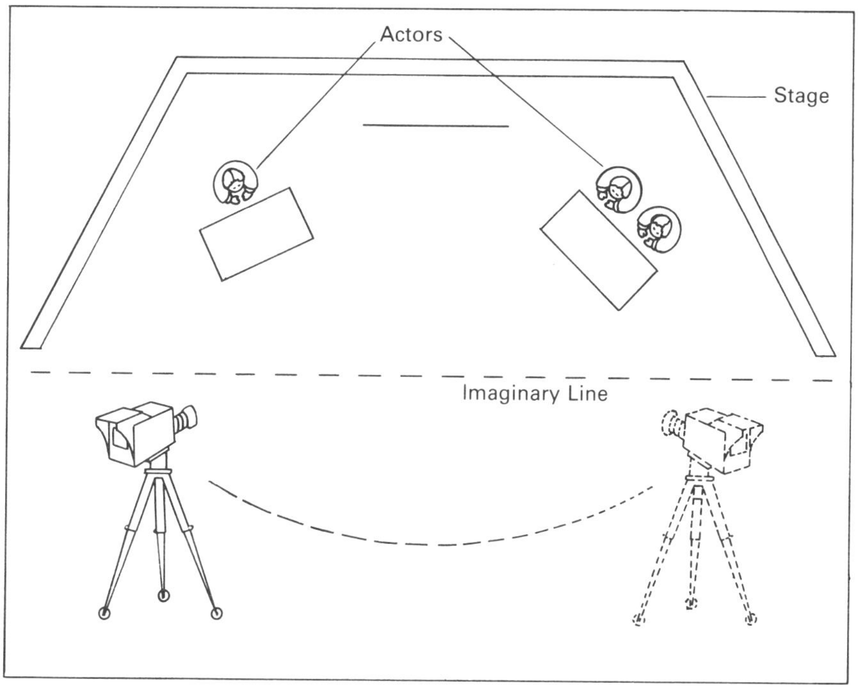

The camera can move anywhere behind the imaginary line and action will cut together correctly.

Scene No. 1

Scene No. 2

Scene No. 3

The cutaway shows some action that differs from that shown in the two scenes between which it is inserted.

Scene No. 1

Scene No. 2

Scene No. 3

A cut-in is an extreme close-up of some action in, or part of, the scene being recorded.

One of the best ways to provide for a variety of combinations during editing is to shoot *insert scenes.* Inserts are also called transition or protection shots. Inserts are used between scenes that do not match. They are also used between sequences to tie them together. Title scenes work well as inserts.

Cut-ins and *cutaways* are two commonly used inserts. A cut-in is a CU or ECU of some part of a scene being recorded. Faces, hands, and set furniture are often used for cut-in inserts. Cutaways are different from cut-ins because they show some action or object different from the two scenes between which they are inserted. Crowd shots, and shots of objects being discussed but not actually present, are examples of subjects for cutaways.

One additional use of the insert is to compress time. For example, if a sequence of scenes was scripted to show a man walking down a street, the time to visualize this action could be compressed by first showing the man at the beginning of the block. A cut-in of the man's feet would be the next scene, and this would be followed by an LS scene of the man at the end of the block. Because of the insert of the man's feet this sequence would seem logical when viewed, and for a motion medium sequence, screen time would be reduced dramatically.

Connectives between Scenes

Since film scenes are physically attached to one another, and since video scenes are electronically sequenced during recording and editing, the way they are connected is important to the film maker. The three most commonly used connectives are the *cut,* the *dissolve,* and the *fade.*

A *cut* is accomplished by connecting two scenes directly to one another. When the first scene ends, the next begins immediately, with no visual transition. A motivated cut means that the motion or sound in a scene leads the viewer to the action in the following scene. A person reaching for a ringing telephone in scene No. 1 would provide a motivated cut to scene No. 2, which might be a close-up of the telephone receiver being picked up. Generally, cuts are used when action is occurring at the same time and in the same place from scene to scene.

Dissolves show one scene slowly fading out while another scene slowly fades in. At the midpoint of the dissolve, both scenes are on the screen. An adaptation of the dissolve is the *wipe.* During a wipe it appears that the new scene is replacing the previous one by pushing it off the screen, usually either vertically or horizontally. In video productions, dissolves and wipes require at least two cameras and a video switcher. Filmed dissolves are done during printing. Dissolves impart, in the mind of the viewer, a feeling that a short time has passed between scenes, or that two events are occurring in different locations but at about the same time.

The *fade* or, more correctly, the fade out, fade in, is used to show a fairly long passage of time between scenes. The longer the fade to dark lasts, the longer the perceived passage of time between scenes. Fades can be easily accomplished in a one-camera video or film system by closing the camera aperture at the end of a scene, or by placing something opaque over the lens.

When a time-consuming sequence, such as a person moving down a sidewalk, must be depicted, it is possible to use an insert, . . .

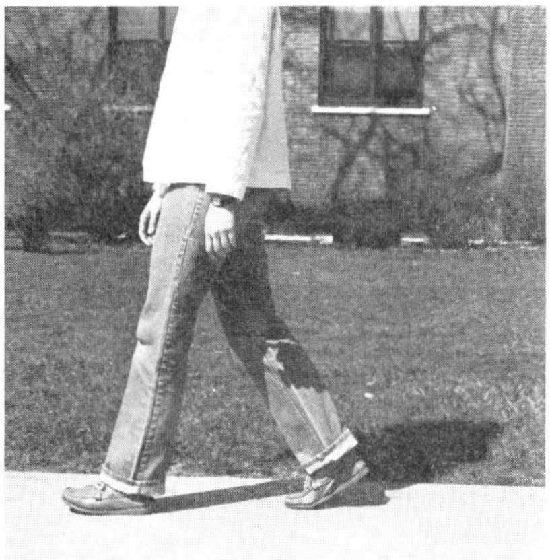

such as this close-up of the person's feet, . . .

between two scenes of the person at the beginning and the end of the walking sequence. Using an insert in this way reduces screen time dramatically.

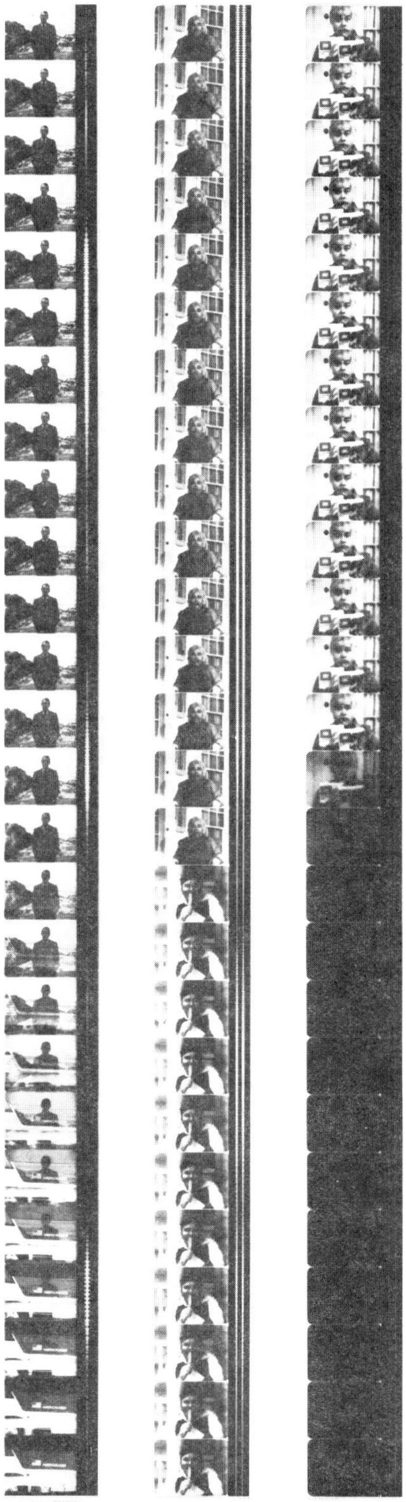

Dissolve Direct Cut Fade

Connectives

Four types of camera movement.

When scripting a motion program, it is a good idea to also script the connectives between scenes. This is even more important when postproduction editing is not anticipated. "In-camera" editing, the recording of scenes in the exact order and for the exact length of time planned for the script, requires a great deal of prior thought and planning on the part of the program producer. When a program is to be recorded once, and then used, the importance of a carefully written, complete script with connectives included is accentuated.

Camera Movement during Scenes

During filming and videotaping, it may be necessary to move the camera. While camera movement is normally undesirable and to be avoided (your eyes do not move a great deal as you watch some activity unless it moves), there are times when it is necessary to move the camera to follow the action of the scene.

There are four types of camera movement:

1. *Tilting* Vertical movement of the camera.
2. *Panning* Horizontal movement of the camera.
3. *Zooming* A change in the camera zoom lens's focal length during a scene that moves the viewer closer to or farther from the action.
4. *Dollying* The movement of the camera toward or away from the action.

Pans and tilts can be used if the subject moves, and you want to follow the movement. Following a skier down a ski slope might require panning and tilting. Adjacent scenes can be connected to each other visually by using pans and tilts. For example, if the first scene in a sequence shows a dog jumping up, the next scene could be of a child climbing a jungle gym to get away. When these scenes are spliced together and viewed, they will appear to be related because of the visual symmetry between them.

Zooms allow the photographer to get closer to, or farther from, action without moving the camera. However, since our eyes do not zoom it is usually not a good idea to zoom during a scene. It is acceptable to zoom to follow action as it occurs, but try to avoid it otherwise.

Dolly shots are difficult to execute effectively. When dollying, the camera is physically moved to follow the action in the scene. The dolly shot gets its name because motion picture cameras were originally placed on a wheeled cart, called a dolly, that was pushed around the set to follow action during filming.

When dollys, zooms, pans, or tilts are used, you should:

1. Use a tripod to hold the camera.
2. Make several takes of each scene.
3. Anticipate the direction of action and keep it slightly behind the middle of the frame as you move the camera in the direction of the motion.
4. Start and finish each scene with a few seconds of recording without camera motion, unless you plan to use camera motion to connect scenes visually.

A FINAL WORD

Scripting a media presentation is important whether you are an amateur making your first super-8 film, or a professional with many shooting sessions behind you. There are five steps to consider when media are to be used in teaching. The first is the development of a story line, which is a sequential listing of all instructional media to be used in a lesson. If some materials needed to complete the story line are not already available and must be locally produced by the teacher, then a treatment description, production script, storyboard, and teaching script are written.

A treatment is a general outline, in narrative form, of what the presentation is to show. A production script matches the visuals and narration of the presentation. It is the guide for most production activities. Sometimes, for complex presentations, a storyboard must be developed. A storyboard is a sequence of cards, each showing an individual scene or shot. Storyboard cards are more detailed than script blocks, and are often given to media technicians by the script writer to assist in production.

It is also important to consider how scenes in a script are related to one another. Directed and chronological sequences are often used to show action or events in a presentation. The LS, MS, CU sequence is probably the one most widely used in media presentations. In motion media productions, cuts, dissolves, and fades are usually used to connect scenes to one another.

If you remember nothing else from your study of this chapter, remember this: The more planning that occurs before a production is begun, the more likely it is that a successful presentation will result.

LOOKING AHEAD

There are four media that usually require scripting—audiotapes, slide presentations, films, and videotapes. The next four chapters of *Media Planning and Production* will discuss techniques for producing these media. Be sure to refer to this chapter during the study of these media, because the information presented in this discussion of planning and scripting relates to them all.

REVIEW QUESTIONS

1. Describe the relationship between a story line, a treatment description, a production script, and a storyboard.
2. What are the three components of a script block?
3. What is an average length of time a slide should be on the screen during a presentation?
4. What is added to a storyboard card when it is written from the production script?
5. Name the form used to keep information about what occurs during production.
6. Is it generally considered legal for a teacher to make one copy of some copyrighted materials for his/her own use in the classroom?
7. What is the basic sequence?
8. What is the name of the rule that states "when shooting a new

scene you should change the size of the image recorded, or the camera angle, or both"?

9. The over-the-shoulder shot is called what kind of camera position?
10. If you wanted to make the actor in a scene appear dominant and powerful, what camera angle would you be likely to shoot from?
11. Define *imaginary line.*
12. Protection scenes are used to ________________.
13. If you were going to film a board being painted in a scene and then have the next scene show it dry and being used in a construction project, what connective would you be most likely to use between these two scenes?
14. The physical movement of the camera closer to or farther away from the scene is called ________________.

ANSWERS TO REVIEW QUESTIONS

1. A story line is a sequence of media used during teaching to reach a course goal. It is similar to a lesson plan. A treatment description is a narrative describing what a media presentation is supposed to include. A production script is written based on a treatment description and includes a sequential listing of visuals and narration that are to be produced. A storyboard is a detailed sequence of cards that shows each scene or shot to be produced. The most general planning tool is the story line, the most specific is the storyboard.
2. Media used, visual, and narrative
3. 6 to 10 seconds
4. Script sequence number and detailed production directions
5. Log sheet
6. Yes, this is called "fair use."
7. LS, MS, CU
8. The General Rule
9. Subjective
10. Low-angle
11. Also called the stage line, the imaginary line is the line across the front of the scene that the camera should not cross.
12. Match action between scenes that would not match if they were connected directly to one another.
13. A fade out, fade in, because a fade relays to the viewer the impression of a long passage of time.
14. A dolly

REFERENCES

Basic Production Techniques for Motion Pictures, Pamphlet No. P-18 (Rochester, N.Y.: Eastman Kodak Company, 1976).

Edmonds, R. *Scriptwriting for the Audio-Visual Media* (New York: Teachers College Press, Columbia University, 1978).

Gaskill, A. L., and Englander, D. A. *How to Shoot a Movie Story,* rev. ed. (Dobbs Ferry, N.Y.: Morgan and Morgan, 1970).

Kemp, J. *Planning and Producing Audiovisual Materials,* 4th ed. (New York: Harper and Row, 1980).

Movies with a Purpose, Pamphlet VI-13 (Rochester, N.Y.: Eastman Kodak Company, 1971).

Rowe, M. R., et al. *The Message Is You: Guidelines for Preparing Presentations* (Washington, D.C.: Association for Educational Communication and Technology, 1971).

Sleeman, P. J.; Cobun, T. C.; and Rockwell, D. M. *Instructional Media and Technology* (New York: Longmans, 1979).

Swain, D. V. *Film Scriptwriting: A Practical Manual* (New York: Hastings House, 1976).

Volker, R. P., and Simonson, M. R. *Media for Teachers,* 3d ed. (Dubuque, Iowa: Kendall/Hunt, 1981).

Zettl, H. *Television Production Handbook,* 3d ed. (Belmont, Calif.: Wadsworth, 1976).

GOALS

☐ Gain an understanding of the use of audio materials in teaching.

☐ Acquire skills in planning and producing a variety of types of audiotape.

OBJECTIVES

☐ Identify commercial sources of appropriate audiotapes for teaching.

☐ Operate a variety of equipment to make audiotapes.

☐ Define terms associated with audiotape production.

James Jones: How do you feel about the fact that you may die tonight?

Grade School Child: I'm prepared to die for this family if I have to for freedom. Thank you, Daddy.

NPR: In the aftermath of Jonestown the FBI found in the jungle hundreds of tape cassettes—900 hours of recordings made in Guyana, and some in California. James Reston, Jr., who had visited Jonestown in the days just after the tragedy, filed legal action under the Freedom of Information Act to obtain access to the tapes. Reston has now written "Our Father, Who Art in Hell." This radio program is the story of the last of Jonestown, based on Reston's research for his book.

From *Father Cares: The Last of Jonestown.* A National Public Radio production, 1981.

This two-cassette package, called FATHER CARES, is available from National Public Radio. The package includes a teacher's guide.

Newspaper headlines during the latter part of November 1978 carried the story of the mass deaths of the people of Jonestown, in Guyana, South America. Among the items discovered at the site were dozens of audio cassette tapes, recorded during the "family" meetings and other gatherings. Even though stories were subsequently written by reporters who visited the scene, and photographs and videotapes helped portray the setting, the cassette tapes have proved to be an even more vivid and powerful medium for providing first-hand information about Jonestown.

This chapter is divided into two sections: teaching with prepared, available sounds and producing your own recordings. Appropriate background information as well as explanation of the technology of tape recording is included where necessary. A few assumptions have been made about your knowledge and background. For example, we assume that nearly everyone has had some experience with audio cassette players, either in a home stereo system, at school, at the office, or as a music system in an automobile. Because of the wide availability of equipment, types of tape, and prerecorded tapes, the medium of audiotape is found throughout the world.

10 AUDIO RECORDING

Car stereo with 8-track or cassette players are very popular.

Portable, battery-operated stereo systems can be taken anywhere.

TEACHING WITH SOUNDS

As our previous example indicates, the sense of immediacy provided by hearing an event as it occurs can be a strong motivation for involvement.

Even though visual elements and support are lacking in a situation where content is being relayed exclusively by sound, if the events are realistic enough the audience will become thoroughly involved. When radio first became available, for example, listeners found themselves fascinated by listening to news, music, and voices from elsewhere. Radio shows, dramatizations, concerts, and other events captured the attention of people everywhere. A teacher can use this appealing characteristic of sound by using recordings in the classroom. Some instructional events involving voice only are described below:

☐ *Documentary sounds* History, poetry, political events, and major speeches by significant national and world figures are available on records and tape. Many business leaders routinely listen to motivational tapes. While the information may be combined with mu-

Media in the U.S.

Telephones	169 million
TV sets	78 million
Radios	79 million
Home hi-fi Systems	50 million

A large collection of documentary records is available.

sic and sound effects, it is mainly the verbal delivery that carries the message. The poet Robert Frost, for example, creates the focal point of interest in a reading of his own poetry. While a professional narrator might provide a smoother reading, the fact that Frost is reading his own material provides an air of realism that would otherwise be absent.

You can create your own collection of documentary sounds by recording from radio or television. During your lifetime, for example, there has been an active space exploration program and the first occasion for someone from Earth to land on the moon. Reports of these events could have been recorded from the news media at the time they happened and used later in a teaching setting.

Hints about Guest Speakers

Get permission from your guest speakers to tape record their presentation for playback to other classes.

Try to photograph guests' materials to make a slide set that can accompany their talk.

☐ *Guest speakers* If you have the opportunity to have a special person come to the classroom to talk with the students, or give a "guest lecture," it might be helpful to make an audio recording of the event. The recording can be used later in the day, if the speaker is unable to be with you the entire day—or saved for use at a later time in the year. The remarks can also be reviewed by the students if they choose to listen to them again. Or, if you plan to teach the same subject in the future, you can use the recording. Chances are that you may not be able to have the same person as a guest speaker a second time.

☐ *Teleconferences* Increasingly, the telephone is being used to bring guest speakers, resource people, and celebrities into the classroom. This technique not only saves time and money, it is often easier to arrange for guests to speak to the class by phone (in terms of schedule problems) than it would be if you arranged for them to appear in person. For years, radio talk shows have used teleconferencing hookups to bring prominent guests "to the studio."

Teleconferencing is an inexpensive alternative to traveling to meetings.

The conversation and discussion you and the class carry on with the guest, over the telephone, may be significant enough to warrant recording on tape. The tape recording can be used later, either in the near future or next year, when the topic is again being taught in your classes.

TRY THIS

Select recordings that would be useful in teaching a concept in your subject area. Use a variety of sources, such as a catalog from National Public Radio, the Center for Cassette Studies, or the NICEM Index that lists audiotapes. Consult your media center for details.

TRY THIS

Make a list of people you would like to invite to your class to discuss a concept, idea, or method that has application in your subject area. Indicate whether they would be making a personal appearance or whether you would be having them speak in a teleconference setting.

☐ *Music* The current widespread and prominent use of audiotape (and disk recordings) for music is well known. Since a great deal of music already exists it is relatively easy to purchase good recordings. It is not necessary to be teaching a course in music or music appreciation to have a need for music in teaching. Some teachers use it to inspire students doing creative writing. It can be used to set a mood, create an atmosphere, or suggest a situation. Certain types of music may be related to the speed and depth at which people learn—and the use of music in the teaching of foreign languages and mathematics is particularly meaningful. For the music instructor, of course, on-the-spot recordings of the school instrumental or vocal group are necessary for analysis and critique. Individuals find a recording of their performance to be valuable too, for self-study.

☐ *Sound effects* To gain a touch of realism it is necessary to make the recording as much like the actual situation as possible. Some simple cases, such as a quiet walk around the block or a trip by car from one city to another, become sterile and artificial if they are recorded without the benefit of all the small sounds that are found in the actual situation. Some of the most effective dramatizations and portrayals of events, when produced for listening only, depend on the vital sound effects that suggest a setting, a situation, and the specific conditions that help the listener understand the implications of the actions.

Sound effects not only add to the realism of a situation, but they can be used in teaching certain types of content. The various cries

Sound effects can be made for audio recordings by using what is available around the house or school.

of an infant, for example, have been shown to indicate hunger, pain, restlessness, or other emotional states. Rather than reading about these cries, it is almost essential that a child-development student listen to them in order to become familiar with the variations. Certain types of malfunctions in electric motors, gasoline or diesel engines, or other types of apparatus may be diagnosed by listening to the characteristic sounds associated with the malfunction.

☐ *Drill and practice* Nothing is quite so patient as a tape recorder that can repeat a message over and over. Even though a student's question might not be the one to which the tape recording is giving an answer, an audio recording is nevertheless excellent for drill-and-practice applications.

The incorporation of sounds into computer programs, through the use of special hardware and software packages, makes it possible for the computer program to include responses that students can hear as well as see. Frequently the sound is recorded on the same disk on which the computer program is stored. Usually, the sound is not actually "recorded" on tape or disk; rather it is stored in solid-state devices in educational teaching machines such as *Speak 'n Spell* or *Speak 'n Read*. By changing the programmed chips, it is possible to change the vocabulary of these devices to match the accomplishments of the children using them.

Sound effects that are mixed with narration add a dimension of reality to a classroom presentation.

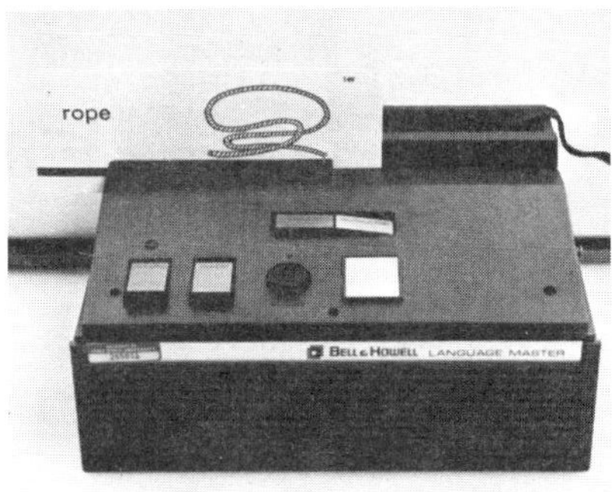

The Language Master is an excellent tool to use for vocabulary-building exercises. The student sees the word, sees a picture of what the word stands for, and hears the word being spoken. It is even possible for the student to record his or her pronunciation of the word.

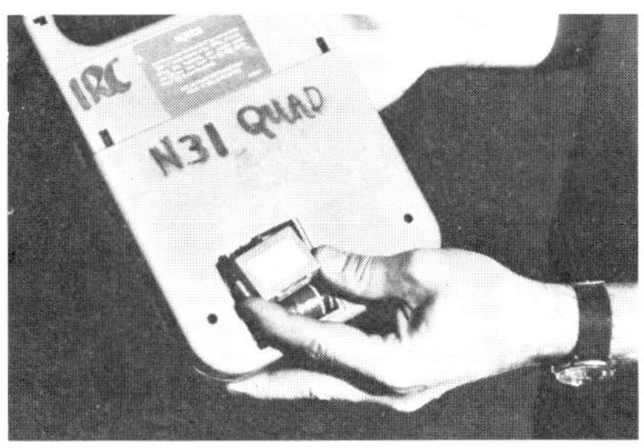

Speak 'n Spell teaching machines have interchangeable memory chips so the machine's vocabulary can be varied.

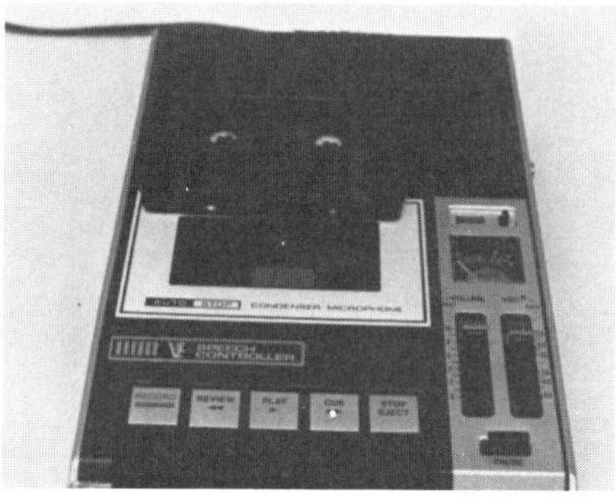

Speech controllers make it simple to play back an audiotape at any speed from one-half the normal rate up to two and one-half times normal without changing the pitch of the recorded voice.

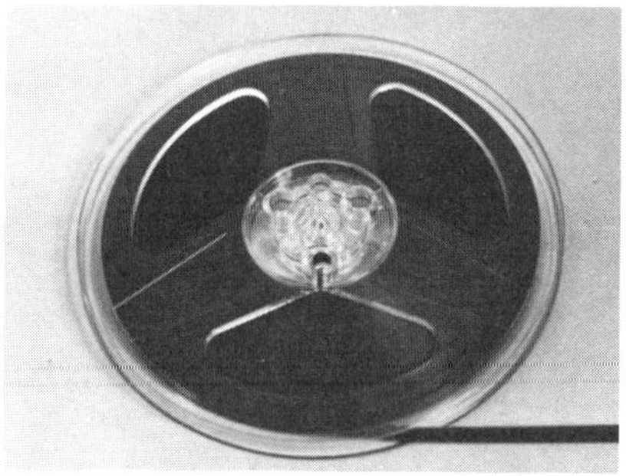

Open-reel audiotapes allow recording times up to several hours.

The standard audio cassette is the most popular format currently being used in schools and training centers.

The 8-track cartridge format is not used extensively in education. It is primarily a consumer format.

☐ *Special needs* A learning situation may call for sound (especially spoken words) to occur faster than normal or more slowly. The ideal device to accomplish this is now available as a built-in feature in some types of audio players, as well as videotape players. One such brand, the VariSpeech, can play an audiotape at half-speed or two and one-half times as fast. This means that the normal speaking rate of around 150 words per minute can be slowed to 75 wpm, or increased to nearly 375 wpm. A number of studies have shown that it is usually possible to more than double the rate of speech without losing the ability to comprehend the message. Using audiotape in conjunction with individualized instruction or learning-package materials means that it might be possible to nearly double the rate of student comprehension if the message is offered on tape, as opposed to delivering verbal instruction via the spoken word, in class.

With these suggestions for using sounds in teaching a variety of subject areas, it becomes clear that there is wide application for audiotape recordings or disk recordings in education. At one time it was possible to use only prerecorded information because the recording process required highly technical, specialized professional equipment. However, for some time, persons without special technical skill or expensive equipment have been able to make recordings. The availability of easy-to-use equipment has greatly increased the usefulness of audiotape in education, and you will find that it is quite beneficial to make your own recordings for your own specific teaching situation.

To provide an appropriate background for the third section of this chapter, in which you will be making a variety of tape recordings, we will take a brief intermission to learn how tape recorders work and to answer some of your questions about the operation of the equipment.

INTERMISSION

The purpose of this section of the chapter is to provide just enough information for you to operate tape recorders and hook them up to each other or various other sources of sound. While we are at it, some of the characteristics of the sound (monaural, stereophonic, and fidelity, for example) will be presented. Then, in the next part of this chapter we can proceed smoothly, building on this background.

The Nature of Magnetic Tape

Audiotape, that plastic strip wound on a reel inside a cassette, a cartridge, or an open reel, is covered on one side with magnetic particles. Viewed under a microscope the particles look like grains on sandpaper, but they are extremely tiny, of course, and feel smooth to your touch—as well as to the head of the tape recorder. When a recording is made on the tape these magnetic particles are arranged in a certain pattern. They are not exactly moved around, but rather each one is magnetized in a specific way, so that the pattern of the magnetic arrangement is locked in the particles on the tape. Unless a magnet or a magnetic field comes close to the tape and "unlocks" the arrangement, the pattern will stay in its recorded form indefinitely. The same tape can be "erased" and rerecorded, however, if the particles are magnetically rearranged into a new pattern. You might record a new song on an old prerecorded cassette tape, for example. When you play the tape the new music is reproduced; the old music is no longer present. Recording and rerecording can be carried out thousands of times and since the pattern is magnetic—not in the form of lines or grooves as it is in a phonograph record—there is nothing to scratch or wear out. Of course, different qualities of the *emulsion* (the magnetic particles glued on the tape) and the *backing* (the plastic strip to which the emulsion is glued) have something to do with how long the recording will last, and the quality of the recording.

Recording Sounds on Tape

To create a system that can make the proper arrangement of the magnetic particles on the emulsion side of the tape, the incoming sound must be treated in a special way. Obviously you cannot hold the tape in front of your mouth as you sing and expect the magnetic particles to jump into the proper arrangement in the tape's emulsion. Magnetic particles are arranged—as iron filings (or nails or thumbtacks) are—arranged by a magnet. The "arranging magnet" in this case is called the *record/playback head*. It is an electromagnet that responds to changes in electricity supplied by a *transducer* (a microphone or another source). The changes show up in the recording head as electrical fluctuations that can be transferred magnetically to the particles in the emulsion.

Transducers and Inputs

If the electrical impulses are to transform the singing (or speaking, traffic noises, dogs barking, for example) into electrical impulses, the sound must be picked up by a device that can change the sounds you hear into electrical energy. Such a device is called a *transducer*. A microphone is a transducer; so is a loudspeaker. Tape recorders receive incoming electrical information through an *input*. Since the information or *signal* may come from a transducer, or microphone, there is a *microphone input*. However, the information may already be "transduced," or in the form of electrical energy, from another source that is not a microphone—such as another tape recorder or a record player. To receive the electricity from these devices the host, or recording, machine has an *auxiliary input*.

Erasing Tape

Please do not think of erase in this case as an act of rubbing out a mistake, such as is done with a pencil eraser, or an eraser on the chalkboard. There is no physical rubbing on the audiotape to remove unwanted signals; the process takes place electronically.

You recall the discussion of the magnetic particles in the emulsion, and the precise arrangement of these particles in order to represent a song, a speech, or a cat's meow. To remove the song, the speech, or the meow it is simply necessary to disarrange the particles in the emulsion. This is called erasing, and it happens when a special head called an *erase head* is activated. It might be easy to forget what you are doing, and accidentally push the wrong button, which would erase a tape. Manufacturers of tape machines know this, and have designed the equipment to require you to push, turn, pull, or move *at least two controls* on the audiotape recorder when you want to erase and rerecord. Erasing is done automatically when you are recording. So the two-controls-to-erase operation applies, in actual practice, to the recording process.

Some Transducers

- ☐ Photo cell in movie projector
- ☐ Phone patch
- ☐ Speaker
- ☐ Microphone
- ☐ TV picture tube

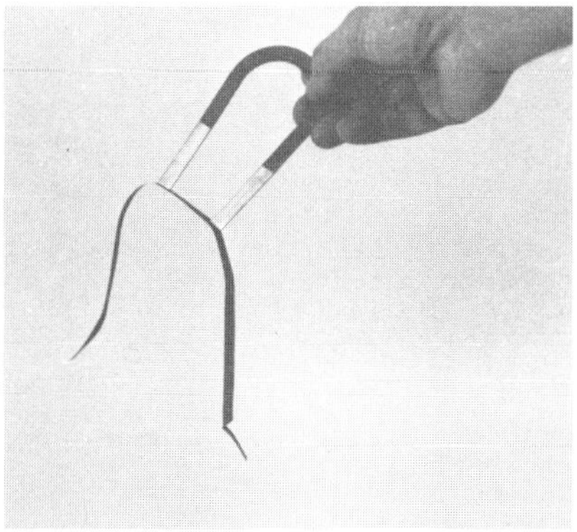

The oxide on audio recording tapes can be magnetized. The recording head is actually a small, but precise, electromagnet.

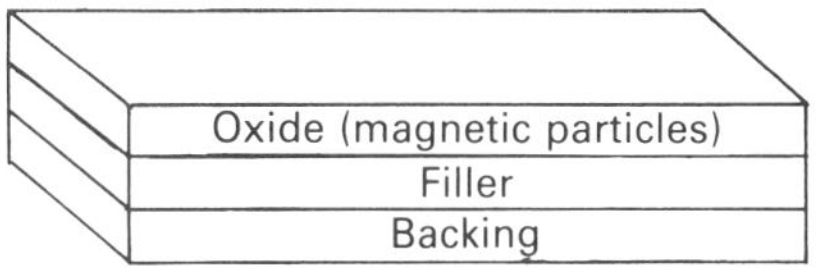

Cross section of magnetic recording tape.

The record/playback head is an electromagnet that acts as a transducer. It places a magnetic signal on tape during recording, or reads a signal from a prerecorded tape during playback.

To put a magnetic recorder in the record mode, at least two controls must be pressed or held at the same time.

Tape Speeds and Frequency Response

Think of audiotape as a writing surface, and the record/playback head as the write/read device. The analogy corresponds to a piece of paper and a pencil. When you write on a piece of paper your hand moves and the paper stays stationary. But imagine that you want to write on paper in the same manner in which a signal is written on tape. The paper would move and your hand would stay in one place. This arrangement would work well if the paper moved at the correct speed so that the letters and words did not pile up on each other. If the paper moved too slowly, though, all of the information would pile up on itself and it would be difficult to read when you were finished. So, for clear definition of the written material the paper would have to move at a certain speed. The same is true with tape. To achieve a clear and realistic sound—a *high-fidelity* sound—the tape must move at a sufficiently fast speed. The faster the tape moves the more clearly the information is recorded. The chart below shows the relationship between tape speed and fidelity:

We mentioned the word *frequency* along with fidelity. When something vibrates or interrupts the air hundreds of times a second, a pitch or frequency is created. Exhaust from a motorcycle, hitting the air hundreds of times a second as the piston fires, creates a pitch. The faster the motorcycle, the higher the whine of the engine. The same is true of a vibrating string, such as on a violin or piano. When it is bowed or struck, the string vibrates a certain number of times each second, creating a musical sound with a certain pitch. These pitches or frequencies are designated by the number of times each second they occur, and the label given to this vibration is *hertz*. Frequencies, measured in hertz, are a way of designating pitch. Low pitches, such as those on the low end of a piano, have frequencies of a few hundred hertz. At the other end of a piano the frequencies are higher. Middle C is 256 Hz, and the highest note on a piano is roughly 4000 Hz.

Tape speed is measured in inches per second, or *ips*. The standard speeds for tape recorders are listed below.

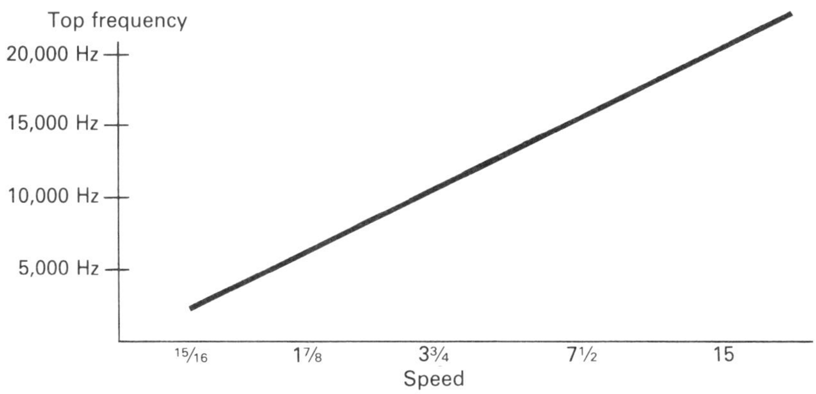

As tape speed increases, so does fidelity.

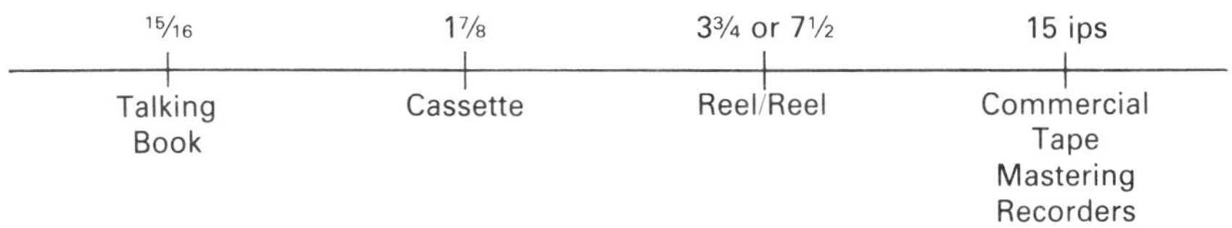

Standard tape recorder speeds.

Obviously, you must play back a tape at the same speed as it was recorded. Otherwise the playback pitch will not be the same as the original pitch. Fortunately, there is a direct relationship between tape speed and frequency. If you double the tape speed you double the frequency. In other words, if you record middle C on the piano (256 Hz) at 3¾ ips and play it back at 3¾ ips you will get middle C. But if you record the note at 3¾ ips and play it back at double speed (7½ ips), you will get the C one octave above middle C. To create a special effect, if you want to make your voice sound like a chipmunk, record your voice at 3¾ ips and play it back at 7½ ips. The technique is particularly effective for creating character voices for children's stories when you want to simulate mysterious creatures. You can try it the other way, recording at 7½ ips and playing back at 3¾. In that case you lower your voice one octave and the words take twice as long. You can become the troll under the bridge in *The Three Billy Goats Gruff* in this way. Try it and find out.

TRY THIS

Record the sound of middle C from a musical instrument at 3¾ ips; play back at 7½. Say a few words, too. Remember—the pitch will be one octave higher during playback at the double speed, and the rate of speech will be twice as fast. So you may want to speak more slowly as you make the original recording.

Tape speed, and its relation to frequency response, is not as important today as it once was, because the machinery, the electronic circuits, and the quality of the tape have been greatly improved. Using good-quality equipment, it is now possible to achieve recordings of very high fidelity at relatively low tape speeds. The standard cassette speed of 1⅞ ips is adequate to reproduce music of exceptionally high fidelity. A decade ago this was not true. In those days a cassette recording was a low-fidelity, poor-quality product.

Other Factors That Affect Fidelity

Tape speed is not the only factor that determines the quality of the finished product. Types of tape (special oxide formulations) should be considered so that appropriate quality can be achieved. Some of the other elements in the process that are related to the quality of recording and playback include transducers, quality of the oxide coatings, and electronic circuits in the tape recorders.

Transducers We discussed the role of transducers—the microphone and speakers—in the tape recording process. It turns out that they are the weakest parts of a system. The recorders themselves are usually capable of a far better job than the transducers. If you want to improve recording fidelity, use a high-quality microphone and good-quality speakers. The inexpensive microphone that comes with many cassette recorders and the small speaker built into the recorder may be thought of as rather narrow "sound windows." A microphone with specifications that indicate it will record from around 50 Hz to 12,000 or 15,000 Hz will capture nearly all of the frequencies that humans can hear. And a speaker that reproduces the same frequencies will guarantee that you will get back, for listening, those sounds you recorded in the first place.

This inexpensive microphone has a frequency response of approximately 150 to 5000 Hz (compliments Califone International, Inc.).

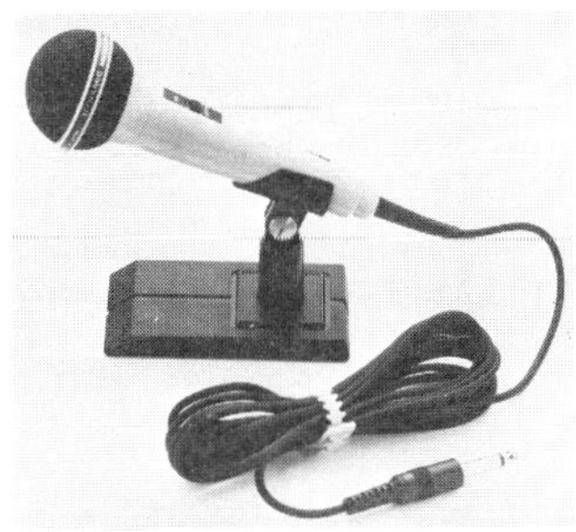

This microphone would be considered a high-fidelity one because it has a frequency response of 70 to 15,000 Hz (compliments Califone International, Inc.).

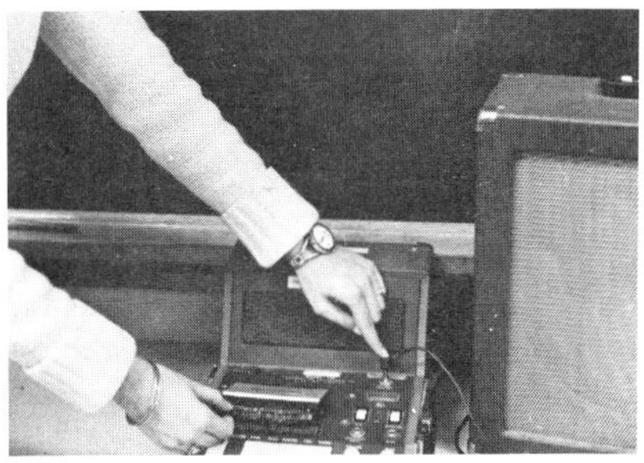

One simple method to improve the quality of recorded audiotapes is to play them back through a higher-quality speaker. Generally, the small speakers that are built in to many recorders are not very high fidelity.

Alligator clips and a patch cord can be used to connect the speaker to the output jack on the audio player.

Audiotape decks are usually high fidelity and are used as part of a recording/playback system that also might include an amplifier, several speakers, and a record player (compliments Ampex Corp.).

TRY THIS

Use two microphones: an expensive one and a very low-priced one. Make a recording with each and compare them.

Tape Quality If you have shopped for audiotape recently, you might have felt confused by the wide variety of brands, types of oxide used in the recording surface, and types of plastic backing. Because these factors are continually changing, the best source of information may be a current audio magazine from a local newsstand. A great deal of consumer information about audiotape and tape recorders is available. As a start refer to the summary below:

The Mechanics of Tape Recorders

While we have described the record/playback head, erase head, and types of tape, we have not looked at the tape recorder itself. Even though tape recorders differ somewhat in physical appearance and design, their mechanical and electrical features are very similar. A generalized tape recorder is shown in the diagram below, with the important features labeled and described.

The portable cassette model is the type you are most apt to find in an educational setting. One other type of recorder, called a *deck*, is also available, but it is more often found in a home stereo system.

Comparison of cassette tapes

Type of Magnetic Surface	Cost	Advantages	Uses
Iron oxide	Low	Low cost	Good for general recording of voice and music, response good to 10 to 12 kHz
Chromium dioxide	Medium	Good high-frequency response, lower noise than iron oxide	Excellent for general recording of voice and music, response up to 15 kHz
Metal	High	Excellent response, wider dynamic range	Excellent in all respects, response up to 20,000 Hz

Open-reel tape decks are most often the format used in commercial recording studios (compliments Ampex Corp.).

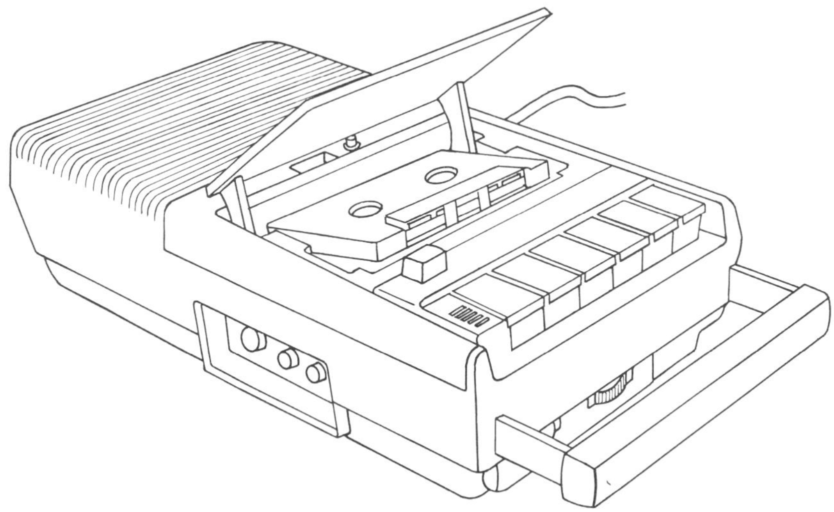

Monaural audio cassette recorder.

OPERATE

Audio Cassette Recorder

1. Insert audiotape, making certain that the left-hand reel inside the cassette is full of tape and the take-up reel is empty.
2. Connect desired components, such as microphone or patch cords, that will be used to make the recording. Patch cords are cables used to connect equipment.
3. Following instructions for the particular model and type of recorder you have, set the controls to record. Determine if the recorder has an automatic level control which will automatically adjust the volume of the sounds you are taping.

This Hitachi model AVA 150 has a built-in condenser microphone to make voice recording easier (courtesy Hitachi Corp.).

OPERATE
Reel-to-Reel Recorder

1. Place the reel of tape on the left spindle and thread tape on appropriate path through machine, attaching tape to empty take-up reel on right spindle.
2. Connect appropriate components, such as microphone or patch cords.
3. Follow instructions for your particular model and type of recorder. Determine if the machine has an automatic level control that does not require manual setting.

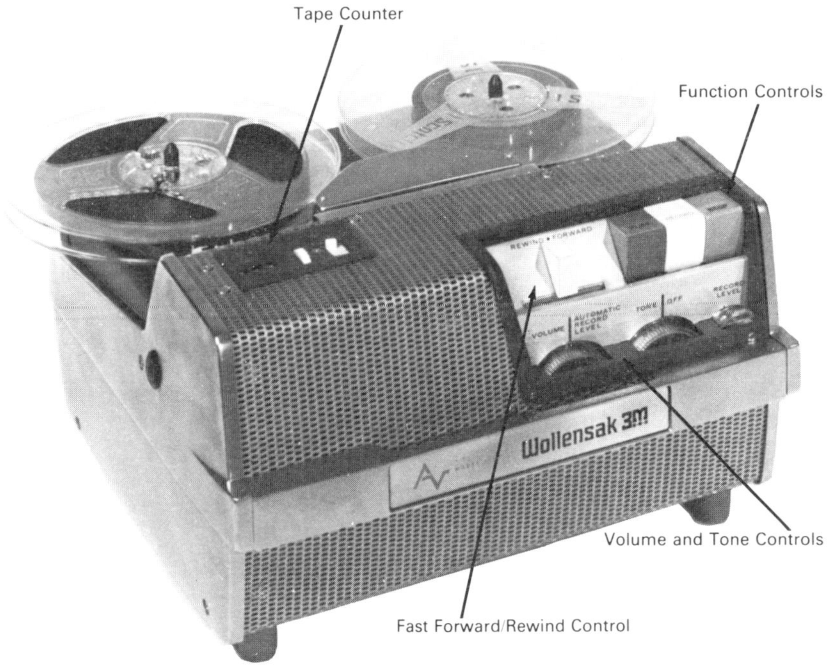

Reel-to-reel recorders usually offer higher fidelity. They often have adjustable tape speeds so recording quality can be improved by recording at a higher speed (compliments Minnesota Mining and Manufacturing, Inc.).

Some of the important features of both cassette recorders and reel-to-reel decks include:

Tape Transport System

1. *Capstan and pinch roller* A rubber wheel (pinch roller) pinches the tape against a revolving metal shaft (capstan) so that the tape is pulled through the machine at a constant speed. It is important to keep these surfaces clean so that the tape moves smoothly.
2. *Feed reel* The reel contains the tape that is being supplied to the capstan/pinch roller. Tape recorders are standardized in this respect; feed reels are on the left, with take-up reels on the right.
3. *Take-up reel* The reel that receives and winds the tape coming from the capstan/pinch roller.

Inputs and Outputs Several receptacles are provided for the tape recorder to receive incoming signals. Microphones and pick-up coils for recording from a telephone are connected to MIC. All other sources, such as record players, tape recorders, radios, television sets, and movie projectors, should be connected to AUX, or AUX IN, or LINE IN. Since the exact label on the receptacle may vary from one machine to another, consult the operating instructions for the unit with which you are working.

In a similar manner, several receptacles may be labeled to guide you in connecting a speaker or patch cord to an output. Typical labels may be LINE, or LINE OUT, SPKR, or MONITOR.

To connect two devices, such as a cassette recorder and a reel-to-reel machine, merely locate the output receptacle of the device supplying the signal, and the input receptacle of the device receiving the signal, and connect these receptacles with a patch cord. Details of the process for transferring sound from one source to a recorder will be described later in this chapter. The process is called *dubbing*.

Special Features A general rule applies to equipment that is designed for the casual user. In the case of cameras, tape recorders, television systems, and other consumer goods, there has been an attempt to create a machine that is easy to use, with few buttons, knobs, and levers that require special settings. However, equipment that is more sophisticated is generally more versatile—but the user must begin to make some adjustments during set-up. Some of the special features described below may not be found on less expensive machines:

1. *Dolby noise reduction* A special electronic circuit is often built into a tape recorder to reduce the high-frequency tape hiss that is an inherent part of magnetic tape recordings. A recorder equipped with a Dolby system requires some adjustment to control the effect of the Dolby circuit.
2. *Tape counter* A type of "odometer" to measure how much tape has been run through the machine. Generally the numbers are not calibrated to correspond to a certain number of inches per second, but merely permit location of program segments that have been recorded on the tape.
3. *Pause control* A control that permits instant stop of the tape transport mechanism. While use of the pause control on inexpensive recorders will often introduce an audible click into the finished recording (if you use it during the recording process), on more expensive machines the design of the mechanism minimizes or eliminates the problem.
4. *Equalizer* A series of sliding knobs that raise or lower certain portions of the frequency range. For example, the lower sounds can be increased in volume to provide a greater bass response, or the lower and higher frequencies can be increased, with the middle frequencies lessened. An equalizer provides adjustment for the acoustics of the room where the tape is being played.

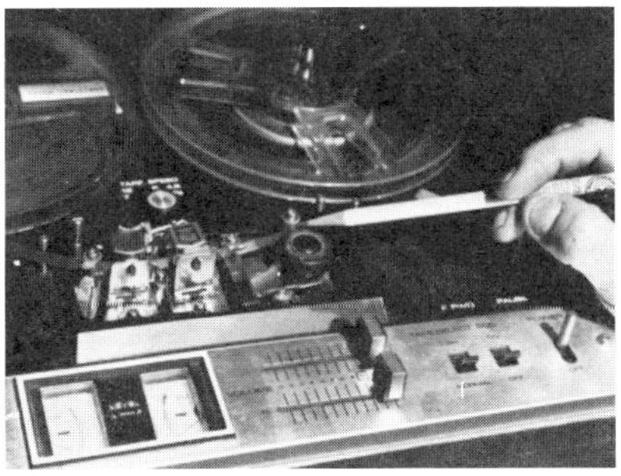

The capstan and the pinch roller squeeze the magnetic tape between them and as they turn the tape is pulled through the machine at a constant speed.

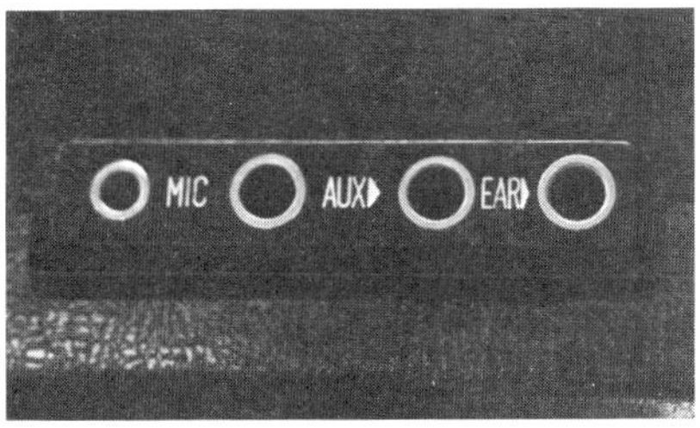

This cassette recorder has two inputs and one output. From the left, the first small jack is where the remote control is plugged in, the second jack is for the microphone. The next jack is an auxiliary input, and the last jack opening is for an earphone or speaker output.

Half-track, monaural recorders place two signals on the tape when it is fully recorded, one in each direction. When recorded on, the tape is similar to a two-lane highway.

Cassette, stereophonic recorders place up to four tracks on a tape. The right and left channels for each signal are recorded side-by-side, then the tape is turned over so another set of signals can be recorded in the other direction. When recorded on, a stereo cassette audiotape is similar to a four-lane superhighway, with two lanes, or tracks, side-by-side going in each direction.

Monaural and Stereophonic

These terms refer to the number of channels or sound sources involved in the recording. Monaural (literally "one ear") recordings result from the use of the microphone, with the resulting playback through one speaker. Even if the recording is listened to with headphones or is played through two speakers, it will be monaural because of the method by which it was recorded. Monaural recordings were the only type available for many years, and still have important applications, even though stereophonic recording and playback systems are now in common use. For the typical educational use, monaural systems are quite adequate.

In the late 1950s it became possible to produce equipment for consumer use that could permit the recording and playback of two sound sources at the same time. Two microphones are used in stereophonic, or stereo recordings. A stereo tape or disk recording contains two simultaneous "impressions" of what is recorded, except that each differs from the other very slightly. Just as you hear sounds with two ears, a stereo recording is made by "listening" with two microphones. While you may believe that each of your ears gets an identical impression from the sound source, each actually hears a slightly different signal. Such minute differences enable you to perceive direction of the sound, as well as other characteristics.

When a stereophonic recording is made, then, two tracks are magnetized on the tape. When the reel or cassette of tape is fully recorded on one side (when it has run from the feed reel to the take-up reel), the tape is replaced on the recorder feed reel and recorded by being passed through the machine again. Thus, four tracks are recorded; two the first time, and two more when the tape is reinserted and recorded the second time.

The patterns for stereophonic reel-to-reel recordings are not the same for cassette recordings, as the diagram below indicates.

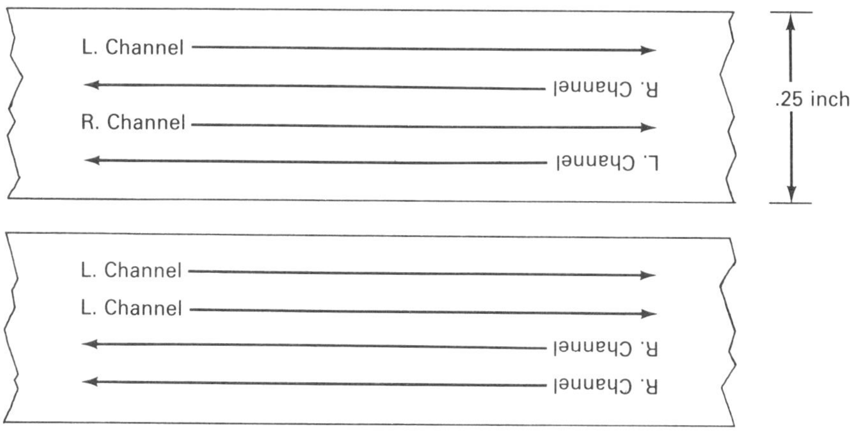

Reel and cassette audio recording tracking patterns (magnetic tape).

MAKING TAPE RECORDINGS

Even though the most common method of making tape recordings is to use a microphone and a portable cassette recorder, a wider range of techniques is available. Not only can you produce finished recordings that are more effective, but the recordings can be of better quality. Several suggestions are described in this section of the chapter.

Recording from Other Recordings

You may be tempted to use a microphone to record sounds from other sources, such as radio, television, or record player. Unfortunately, the result may not be particularly pleasing because, as you recall, transducers are involved. The speaker that produces the sound is one transducer, and the microphone you use is the other. If you can bypass one or more transducers you stand a better chance of preserving quality, and the result will more likely be a higher-fidelity recording (if your original source was high fidelity). Instead of using a microphone, transfer the signal by means of a wire that connects the source with a recorder. The process, called *dubbing*, requires a wire called a *patch cord*.

In patch cords the wires are alike, but the ends are different. The electrical requirements of different sorts of patch cords, recorders, and sources of sound are essentially alike; all that differs is the sort of receptacle and plug needed to hook the two together.

The two main categories of ends for patch cords are plugs and jacks. Plugs insert into jacks. Usually machines have jacks and patch cords have plugs that fit into them. The four types of connectors most used in audio patch cords are mini plugs or jacks, phone plugs or jacks, RCA (phono) plugs or jacks, and Canon plugs or jacks.

When you patch two units together remember to locate the appropriate *output* on the source of the sound, and the correct *input* on the recorder. Some of the labels for outputs and inputs are shown in the illustrations below:

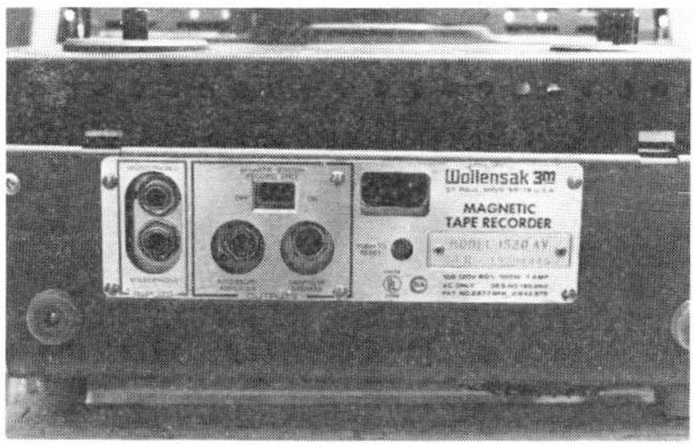

Input and output jacks on a reel-to-reel audio recorder.

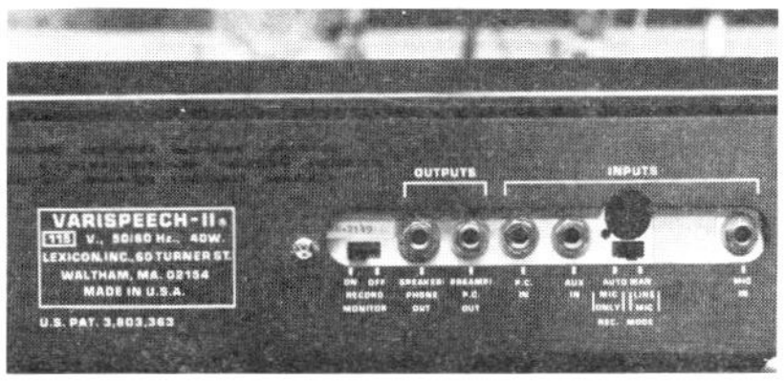

Inputs and outputs on the back of a speech compressor.

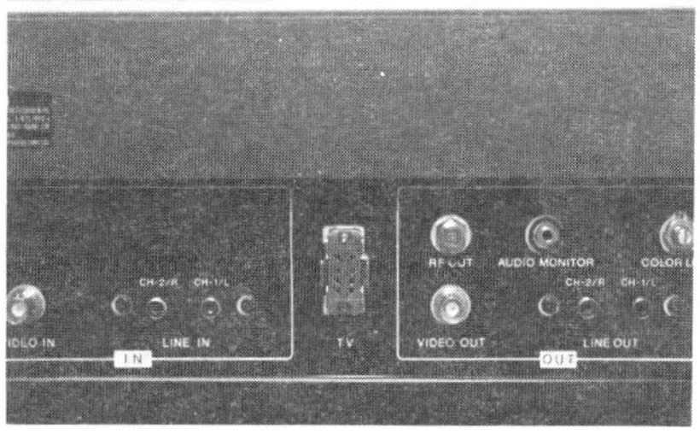
Inputs and outputs on the rear of a video cassette recorder.

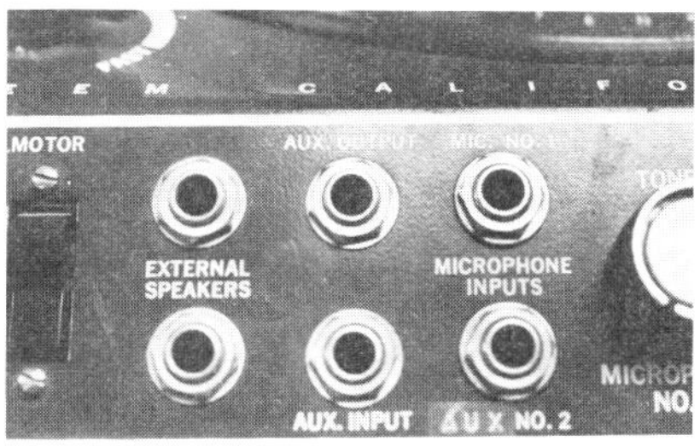

Inputs and outputs on a record turntable.

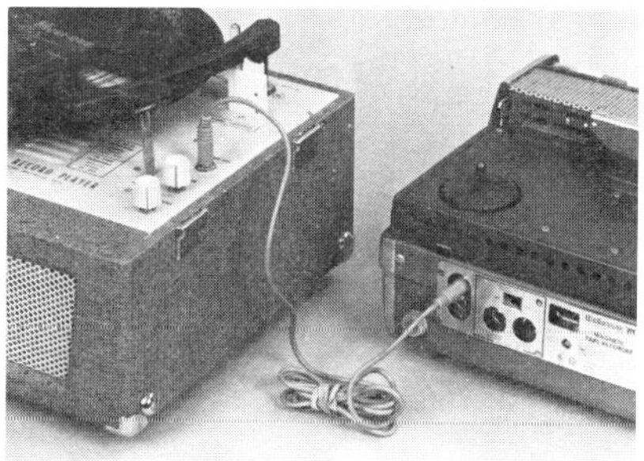
To dub from a record player to a reel-to-reel tape recorder, use a patch cord and connect the output jack on the player to the input jack on the tape recorder.

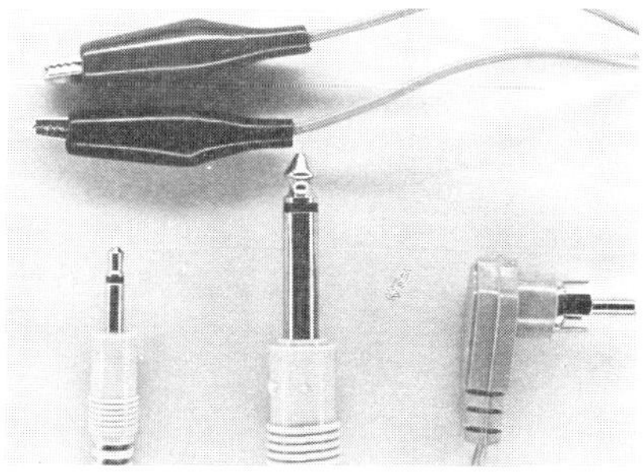
The most commonly used plugs found on audio patch cords are the miniplug, the phone plug, and the RCA plug (sometimes called a phono plug). Alligator clips can be used if the correct patch cord plug is not available.

Don't worry about ruining the equipment as you experiment in making a dub. The electricity carried by a patch cord is very weak. You will note, however, that certain combinations may result in copies that are not usable. For example, if you connect the speaker output of a movie projector to the microphone input of a tape recorder, the resulting dub will be distorted and unclear. An important rule to remember is to use the *mic input* for microphone only. All other sources are usually fed into the *aux input*. Be sure to adjust the output volume on the source and the input volume on the recorder so that each is approximately 50 percent of the maximum. Make a test tape, and then increase or decrease each the same amount, as needed.

OPERATE

A Record Player

While there are variations in the design of classroom record players, each has certain features in common. The instructions here should be usable for nearly any type of record player you use.

1. Locate the off/on switch and volume control. Usually the knob that controls volume also serves as the off/on switch, similar to the knob found on radios and television sets.
2. Locate the speed control. This permits a setting for the turntable speed of 33⅓, 45, or 78 rpm. Usually this control is on the face of the record player near the turntable.
3. The output receptacle is the place to plug in a patch cord or headphone jack. The receptacle will probably be labeled OUTPUT or PHONES or HEADPHONE, and can be used for listening with headphones or to patch the output from the record player into the input of a tape recorder, an external amplifier, or an external speaker. Inserting a plug into the receptacle probably will shut off the speaker of the record player, but the volume control will continue to operate, raising or lowering the power of the signal being sent to the receptacle.

Audiotronics record player components (compliments of Audiotronics Corp.)

PRODUCE

A Dub from a Record Player to a Tape Recorder

Although various models of record players and tape recorders appear to be designed differently, they have certain common features, listed here:

Record Player*

Output: a receptacle labeled with one or more of these notations:

- ☐ Ear
- ☐ Earphone
- ☐ Speaker
- ☐ Out
- ☐ Line out
- ☐ Headphones

*Record players have no input receptacles

Tape Recorder

Output: one or more receptacles

Input: receptacles labeled:

- ☐ Input
- ☐ Line in
- ☐ Auxiliary
- ☐ Mic or microphone

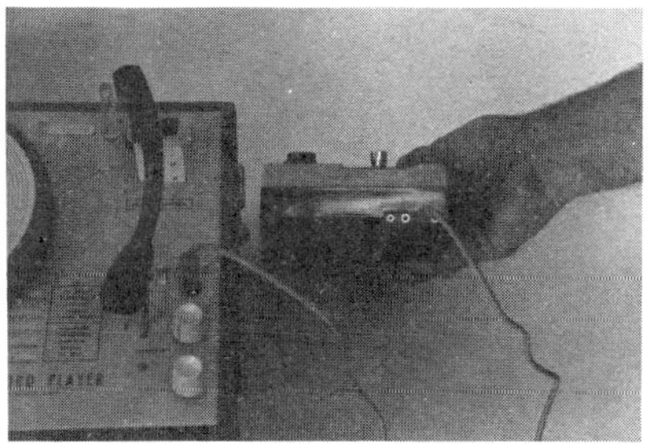

Here a record player is connected to a cassette recorder so a copy of the record can be made on tape. The headphone output on the record player is patched to the auxiliary input on the cassette tape recorder.

To transfer the sound from a record player to a tape recorder—a process called dubbing—use a patch cord. Connect it from the output of the record player to an input of the tape recorder. (*Note*: Avoid using the *microphone input* of the tape recorder when dubbing, because the quality of the resultant recording will be very poor.)

TRY THIS

Set up a dubbing system to transfer sound from a record player or another recorder to made an audiotape on a recorder. Determine the type of patch cord you will need.

PRODUCE

A Dub from One Tape Recorder to Another

One method for making a copy of a tape recording is to use a tape duplicator (see instructions for operating a cassette tape duplicator elsewhere in this chapter). Tape duplicators run several times faster than the normal speed of a tape recorder, and duplicate all tracks of a tape at the same time.

Another method of obtaining a copy of a tape is to patch, or connect, two tape recorders together. One advantage of producing a copy from another tape recorder is that only certain portions of the tape need to be copied. In other words, it is possible to *edit* the tape as it is being dubbed.

1. Determine which tape recorder is to be used for playback, and which will be used to record the copy. Attach a patch cord with an appropriate plug on either end to an *output of the playback machine* and an *input of the recording machine*. Since inputs and outputs have a variety of names, use the lists below for help.

Names of Tape Recorder Inputs	Names of Tape Recorder Outputs
Aux	Ear
Input	Output
Tape in	Speaker
Radio/Phono	Monitor
Mic*	

*Mic is an input that should be used only with a microphone.

2. Adjust the volume level on both the playback machine and the recording machine. Each volume level should be about one-half of the maximum. If the recording machine has an automatic level control it may not be necessary to use any particular setting of its volume control.

High-speed cassette tape duplicators can copy a C-60 cassette in two minutes.

OPERATE

A Tape Duplicator

Instructions will vary, depending on the design of the machine. Read the specific operating instructions for the tape duplicator you are using.

1. Place the master tape and the blank tape into the machine in the appropriate positions. (If the blank tape is not blank, but has been recorded at an earlier time it may be necessary to demagnetize it to erase the prerecorded signal. Some tape duplicators do this automatically and others do not. Consult the instructions for the machine in use.)
2. Adjust volume controls for each track that is being recorded, as needed. (There may be automatic recording level capabilities built into the duplicator.)
3. When the duplication process is completed, check the copy for quality by playing back certain portions on a standard tape player.

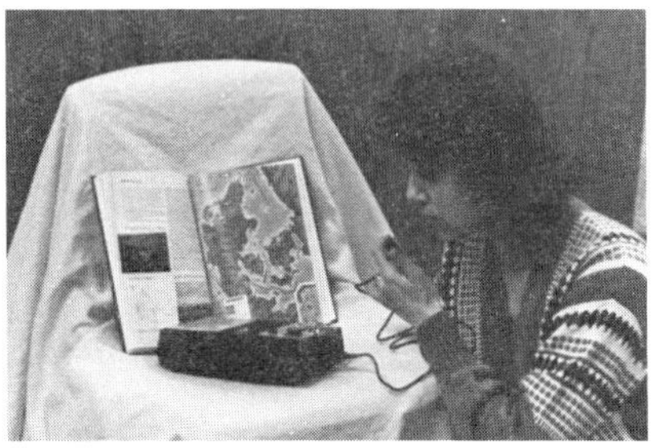

If you are recording at home, muffle echos by covering chairs and furniture with blankets; also, sit on the carpet.

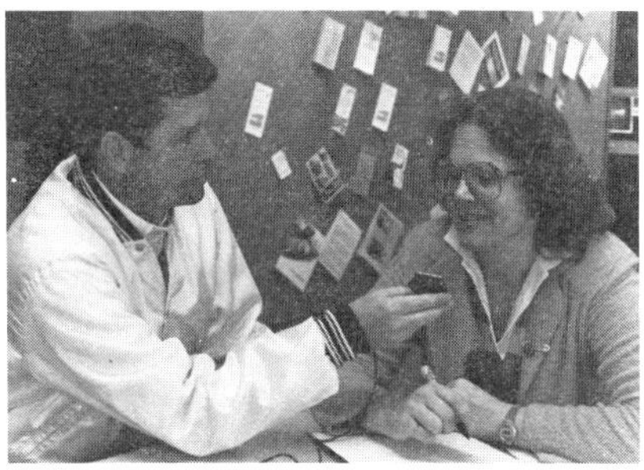

Talking to another person is an effective way of making an audio narration.

Recording Your Voice

If your recorder has a detachable microphone, hold it about 6 to 8 inches away from your mouth, and speak at a normal volume level. You will note that an ordinary room creates a great deal of echo, making the resulting recording sound as if you were in some sort of chamber. Try to make live recordings in a deadened room, which can be created by putting blankets or padding on chairs.

TRY THIS

Set up a home recording studio and make a brief audiotape. Then make the same recording without benefit of the set-up. Compare the results.

Voice inflection is important in order to create an interesting tape. Many teachers have difficulty making recordings that sound interesting, especially when they read from a script. One means of avoiding the monotonous intonation pattern that may occur if a script is read is to consciously attempt to make the voice rise and fall. It may be easier to make an effective tape recording if you have a friend with you as you read the script. Talking to another person is more effective, and more natural, than talking to a microphone. It is also possible to ad-lib a tape recording. In other words, work from notes instead of a word-for-word script. The natural and conversational voice pattern that results will make for more pleasant listening.

If your script calls for character voices that might make the resulting production more realistic, you may wish to ask a friend or two to help you read the script. If very young voices are needed, choose a child. For the voice of an older person, ask someone of the appropriate age to help, unless you can dramatize the effect yourself. It is usually more effective to use people of the appropriate age, unless you are particularly good at dramatizations.

Mixing Your Voice with Other Sounds

Up to this point we have discussed the production of single-source tapes: dubs from other recordings, or live narration. While these tapes can be effective and useful, they lack the versatility and interest that are needed to create effective instructional settings. As you listen to the radio or television you will note the use of multiple sounds, each mixed into the total production to produce a maximum effect. The same technique can be practiced in the recording of instructional tapes. Some of the types of sounds you may wish to consider are:

- ☐ Voice narration
- ☐ Music background
- ☐ Sound effects
- ☐ Audible pulsing signals as an aid to synchronizing slide sets with the tape
- ☐ Inaudible signals to synchronize slide sets

TRY THIS

Make an audiotape that incorporates a variety of sound effects or character voices. Use a mixer if you have one; if not, play the sound effects on a second tape recorder and use your microphone to read the sound effects as needed.

Since each sound source needs its own volume setting, and should be faded in or faded out as required by the script, it is necessary to use two production techniques: mixing and editing.

Mixing of various sounds from various sources is accomplished by using a multichannel device called a *mixer*. Large mixers that can handle twenty or more channels are used in certain musical productions or concerts. Each performer, each instrumentalist, and each vocalist might have a microphone. All microphones—or electronic pickups on the instruments—are fed into the mixer console. The operator of the console adjusts the volume of each component with the aid of headphones and VU meters (meters that register volume).

The same technique is used, on a smaller scale, when you make an audiotape into which music and sound effects have been mixed. One example of the use of a mixer common to schools or training centers is one in which a turntable, a microphone, a cassette tape player, and an audible beeper are patched into a mixer. The narrator controls the volume of all of the elements as the final mixed signal is recorded on tape.

A commercial recording studio will often have the capability of mixing a number of sound sources together.

Commercial multitrack recording systems, such as this one, make it possible to mix up to 24 channels on one piece of magnetic recording tape (compliments Ampex Corp.).

An audio recording studio makes the mixing of several sound sources to make a recording much simpler.

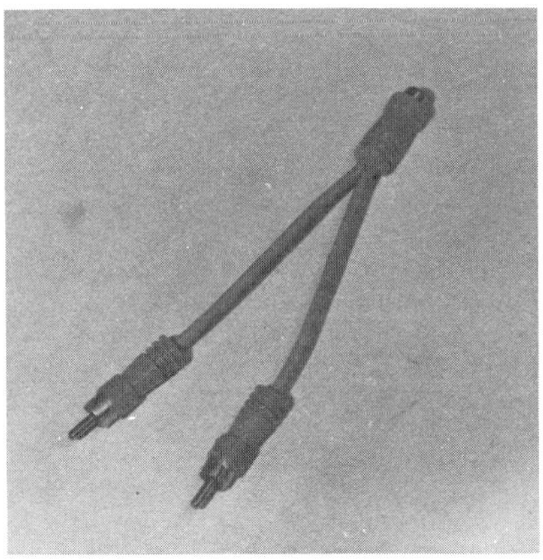

A patch cord makes it possible to dub from a two-channel stereo player to a monaural recorder.

You can understand how busy you might become if you had to narrate, operate the record player, cue up sound effects on the cassette playback deck, and add the beeps for slide synchronization—all at the same time. With a little practice it can be done, but there is an alternative way to accomplish the task, using a stereophonic tape recorder. Since it is possible to record first one channel and then the other you can add sources one at a time. The procedure goes like this:

1. Narrate on the left channel only. Don't record anything on the right channel.
2. Play the narration back and as you listen to it set the tape recorder to record on the right channel only. (You may now add music or sound effects on the right channel.)

When you play the resulting tape your voice will show up on the left speaker and the sound effects or music will come out of the right speaker. If you wish to convert the two-channel tape to a monaural track, patch into a monaural tape recorder using a Y patch cord.

This technique is fine if you have only two sources, such as your voice and some background music. However, if you wish to add sound effects, and then a synchronizing pulse, the process gets a little more complicated. The easiest way to accomplish the production is to use a four-channel tape recorder using the separate channels for voice, music, sound effects, and synchronizing signals.

Editing audiotape may be necessary if you wish to correct mistakes or rearrange material that is already recorded. Since you cannot see the signal on tape it is necessary to play the tape slowly—spinning the reels by hand, actually—past the playback head. When this is done each word, each syllable, each pause shows up in "slow motion." It becomes fairly easy to identify which words are mispronounced or which phrases you need to mark for removal. Marking is done with a felt-tip pen so that the tape can be cut and spliced together at the precise points. The splicing process is shown at right.

TRY THIS

Editing Reel-to-Reel Audiotape*

This exercise will provide experience in removing unwanted portions of an audiotape and splicing the remaining segments together in such a way that the resulting gap in the material is unnoticed.

1. Record some narrative on reel-to-reel tape at 7½ ips. For example, record: "I really do not like cold weather."
2. Replay the recording you have just made and freeze the tape at the precise moment the words *do not* are played. (Freeze by holding the feed reel so that the tape does not move.) You may have to rock the tape back and forth across the playback head to locate the precise point at which the word *do* begins.
3. At the point where you hear the word *do* make a mark on the tape with a felt pen. The mark should be on the portion of tape in the center of the playback head area
4. Allow the tape to continue until the word *not* passes over the playback head. Mark the tape after the word *not* passes over the head.
5. Cut out the section of tape between the two marks and splice the ends together, following instructions for splicing tape.
6. Replay the edited tape. It should say, "I really like cold weather."

*Requires a reel-to-reel tape recorder that has a playback head exposed so that you can made a precise mark on the tape with a felt pen.

OPERATE

Audiotape Splicer

1. Lay tape in splicer, overlap ends for 45° cut.

2. Cut tape.

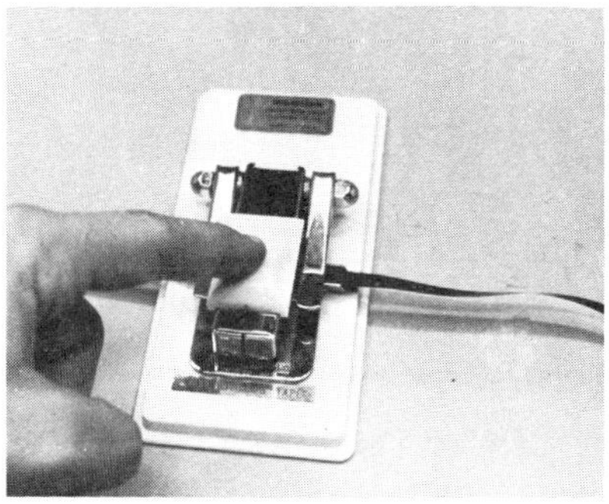

3. Apply special splicing tape.

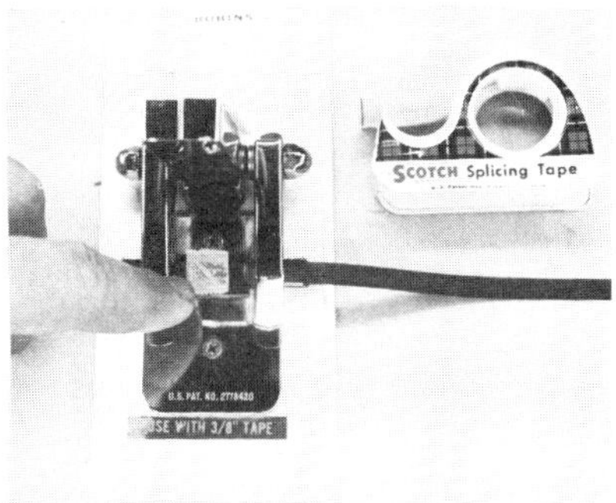

4. Trim splicing tape so edges are even with the audiotape.

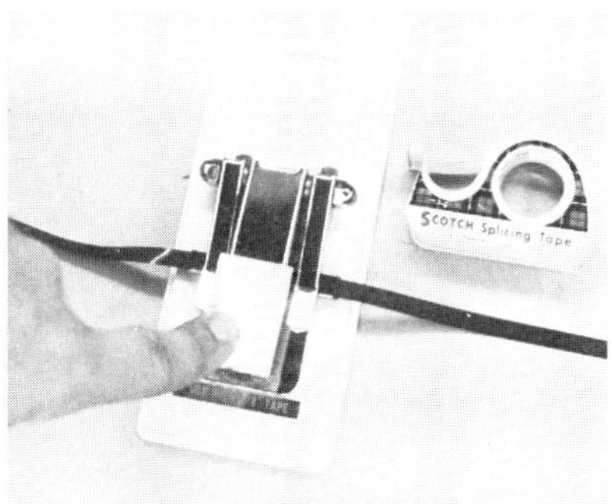

A FINAL WORD

You should not underestimate the power of materials that communicate ideas exclusively through sounds. In the early days of the radio and telephone, a natural fascination developed as the technology grew and matured. Copies of radio serials and important historical broadcasts are available today in retail stores, and sales are good. By relying on a person's imagination, these productions are able to capture and hold the attention of a wide variety of listeners.

LOOKING AHEAD

You probably are aware that the two organs through which we acquire most of our information are the ears and eyes. When both sights and sounds of a subject are combined into a teaching strategy it is possible to maximize learning and achieve instruction that is particularly effective. With the background you now have in the production and use of audiotape it is appropriate that we turn our attention to the production and use of projected materials created with a still camera. The types of cameras, their uses, and the products produced using them, such as slides and photographs, are the topics in the next chapter.

REVIEW QUESTIONS

1. List some general uses of audiotape in teaching your subject area.
2. Describe how magnetic tape records sound.
3. Describe: transducer, input, output, patch cord, dubbing, frequency, fidelity, capstan, monaural, stereophonic, mixer
4. Why is a patch cord used in making a copy of a tape or record rather than a microphone?
5. Describe the relationship between tape speed and frequency response.
6. List the elements in the tape transport system.
7. What is the difference between a microphone input and an auxiliary input?
8. Describe some situations in which it might be important to edit an audiotape.

ANSWERS TO REVIEW QUESTIONS

1. Use of documentary sounds, guest speakers, music, sound effects, drill and practice, varying playback speed.
2. Particles of magnetic material are magnetized in a pattern that corresponds to the sound. When the pattern of the particles is read from the tape, the original sound is reproduced.
3. Transducer: a device that converts energy from one form to another
   Input: receptacle that receives incoming signal, either through a patch cord or microphone
   Output: receptacle that furnishes an outgoing signal, to a patch cord or speaker
   Patch cord: wire through which a signal can be transmitted
   Dubbing: the process of transferring signal through a patch cord
   Frequency: a measure of the pitch of a sound, designated in units called Hertz
   Fidelity: the degree to which a recorded sound matches the original sound
   Capstan: the rotating shaft that pulls audiotape through a recorder as the tape is pinched against the shaft by a rubber wheel
   Monaural: sound that is recorded with a single microphone and generally played back through a single speaker
   Stereophonic: sound that is recorded with two microphones and played back through two speakers
   Mixer: an electronic device that permits the volume of multiple inputs to be controlled so that they have the proper balance to one another in the final mix.
4. The patch cord transmits the signal without loss of fidelity.
5. As tape speed is increased, the frequencies it is capable of recording are increased.
6. Feed reel, capstan and pinch roller, take-up reel.
7. A microphone input adds considerable amplification to the incoming signal, but an auxiliary input is designed to receive an incoming signal that is already amplified.
8. Editing is important if a recorded speech requires shortening, if words are mispronounced, or if it is necessary to rearrange prerecorded material.

REFERENCES

Clifford, Martin. *Microphones: How They Work and How to Use Them* (Blue Ridge Summit, Pa.: TAB, 1977).

Jacobson, T., and Hardin, J. "Education by Telephone," *Phi Delta Kappan* (June 1977): 769–771.

Tremaine, Howard M. *Audio Cyclopedia* (Indianapolis, Ind.: Howard Sams and Company, Inc., 1978).

GOALS

☐ Understand the uses of photography in teaching.
☐ Understand how to produce still pictures.

OBJECTIVES

☐ Describe the unique uses of still pictures for teaching.
☐ Describe how film works to capture images.
☐ Describe how cameras work to control light.
☐ Relate the type of photographic product desired, the format of film, and the location where the pictures will be taken to film and camera selection.
☐ Describe the four major types of cameras.
☐ Describe the operation of slide projectors, filmstrip projectors, and still cameras.
☐ Explain the sequence to follow when taking still pictures.
☐ Process exposed black and white film into negatives.
☐ Print and enlarge black and white pictures from negatives.

Imagine that the room you are in is full of ping-pong balls. It is jam-packed. It would be impossible to cram another ping-pong ball into the room. Now, suppose you decide that you want to remove exactly 100 of the ping-pong balls from this room and put them in the room directly below you (if you are sitting in the basement, imagine a little harder for a few minutes and conjure up a subbasement room).

In order to get the 100 ping-pong balls out of your room and into the room downstairs you are going to have to cut a hole in the floor and let them drop through. First, you must decide how large a hole you should cut in the floor. Naturally, the hole will need to be at least slightly larger than one ping-pong ball, but what about other sizes? Actually, an entire series of hole sizes will work satisfactorily. However, the smaller the hole, the longer it will take for the ping-pong balls to drop through it, and if you decide to cut a hole larger than about 18 inches in diameter too many balls will drop through at one time. Any selection in between will work all right. Remember that the time the hole should be left open will vary depending on its size. Since your task is to drop 100 ping-pong balls into the room downstairs, any combination of hole size and time that will give this result is acceptable.

While it may not seem like it, we are talking about a concept directly related to photography. Instead of ping-pong balls, photographers talk about light. Empty rooms are cameras (as a matter of fact, the term *camera* comes from the Latin *camera obscura,* which means "lightproof room" or "box"). Holes in the floor are apertures, and opening times are shutter speeds.

The primary responsibility of the photographer is to allow the correct amount and type of light to enter the camera so that a light-sensitive substance in the camera called film is exposed. The controls on the camera body, such as aperture ring, shutter speed dial, and focusing knob, permit light to be regulated. That is basically what photography is all about.

This chapter will present information on uses of photography in teaching, how film and cameras work, how to produce the kind of photograph you want, and how to process and print your own black and white pictures.

11
STILL PHOTOGRAPHY

TEACHING WITH PRINTS, FILMSTRIPS, AND SLIDES

The simplest way to use still photography is to teach with pictures produced by others. Prints, filmstrips, and slides are available from a number of commercial and educational sources. These visuals can be easily incorporated into story lines and lesson plans.

Study Prints

Because no equipment is needed, probably the easiest type of photography to use is the study print. A study print is a black and white or color picture. Usually, study prints are mounted onto a cardboard or posterboard backing. For individual use, small prints, like those found in Kodak's *Photo Discovery Sets,* are appropriate. Most often, however, the 8 × 10-inch or larger print size is best for teaching situations.

Photo essays are another use of prints. A photo essay, as the name implies, is a picture story. They usually have directions that accompany the pictures. These directions are either written or on an audiotape recording. Photo essays are normally used to visually and verbally describe a sequential pro-

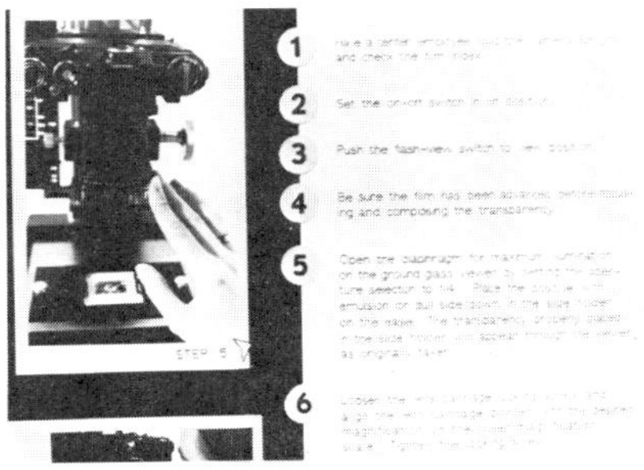

Photo essays can be mounted on the wall near equipment . . .

These photographs are used by students to illustrate a story they create.

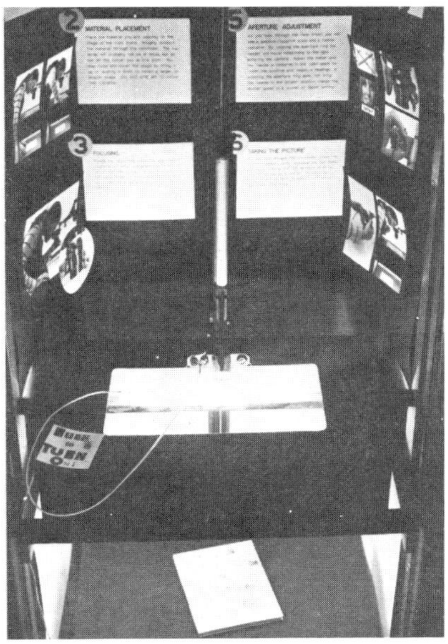

so students can follow the directions when operating the device.

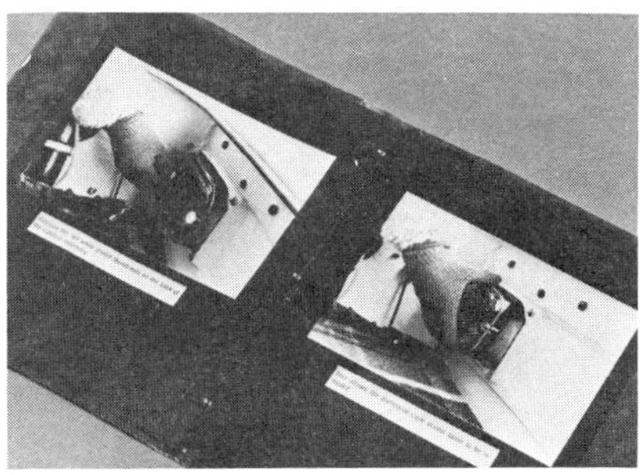

Photo essays can also be placed in notebooks so they are portable and can be used anywhere.

Slides enable the teacher to project large, colored images easily and inexpensively.

cess, such as equipment operation. As you have certainly noticed, this text has several photo essays in it.

Bulletin boards are another way to use study prints effectively. Slightly larger photographs are needed for bulletin boards so that they can be seen easily by all students, even those sitting quite a distance from the board. The 11 × 14-inch size is often used when photographs are mounted on bulletin boards.

Sometimes teachers like to explain a process visually. Mounted photos can be held up by the teacher, described, and then passed around the class. Study prints are popular because no equipment is needed and the classroom lights can be left on.

Slide Sets

Commercially produced 2 × 2-inch slide sets are available for almost all subject areas and grade levels. The primary advantages of slides over prints is that they are less expensive, are in color, and can be projected to fill a screen so that the entire class can see them easily. However, a projector is needed to show slides and the classroom must be darkened.

OPERATE

Projecting Slides with a Carousel Slide Projector

You need: slides, a carousel slide tray, a carousel slide projector

Procedure

1. Load your slides into the carousel tray in sequential order. Position each slide as you want it to appear on the screen with your hand holding on to the lower right-hand corner of the slide. Place the slide in the tray by rotating your hand so that when you are done you are holding the slide in the top left-hand corner. Drop the slide straight down into the tray.
2. Load the tray on the projector. Hold it over the top of the projector with the zero slot in the 3 o'clock position. Align the zero slot in the tray to the guide indicator on the projector and place it on the projector.
3. Turn the projector on. The on/off switch is located at the rear of the projector.
4. Advance the tray to the first slide. Use either the remote-control slide advance button or the slide advance button on the right side of the projector. The remote cord is connected at the rear of the projector just above the on/off switch.
5. Fill the screen with the projected image. Use the zoom-lens control and the focusing knob to project a screen-filling and focused image. The elevation control on the front of the projector can be used to position the projected image correctly.
6. Advance the slides using the remote-control button. Focus can also be adjusted with the focus button on the remote unit.
7. Remove the tray, or spin it ahead, by depressing the select button on the right side (3 o'clock position) of the projector and spinning the tray with your other hand. Normally, for most models the projector motor must be on to do this.

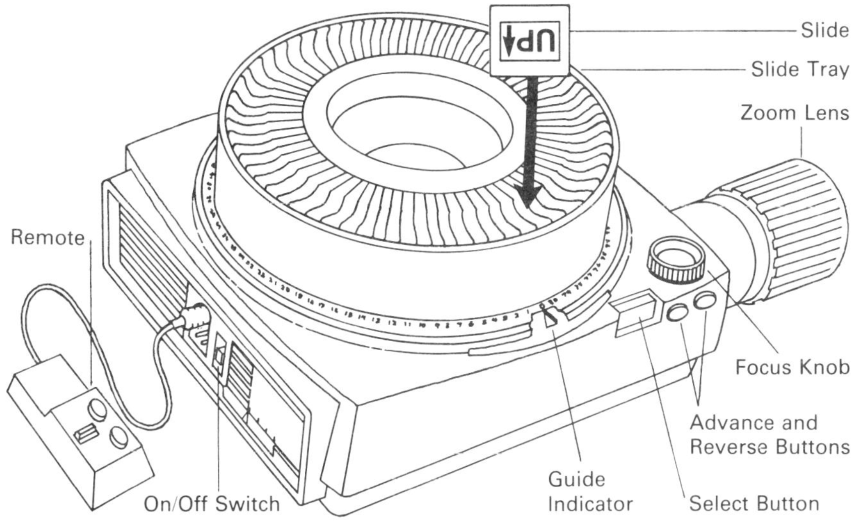

Parts of a Kodak carousel slide projector.

TRY THIS

Load a carousel slide tray with slides and project them on a screen using a slide projector such as the Kodak Carousel. Be sure the image projected is in focus and fills the screen.

When loading slides in the tray be sure they are positioned correctly.

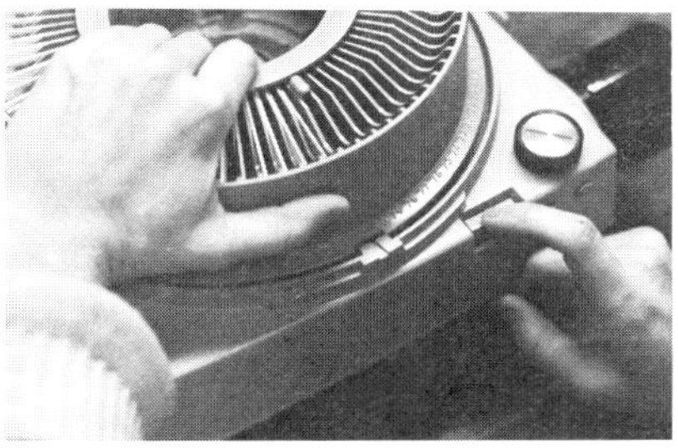

Place the loaded tray over the projector, and align the zero slot with the guide indicator on the projector.

Connect the remote control to the back of the projector next to the on-off switch. Align the dot on the remote cord to the dot on the projector case.

The zoom lens is controlled by rotating the front of the lens housing. This allows you to change the size of the image being projected.

The image is focused by rotating the focusing knob on the front of the projector.

To spin the tray ahead, turn the projector on, depress the select button, and rotate the tray with the other hand.

Filmstrip projectors are available in models with automatic advance capability, or with manual advance, such as this popular brand.

Filmstrips

Many teachers like to use commercial filmstrips instead of slides. In actuality, the two techniques give approximately the same effect on the screen. The filmstrip format is favored by some because the visuals are all in sequence so they cannot be projected out of order, and the projectors used to show filmstrips tend to be a little less expensive than slide projectors. Also, commercially produced filmstrips usually cost less than slide sets. Storage of filmstrips is easy, too, because when they are placed in their storage containers they are only a little more than an inch deep.

Filmstrip and slide sets are often synchronized with audiotapes to form complete presentations. Some projectors even have built-in audiotape players that have the capacity for using audiotapes recorded either with audible or with inaudible synchronization pulses. The inaudible pulses advance the projector automatically so the narration on the audiotape stays in synchronization with the image being projected. Many commercially prepared slide and filmstrip sets are made with this automatic sequencing.

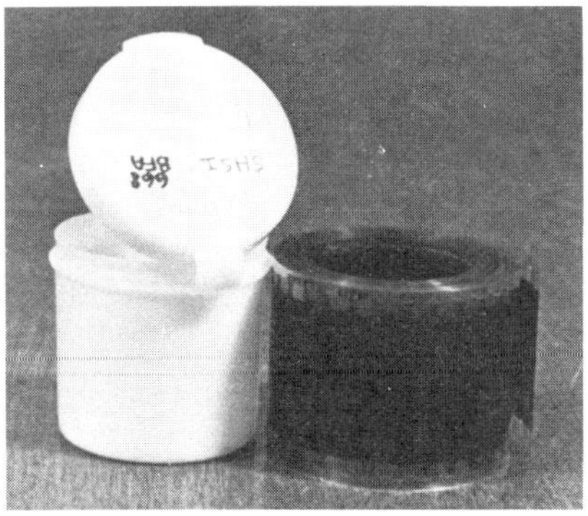

Filmstrips are popular because they are small, easy to store, inexpensive, and locked in order.

OPERATE

Projecting a Filmstrip with a Filmstrip Projector

You need: a filmstrip, a filmstrip projector

Procedure

1. Prefocus the projector. Turn on the lamp and fan switches. Adjust the elevation control so that the white light image is projected to fill the screen. Focus using the focus knob or by turning the objective lens.
2. Load the filmstrip. Filmstrips unroll counterclockwise. Thread the filmstrip into the top of the filmstrip carrier by hand until the sprocket holes on the filmstrip mesh with the sprockets under the film gate.
3. Advance the filmstrip using the advance knob on the right-hand side of the film gate under the lens to show the first full image.
4. Frame the image so just one image is seen at a time. The framing control is usually located on the right side of the filmstrip carrier, near its bottom, alongside the advance knob.
5. Advance the filmstrip and show the entire sequence.
6. Remove the filmstrip from the projector and rewind it by hand. Remember, it must fit into its storage container. Handle the filmstrip by its edges only. Do not pull on the end of the filmstrip to tighten it because this may scratch the emulsion or break the filmstrip itself. Wind it tightly, starting with the end and working toward the beginning.

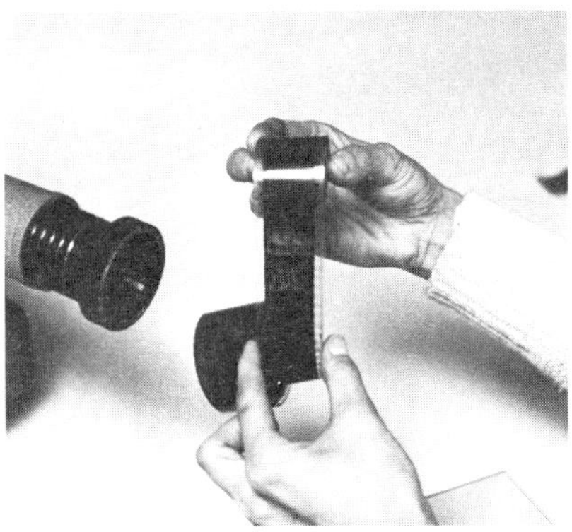

When rewinding a filmstrip, hold it by the edges and wind from one end toward the other. Do not pull on the filmstrip to tighten it.

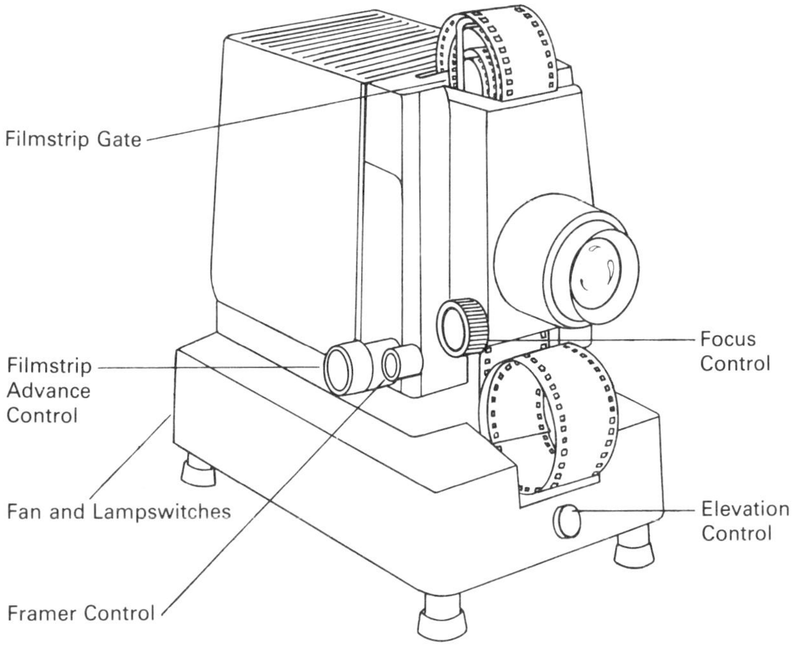

Parts of a filmstrip projector.

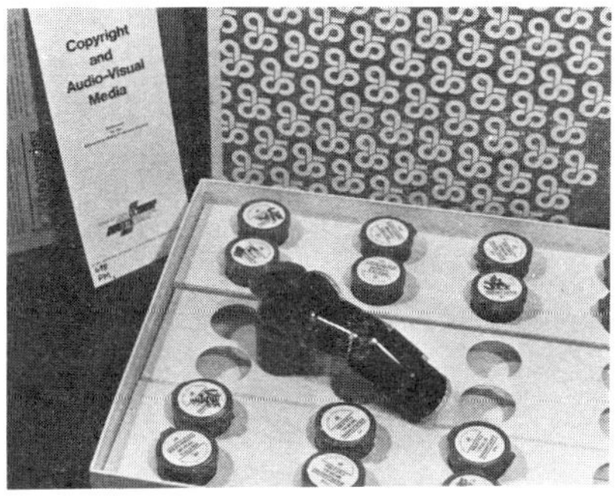

Many commercial filmstrip kits are available.

TAKING YOUR OWN PHOTOGRAPHS

Once you begin using prints, slides, and filmstrips in your classes you will soon see how helpful they can be to your teaching. The next step you will probably want to make will be to take your own photographs. Two fundamental questions are usually asked by teachers who want to use photography for the first time. They are: How do cameras work, and how does film work? In this section of Chapter 11 the basic function and operation of film and cameras will be presented.

How Film Works

Unexposed film stock contains a light-sensitive substance. When light strikes it, a chemical reaction occurs. The results of this chemical change can be made permanent and visible by another chemical process. The final product of these chemical reactions is either a positive or negative image that can be viewed directly (positive slides) or used in another process to produce photographs for teaching (color and black and white prints).

Light is a key ingredient in photography. Light is reflected off of objects. That is why it is possible to see them. The amount and type of light reflected from someone or something vary because of the type of reflective surface. Brightly colored, shiny, smooth objects reflect a great deal of light. Dark, dull, and rough objects reflect less light. Since different objects and parts of objects reflect different amounts of light, a sheet of film that is struck by this reflected light reacts differently and records an image, depending on how much light strikes it.

Light Sources

- ☐ Sun
- ☐ Light bulbs
- ☐ Fluorescent lights
- ☐ Electronic flash
- ☐ Fire

While basically snapshot cameras, the easy operation of the disc camera format makes them attractive for educational and training uses, too (compliments Eastman Kodak Co.).

People and objects can be seen because light is reflected off of them.

Film is made up of two basic components—emulsion and backing. Film backing is a clear plastic. The emulsion is a gel-like substance containing suspended silver crystals. The silver crystals suspended in the film emulsion are the sensitive substance that reacts when struck by light.

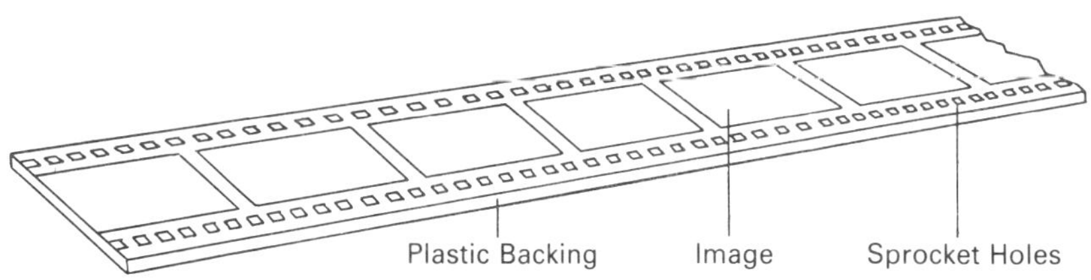

Components of exposed film.

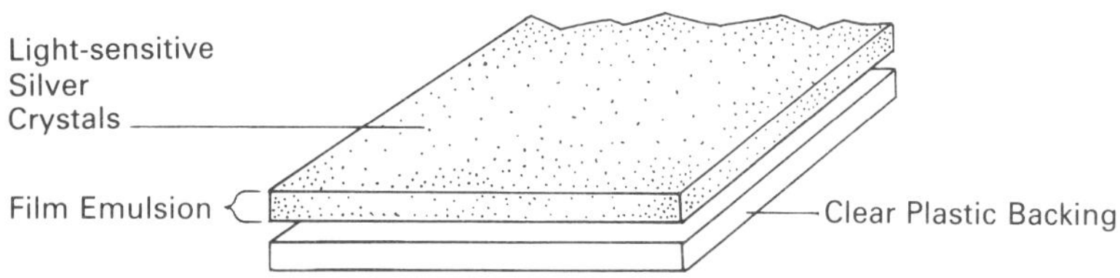

Cross section of film.

Pocket cameras make photography simple and less formidable than carrying around a larger-format camera.

Stated another way, film works because light strikes the silver crystals suspended in the emulsion. When this happens a reaction begins. If the correct amount of light (too much would give an overexposure, and too little, an underexposure) is allowed to fall on the film, a latent image is formed. If this latent image is processed in the appropriate chemicals, a visible, permanent photograph will appear. The camera is the mechanical device used to control how much light will strike the film to produce a latent image.

How Cameras Work

Cameras have three primary components—a light-proof container (the camera body), a hole in the container (the camera aperture), and a mechanical device for opening and closing the hole (the camera shutter). All cameras, from the simplest to the most complex, have these three components.

As cameras are made more versatile—and more complex—additional controls are added to those found on the basic camera. For example, a viewfinder permits the photographer to "sight-in" the camera accurately on the subject of the photograph. The lens makes it possible for the photographer to focus the light that enters the camera body, so that a crisp, sharp image falls on the film. Variable aperture openings and shutter speeds allow the photographer to more accurately control how much light will enter the camera when a picture is taken. All these components are adaptations of, or additions to, the basic camera's three components—the light-proof container, the hole, and the shutter.

Apertures, f Numbers, and f/Stops

An *aperture* is an opening in the camera body through which light passes. Aperture sizes can be changed in many cameras. An index of the size of an aperture is determined by using the ratio of the diameter of the aperture opening to the focal length of the camera lens. Focal length is the distance from the optical center of the camera lens to the film in the back of the camera when the lens is focused on infinity.

The result of this ratio is called an *f number*, and is used to designate a certain aperture size. Actually, the f number is just the denominator of the fraction of the ratio. Because only the denominator is used, f numbers like f/1.4 or f/2.0 indicate fairly large openings, while f numbers such as f/16 or f/22 indicate quite small apertures.

Sometimes photographers use the term *f/stop* to refer to the lens opening size. The f/stop is the mechanical device that changes the physical size of the aperture. Before variable apertures were invented, photographers had metal plates with holes of various sizes cut in them. Depending on the amount of light in a situation they would insert one of these "stops" behind the lens to control how much or how little light would enter the camera when the shutter was opened. Because these plates were used to stop light they became known as f/stops. Many modern cameras have variable aperture leaves that allow the photographer to adjust the amount of light that will enter the camera by merely turning a dial. However, the two terms *f number* and *f/stop* are still used.

Shutter Speeds

Shutter speeds are listed in fractions of a second. The number listed on the camera's shutter speed dial is the denominator of that fraction. Commonly, cameras with variable shutter speeds will have ones that range in speed from one full second up to 1/1000 of a second.

There is a direct relationship between aperture size and shutter speed. As the shutter speed is adjusted one step in either direction, the amount of light that enters the camera is halved or doubled, depending on the direction of the change. Similarly, as aperture sizes are changed from one f number to another, the amount of light that enters the camera is halved or doubled. Thus, a shutter speed of 1/60 and an f number of f/8 indicate that the same amount of light would enter the camera as settings of 1/120 and f/5.6, or 1/30 and f/11.

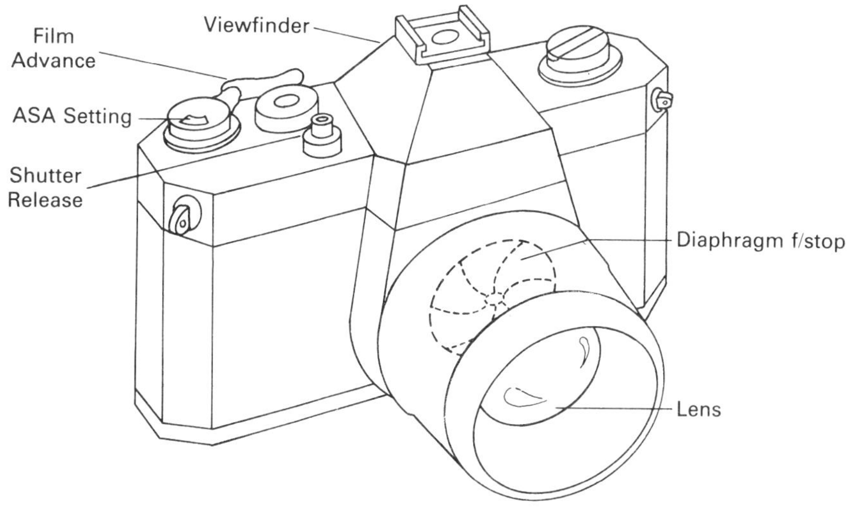

A simple camera.

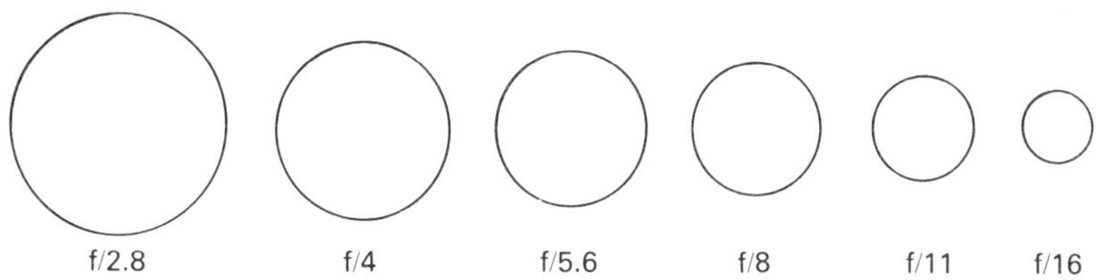

Relative size of various apertures.

If you want to freeze the action of people and animals, you should use a shutter speed of at least 1/125 of a second. It is also a good rule to never hand-hold a camera when you plan to use a shutter speed below 1/15 of a second. At a shutter speed slower than this it is likely that you would move the camera while the shutter is open, producing a blurry picture.

There are two types of camera shutters: in-the-lens and focal plane. Inexpensive, amateur cameras are most likely to have in-the-lens shutters. Often the leaves of the variable aperture serve double duty as both the f/stop control and the shutter. When the shutter release button is pushed, the aperture leaves of the lens open to the preselected aperture size, then close. The aperture remains open for as long as the shutter speed dial was set for. In short, in-the-lens shutters are just that; the shutter is found *in the lens*.

Typical Shutter Speeds

1
2
4
8
15
30
60
125
250
500
1000

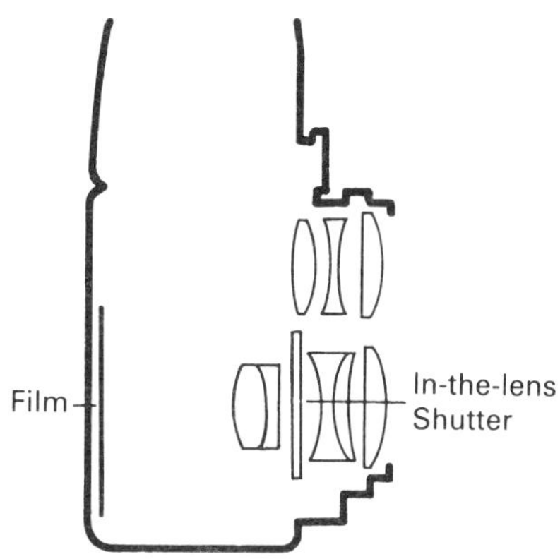

Camera with an in-the-lens shutter.

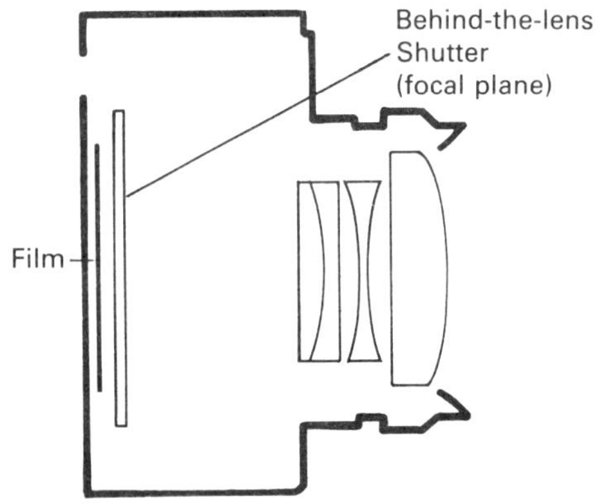

Camera with a focal-plane shutter.

Focal-plane shutters are used in single-lens reflex cameras (SLRs). In an SLR, the photographer views the subject to be photographed directly through the lens of the camera, so the shutter is placed behind the lens and directly in front of the film. This makes it possible to look out of the lens without opening the shutter. The area directly in front of the film is called the film's focal plane; that is how this kind of shutter gets its name. Either shutter type works satisfactorily for most photographic jobs.

Focusing

There are three techniques for focusing the light that passes through the lens of the camera. Preset focusing is used in the most inexpensive, amateur cameras. Rangefinder focusing is used in medium-priced cameras. Ground-glass focusing is used in single-lens reflex cameras and in special types of professional cameras.

Preset focusing is found on only the simplest and least-expensive consumer cameras. Usually focusing for this type of camera will be preset so a picture can be taken of any subject that is more than five feet from the camera. Cameras with this type of focusing cannot be used for close-up photography.

The Kodak disc 6000 camera has preset focusing, one setting for close-ups and one for long shots (courtesy Eastman Kodak Co.).

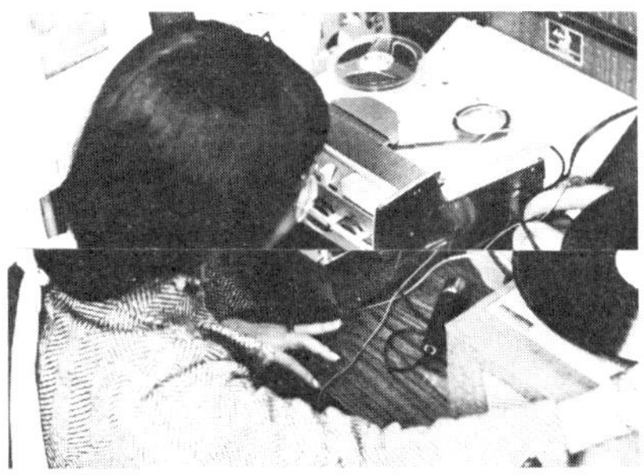

For split-image focusing systems, the scene in the viewfinder is split in half when the camera is out of focus.

A popular technique for focusing many of the medium-priced amateur and professional cameras is called *rangefinding*. There are two types of rangefinder systems—double image and split image. Both work in basically the same way.

With double-image rangefinders, a double image is seen in the center of the viewfinder when the camera is out of focus. As the camera focusing knob is rotated, the two images come together and become superimposed. When this happens the camera is in focus. Actually, a built-in measurement system with two lenses has found the exact camera-to-subject distance. That is where rangefinder focusing gets its name—finding the range.

Split-image focusing is a variation of the double-image system. Instead of seeing a double image in the viewfinder when the camera is out of focus, the photographer sees an image that is split into halves. As the camera is focused, the two halves of the image move together to come into correct orientation to one another.

Ground-glass focusing is used in single-lens reflex, twin-lens reflex, and view cameras. In this system, the light passing through the lens falls on a translucent ground glass that the photographer can see. The image is focused by the photographer so that what is seen on the ground glass looks clear and sharp. A Fresnel lens is often used in ground-glass focusing.

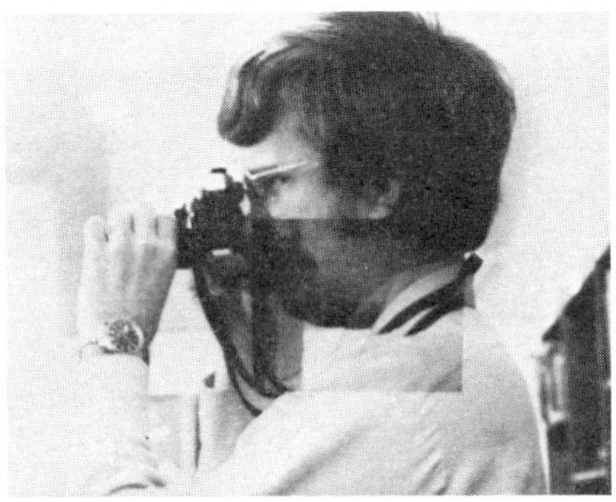

When focusing a camera with double-image focusing, two images are seen through the viewfinder when the camera is out of focus.

Fresnel lens or ground-glass focusing systems require the camera operator to focus the camera so a clear, distinct image is seen in the viewfinder. If the image is fuzzy, the camera is out of focus (compliments Pentax Corp.).

TYPES OF CAMERAS

There are four major types of cameras—viewfinder cameras, single-lens reflex cameras, twin-lens reflex cameras, and view cameras.

Viewfinder Cameras

The photographer observes the scene to be photographed through a small peep hole (the viewfinder) when using a viewfinder camera. Generally, viewfinder cameras are less expensive, less complicated, and less likely to malfunction than other camera types. They can be very small, lightweight, and simple to operate. However, because the viewfinder is in a different location than the lens that the light passes through to expose the film, viewfinder cameras have an inherent defect called *parallax*. Parallax occurs because the photographer views the subject to be photographed through a different opening than the one exposing the film. It is difficult to take close-up pictures with a viewfinder because it is never possible to be exactly sure what image will strike the film when the exposure is made.

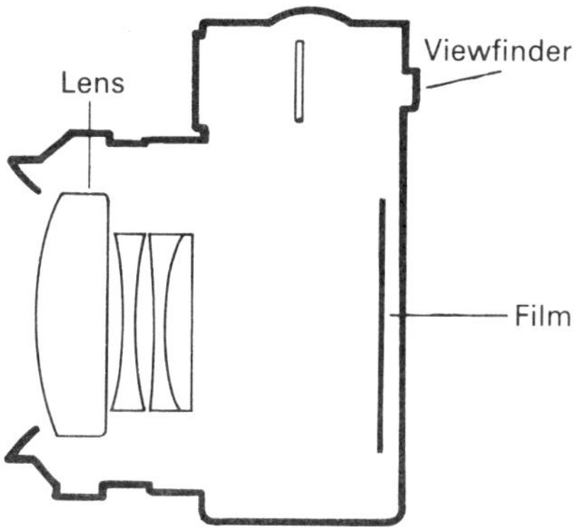

Cross section of a viewfinder camera.

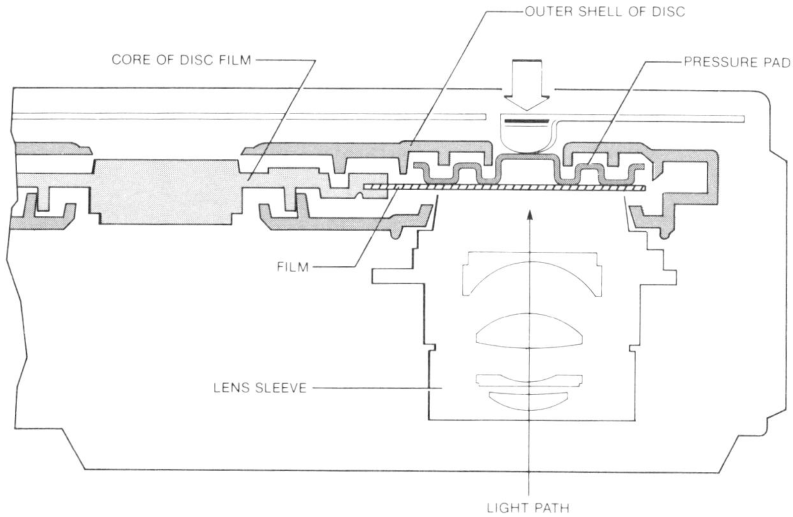

Cross section of a disc camera.

Hints

☐ Action shots require a shutter speed of at least 1/125.
☐ Flash units use a shutter speed of 1/30–1/60
☐ Never hand-hold a camera at a shutter speed below 1/15.

A disc camera.

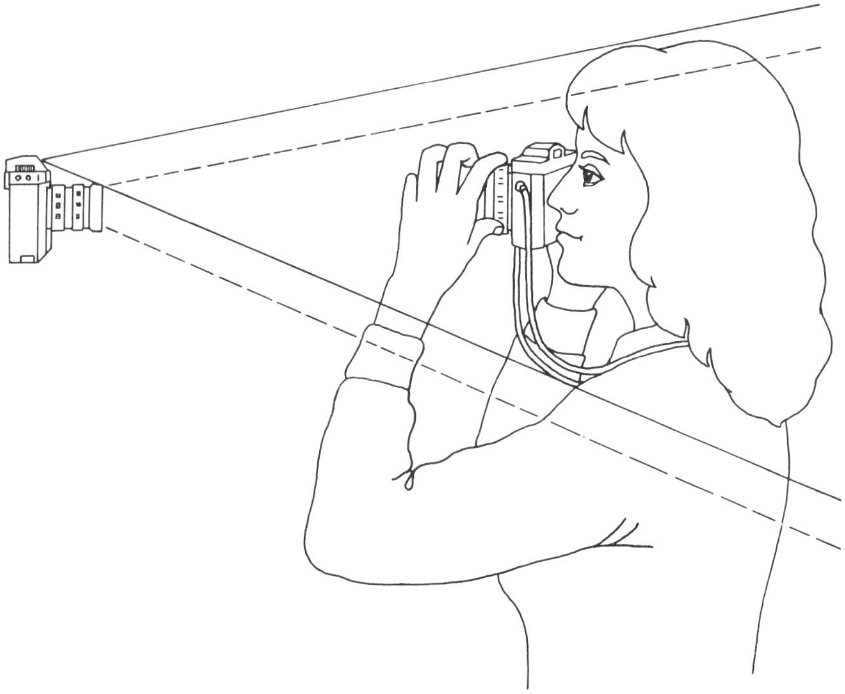

Parallax occurs because the viewfinder and the lens are not in the same spot, so two different images are seen. Parallax is a problem when viewfinder cameras are used (compliments Eastman Kodak Co.).

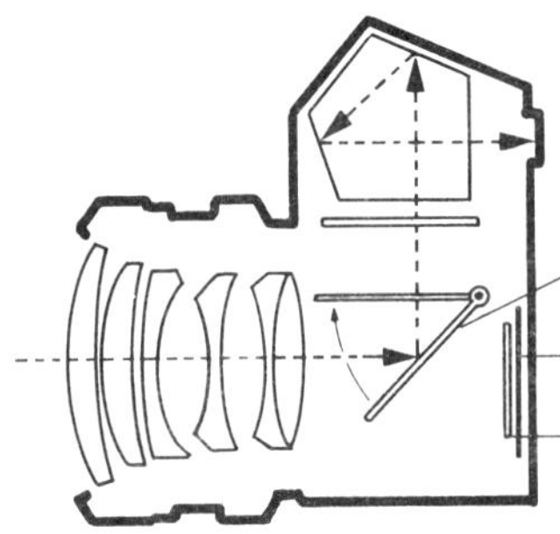

Single-lens reflex camera.

The Pentax is a popular single-lens reflex camera.

Single-Lens Reflex Cameras (SLRs)

A prism and lens make it possible for the photographer using a single-lens reflex camera to look directly out the camera lens and through the aperture when focusing and framing. This makes close-up photography simple. Whatever is seen through the lens will be the same image that falls on the film when the shutter is released. Because SLRs have focal-plane shutters and through-the-lens focusing, they are the most versatile camera type. It is possible to change lenses, take extreme close-ups, and even take pictures through a microscope or telescope with a single-lens reflex camera. Parallax is not a problem with an SLR. In the foreseeable future, the single-lens reflex camera that uses 35mm roll film will probably continue to be the most popular camera format for educational uses.

Twin-Lens Reflex Cameras (TLRs)

The twin-lens reflex cameras, like the SLR, use a mirror that reflects an image upward onto a ground-glass screen. However, the twin-lens reflex camera has two sets of lenses, one for focusing and one for exposing the film, and its mirror is fixed. Because its mirror is fixed the TLR has few moving parts and is quite reliable. However, parallax is a problem, so close-up work with a TLR is difficult. Another disadvantage of TLRs is that the image that falls on the ground-glass plate

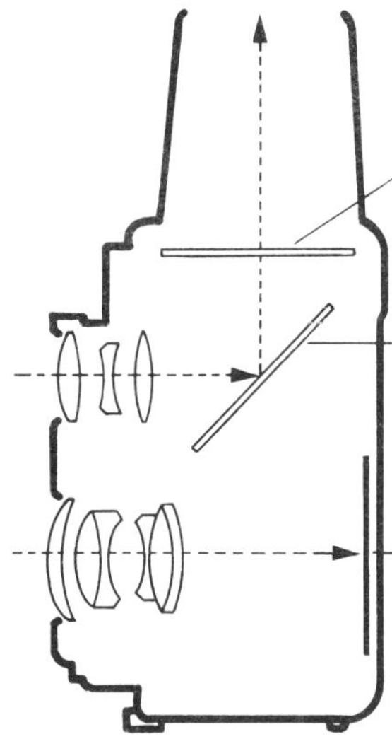

Cross section of twin-lens reflex camera.

Twin-lens reflex cameras, sometimes called "press cameras," have two lens systems—one for focusing and one for light to pass through to expose the film.

is reversed left to right. This tends to make composing and focusing a little difficult. Lenses are not interchangeable on most TLR cameras.

View Cameras

The oldest, simplest technique for focusing a camera is to look directly from the back of the camera through the lens. View cameras have a ground-glass plate at the rear of the camera body. The photographer looks at the plate to compose and focus the image. View cameras are still used today, but usually by professional photographers.

Focusing is made possible in a view camera by moving the camera lens backward and forward. While a view camera can be used for close-up work, its major advantage over other camera types is its large film format size. Large sheet film is used in view cameras rather than roll film. This usually means that better-quality photographs result because of the general rule that the larger the negative area the better the quality of the printed picture. Of course, the cost of the film is proportionally higher, too.

FILM PRODUCTS, FORMATS, AND TYPES

There are three questions that you should ask yourself everytime you plan to take photographs.

Question No. 1: What kind of a photographic product do I want?

Question No. 2: What format of film do I need to fit my camera?

Question No. 3: How much, and what type of, light will there be where I plan to take my photographs?

Question No. 1: Film Products

As far as most educators are concerned, there are three final products of the photographic process. These products are colored 2 × 2-inch color slides, color prints, and black and white prints. There is a film type for each product.

Film with the suffix *-chrome*, such as Kodak films Ektachrome and Kodachrome, normally give transparent slides from the processor that are ready for projection. Slide film is called positive film because a positive of the image photographed is on the film. This image is on transparent plastic and during processing is mounted into a 2 × 2-inch cardboard or plastic frame or mount so it can be placed into a projector for viewing.

Color-print film in roll form will have the suffix *-color*, such as Kodacolor films from Kodak. Color-print film is a negative film. When you expose the film in your camera and this film is processed chemically, a negative image is formed on transparent plastic. This negative is then used to print positive color photographic prints.

Black and white roll film is a negative film, also. Normally, black and white films have the suffix *-pan*, which is short for *-panchromatic*. Tri-X pan is an example of a popular black and white film.

Question No. 2: Film Format

Format refers to the size of the film software and the way it is packaged for use. There are five major amateur film formats. They are the 110 cartridge, the 126 standard cartridge, the 135 cassette, disc film, and the instant sheet-film package.

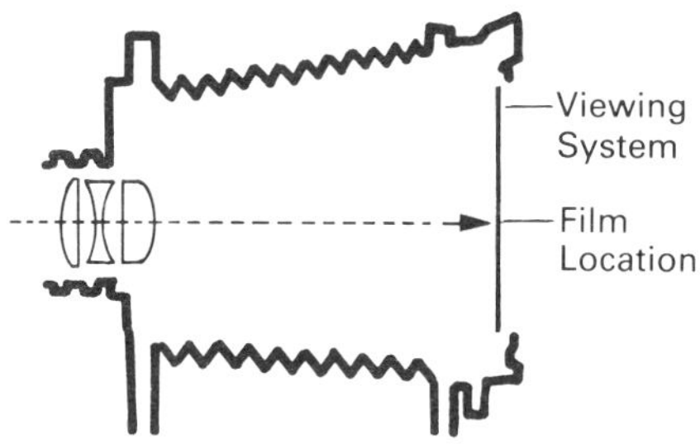

Cross section of a view camera.

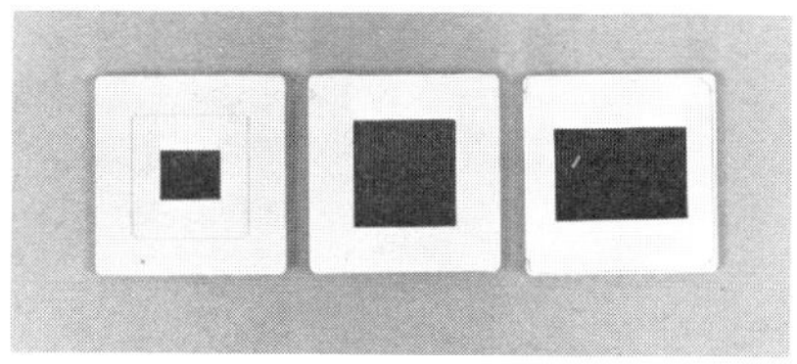

2 × 2-inch slides are available in the 110, 126, and 135 formats.

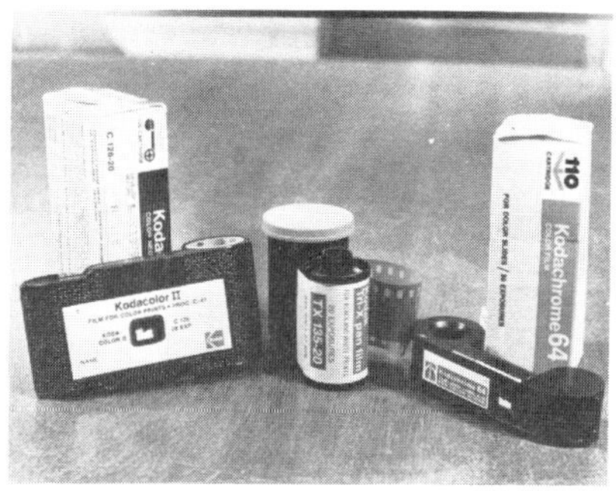

The three most popular roll film formats are the 110 pocket instamatic, the 126 standard instamatic, and the 135 35mm format.

Disc-format film gives 15 exposures . . .

and the disc drops easily into the camera. It will fit one way only (compliments Eastman Kodak Co.).

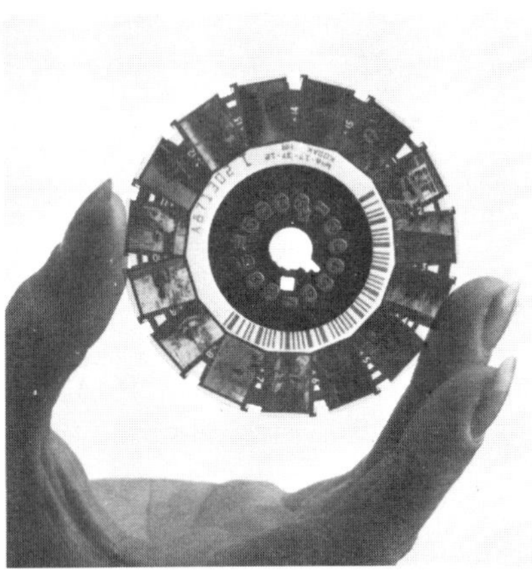

After processing, the exposed disc gives 15 negatives that can then be enlarged into positive prints (compliments Eastman Kodak Co.).

Film packaged in a 110 cartridge is the smallest roll-film format, and is used in the various popular pocket cameras available. It is primarily a snapshot format. The 126 cartridge contains film that is larger and produces a square photograph. Both the 110 and the 126 formats of film are packaged in a light-proof cartridge that is just popped into the back of the camera. When the picture taking is completed, the entire cartridge is sent for processing.

The 135 cassette contains film that is 35mm wide. The largest variety of film types can be purchased in the 35mm format. During the last decade the greatest growth in photographic equipment has been in the 35mm format.

It is important to remember that film of a particular type packaged in either the 110, 126, or 135 format is identical. The film is just cut to different widths and lengths and packaged in different containers in order to fit various types of cameras.

The disc film format is another amateur format. Instead of film stored in a roll it is on a disc that rotates as exposures are made. The disc format is primarily for snapshot photography.

The only sheet film widely used by amateur photographers is the instant film format. Color and black and white prints can be produced using a camera such as the Polaroid, along with instant sheet film. The main advantage of the instant film format is that it is possible to take one picture at a time and see the results immediately. However, instant film photography is limited to snapshot picture taking because it is difficult to make enlargements or copies of instant film prints. Also, slide film is not available in the instant format.

Polaroid instant film is popular for snapshot photography because a picture is available almost immediately.

Question No. 3: Light

A third, and very important question you should ask before buying film is, "What will the lighting be like where I plan to take my photographs?" In other words, how much and what kind of light will be available?

The amount of light available where you plan to take your pictures is a critical concern if you are going to get properly exposed photographs. To provide for various lighting situations, films with different degrees of light sensitivity are available. This sensitivity is indicated by an index called an ASA or ISO rating. ASA is short for American Standards Association. The ASA rating has been replaced with the letters ISO. Often referred to as film speed, the ASA/ISO rating of a type of film is a numerical index of its sensitivity to light. The higher the ASA/ISO rating number, the less light the film needs in order to be correctly exposed. Films such as Kodacolor 400 are called fast films because they require very little light. Films like Kodachrome 25 are called slow films because they need a large amount of light in order to be exposed correctly. Doubling or halving ASA/ISO numbers indicates that the film is twice or half as fast. Stated another way, if you select a film with an ASA/ISO rating of 200 instead of 100, you can use one smaller aperture size or one faster shutter speed setting, and still have a correctly exposed photograph.

While the widest selection of ASA/ISO ratings for films can be found in the 135 format, a variety is available in the other formats, also. Before you purchase film for your camera, try to anticipate where you will be taking your pictures. If you are going to be out of doors on a sunny day where there is plenty of light available, a slow-speed film would be satisfactory. On the other hand, if you are going to be taking photos in a dimly lighted room you will probably need a much faster film.

The type of light present in a place where you plan to take pictures is also important and should be considered before you purchase film. White light from different sources is not the same. The light from the sun and from a light bulb seem the same to the eye, but to film they are dramatically different. The difference is the color temperature of the light. Light temperature is measured using the Kelvin (K) scale. The Kelvin scale indicates the whiteness of a light. A low color temperature of less than 3200 K, such as that produced by a regular incandescent lamp, is yellowish. Higher color-temperature light sources are bluer. Sunlight gives the truest, most complete white light.

Matching film type is important only when using color film. Panchromatic black and white film reacts the same to light of any color temperature.

Stated another way, there are two types of white light: natural, from the sun, and artificial. Similarly, there are basically two kinds of film, each with different color-temperature sensitivities: daylight film and artificial-light film. There are also two artificial-light films, called types A and B. When you buy film for your camera, try to anticipate what the light source will be where you plan to take your pictures, then buy the film type to match your lighting source.

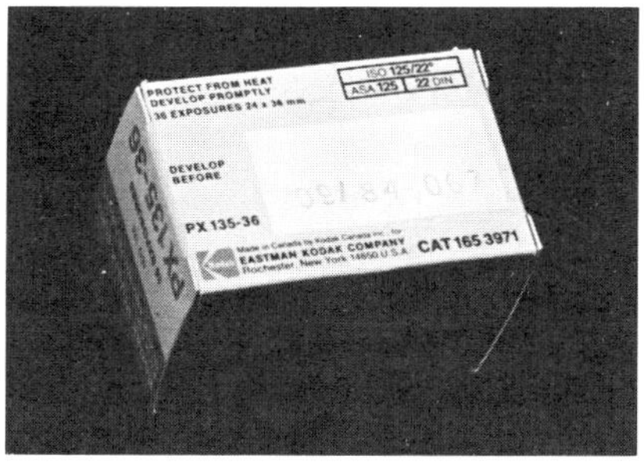

The film box will show its ASA rating.

The ASA rating of the film loaded into the camera must be set on the camera. For single-lens reflex cameras such as the Pentax this is done on the shutter-speed dial.

The ASA is set on the front of the camera for some Polaroid models.

SLOW FILM
(Needs much light)

ASA / ISO	Film Name
25	Kodachrome 25 (Slides)
32	Panatomic-X (B & W Prints)
64	Kodachrome 64 (Slides)
64	Ektachrome 64 (Slides)
75	Polaroid (Color Prints)
100	Kodacolor II (Prints)
125	Plus X (B & W Prints)
200	Ektachrome 200 (Slides)
400	Tri X (B & W Prints)
400	Kodacolor 400 (Color Prints)
400	Ektachrome 400 (Slides)
3000	Polaroid (B & W Prints)

FAST FILM
(Needs little light)

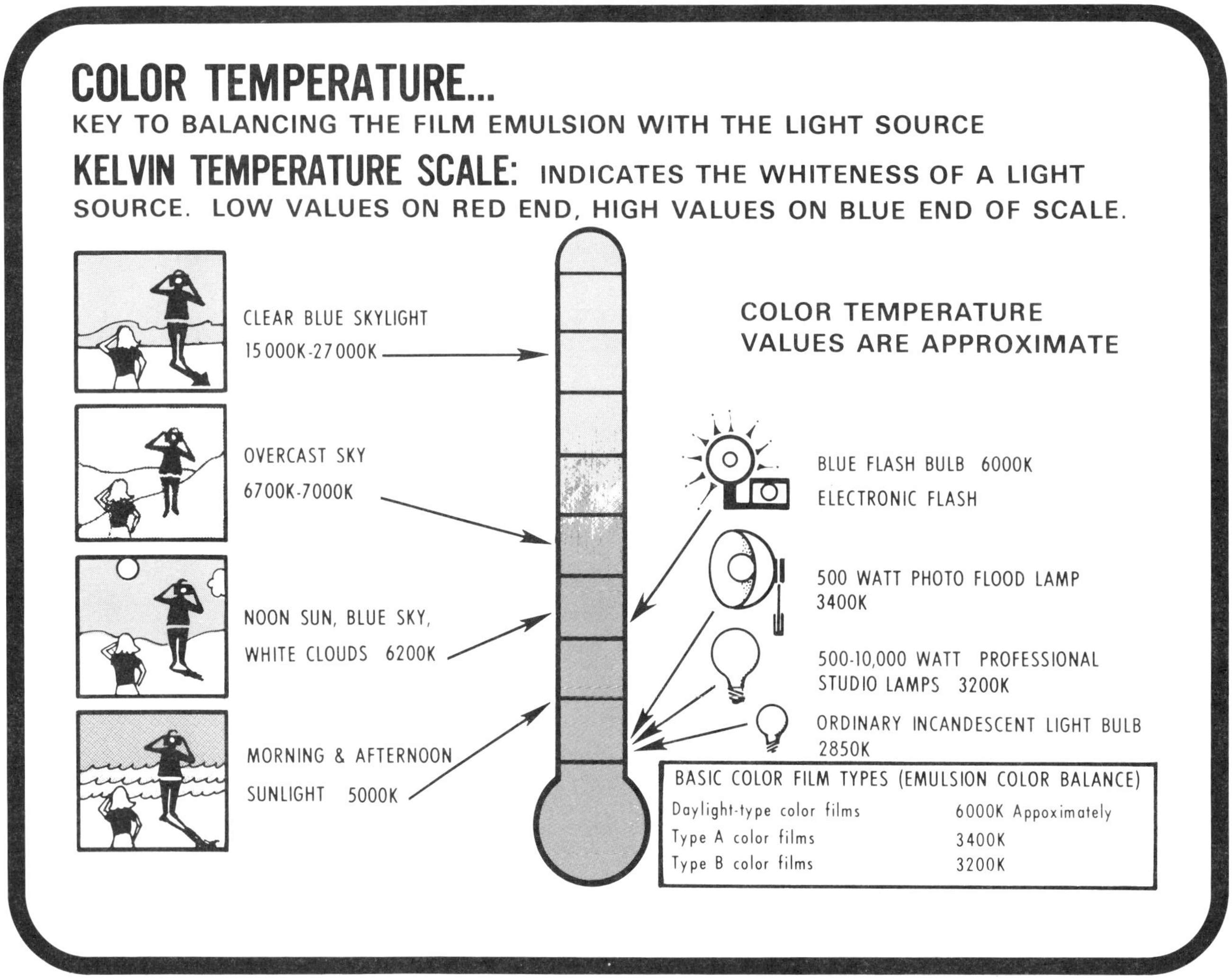

Color temperature is measured in kelvins. Color films are produced that can be used with different light sources (compliments of DCA Educational Products).

Reflected-light meter.

Incident-light meter.

Measuring Light

Light meters are used to determine the intensity of light. Most amateur cameras have built-in light meters that measure light reflected from a subject. These meters give the photographer a visual indication of the amount of light in a scene so the camera's aperture and shutter can be adjusted to produce a properly exposed photograph. Built-in light meters measure *reflected light.* Hand-held reflected-light meters are also available.

A second kind of light meter used by some photographers is called an *incident-light meter.* This type of light-measuring device determines how much light actually falls on a subject. Hand-held incident-light meters have scales that indicate which aperture and shutter speed settings will give a correctly exposed photograph. These settings are then transferred by the photographer to the camera before the shutter is released.

In summary, when you purchase film for your camera you should ask three questions. First, what final product do I want? Do I want slides (buy a -chrome film), color prints (buy a -color film), or black and white prints (buy a panchromatic film)?

Second, what kind of camera do I have? Is it a pocket instamatic (uses 110 film)? Is it a standard instamatic (uses 126 film)? Is it a 35mm camera (uses 135 film)?

Last, under what lighting conditions will I be taking pictures? If I am going to be photographing in low-light situations, I need a high-ASA film. If there will be plenty of light, I can use a lower ASA-rated film. If I plan on taking color photographs I will also need to match the light source to the film type. Daylight films are for daylight or flash photography. Type A and B films are for use in artificial lighting situations.

Remember that there are dozens of films available, one for almost any situation. If you take care to select the right type, your picture taking will be much more likely to be successful. The film you use is the single most important factor in determining the technical quality of your photographic efforts.

CAMERA OPERATION

For most amateurs, the standard 126 instamatic is the best camera to use for gaining experience in picture taking. It is probably the simplest, even though it may not be versatile enough for some photographic requirements. This type of camera gets its "instamatic" name because it is used for taking pictures with few adjustments needed to make the camera work. More expensive instamatics have built-in light meters that automatically set the camera aperture so the correct amount of light will strike the film when the shutter is released.

There is a six-step sequence to follow when operating any camera, including an instamatic.

Steps in Operating a Camera

1. Select film (product, format, ASA).
2. Load the film.
3. Set the ASA.
4. Frame and focus.
5. Adjust the aperture and shutter.
6. Release the shutter to expose the film.

Hand-held light meters are available.

The standard 126 instamatic cartridge is light-tight. To load the camera, the cartridge is just popped in the back. The ASA is set automatically.

The standard instamatic format is rapidly being replaced by the disc-format camera as the standard snapshot format. The new generation of Kodak disc cameras is powered by lithium cells that contain more energy than most users will need for years of camera operation (compliments of Eastman Kodak Co.).

OPERATE

Taking Photographs with an Instamatic

You need: 126 film, an instamatic camera

Procedure

1. Select the correct film. Choose the type that will give you the picture you want. There are several selections of 126 film to choose from.
2. Load the film into the camera. Open the back-door film compartment latch, and pop the 126 cartridge into the back of the camera body. It will fit only one way. Close the door. The ASA will be set automatically.
3. Set the shutter speed. If the instamatic has adjustable shutter speeds, use 1/30 or 1/60 for indoor, dimly lighted situations; 1/125 or 1/250 is good for outdoors. *Note:* Inexpensive instamatics may not have adjustable shutter speeds. Often they are preset for a shutter speed of about 1/50 of a second. If you think you will need more light than this shutter speed will provide to correctly expose your film, place a *burned out* flash cube into its receptacle in the camera. This will automatically slow the shutter speed slightly so more light will enter the camera when the shutter is released. Flash cubes normally are good for snapshot photography only when the subject of your picture taking is about 8 to 10 feet away, so do not use a good cube unless you absolutely need it.
4. Focus the camera. Instamatics have either preset focusing or double-image focusing.
5. Frame the subject in the viewfinder and trip the shutter to expose the film. Advance the film-advance lever to prepare to take the next photo.

TRY THIS

Buy a roll of 126 film, preferably slide film, because it is less expensive than print film. Take pictures with an instamatic. Take your photos in a variety of situations. Remember to keep a log so you can critique them when they are returned from the processor.

Many brands of SLR cameras now have automatically controlled shutters and apertures. In these automatic cameras, the light meter measures the reflected light entering the lens and selects either the appropriate shutter speed or aperture so the film in the camera will be correctly exposed. All the photographer has to remember to do is set the ASA/ISO when the film is loaded, and to focus.

The Pentax SLR camera.

THE SINGLE-LENS REFLEX (SLR) CAMERA

Probably the most versatile camera currently in use by amateurs is the 35mm single-lens reflex camera (SLR). Many film types are available for SLRs, lenses are interchangeable on them, and they are portable and lightweight. In recent years the cost of many SLR cameras has been reduced substantially. They are now within the range of many instructors' budgets. These are some of the reasons that make single-lens reflex cameras so popular.

The most dominant characteristic, however, of the SLR camera type is its viewing system. Because the photographer "sees" the same thing the film "sees," the job of picture taking is fairly simple. Whatever is viewed through the viewfinder is what will be seen on the film after it is processed. Many call the SLR camera the "what you see is what you get" camera.

Most SLRs now have built-in light meters to assist in setting the aperture and shutter speed. There are two types of light-meter operating systems: manual and automatic. Manually adjusted meters require the photographer to set the aperture and shutter controls so the light-meter indicator needle seen in the viewfinder is centered.

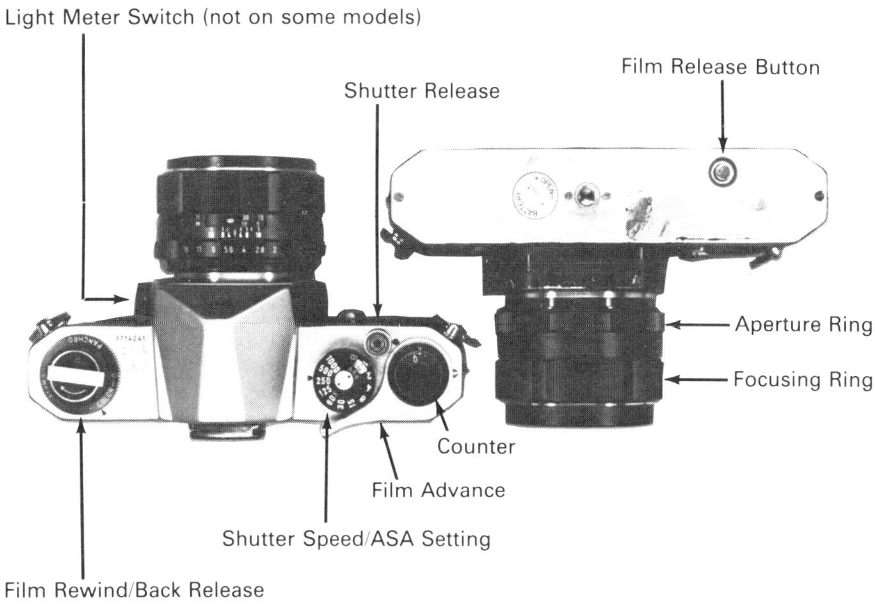

TRY THIS

Expose a roll of film with an SLR. Take your photographs in a variety of settings. Keep a log of your picture taking so you can critique them after they are processed.

OPERATE

Taking Photographs with an SLR Camera

You need: a roll of 35mm film, and a single-lens reflex camera

Procedure

A. Film Loading
   1. Open back; pull back release.
   2. Insert film.
   3. Thread film; attach to take-up reel and match sprocket holes to sprockets on both sides of film.
   4. Close back; advance counter to 1, to move unexposed film behind shutter.

B. Operating Instructions
   1. Set film speed (ASA, or ISO).
   2. Set shutter speed.
   3. Compose and focus.
   4. Turn on light meter, if your model has one.
   5. Adjust light meter by changing aperture size and/or shutter speed.
   6. Press shutter release to take picture.

C. Film Unloading
   1. Press film release button (bottom of camera).
   2. Wind film into light-proof cartridge (film rewind).
   3. Open back.
   4. Remove film.

Set the film speed (ASA).

Set the shutter speed.

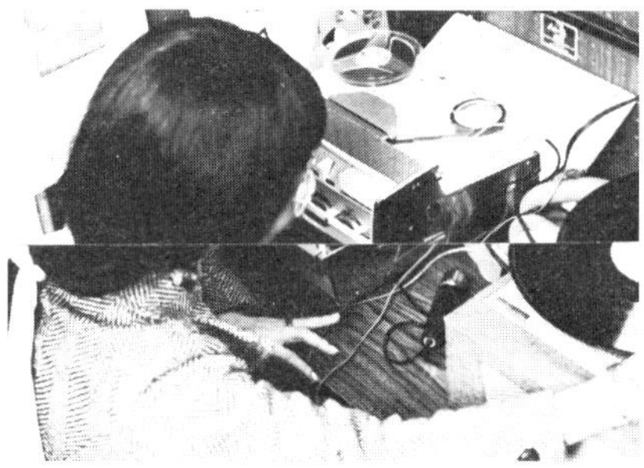

Compose and focus your picture.

If your model camera has a light-meter switch, turn it on.

Change the aperture and shutter-speed controls to center the light-meter needle.

Release the shutter to expose the film.

OPERATE

A Polaroid Instant Camera

You need: a Polaroid instant camera, Polaroid film

Procedure

1. Load film, following instructions on film package. Be sure to clean steel rollers that film passes through. Set ASA according to instructions accompanying film.
2. Set shutter if necessary. On many Polaroids shutter speed is automatically controlled.
3. Focus if necessary. Camera may have double-image rangefinder instead of footage marks on front lens.
4. Cock the shutter.
5. Trip shutter. Picture will not develop until it is pulled entirely out of camera. Follow other developing instructions accompanying film.

TRY THIS

Take a series of prints with a Polaroid camera. Analyze each one immediately after you take it. If there is a problem, try to figure out what it is, and try again.

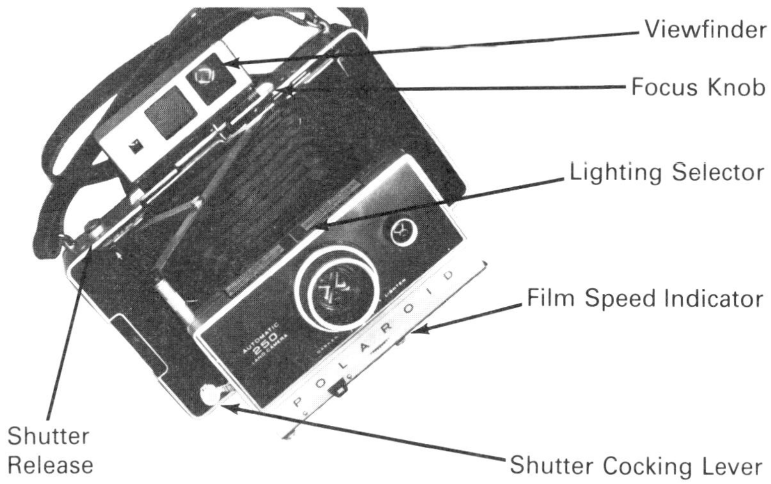

Polaroid ASA setting indicator.

Polaroid lighting selector.

Close-up rings are screwed onto the front of the SLR's standard lens.

A macrolens is a replacement lens that allows focusing to as close as 8 inches.

Extreme close-up photography is possible with a bellows system.

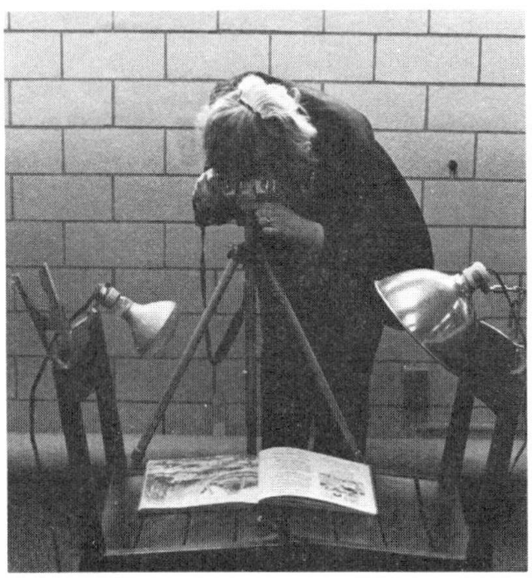

An inexpensive, but effective, copy stand can be made using a tripod and floodlights.

Copy Stands and Close-up Photography

When you use a single-lens reflex camera you have the option of taking high-quality close-up pictures. One commonly used way of doing this is by using a *copy stand.* Copy stands are basically rigid mounts for holding cameras steady. They also commonly have attached light sources and a movable platform, so objects of different sizes can be photographically copied.

Instamatic Copying Some ready-made copy stands are available, including the Kodak Visualmaker. While the device is limited to copying originals of only two sizes (8 × 8 inches or 3 × 3 inches) and uses Instamatic film, results are quite satisfactory.

Copying with a Single-Lens Reflex Camera The SLR camera has two advantages: A variety of lenses can be used, and there are no parallax problems. You must be able to use close-up rings, a macrolens, bellows, or extension tubes so that a variety of sizes of originals can be accommodated. There are four close-up systems and they are described below:

☐ *Close-up rings* Special small lenses can be attached to the front of a standard 50mm lens for close-up photography. These lenses, called close-up rings, alter the path of light passing through them so it is possible to focus the camera on closer objects.

☐ *Macro focus lenses* A replacement lens, made especially for close-up focusing, called a macrolens, is probably the best option for most copying and close-up work. This lens is used with the SLR camera body instead of the regular lens. Because macrolens elements are specially constructed, it is possible to focus on objects as close as a few inches from the camera.

☐ *Bellows attachments* By placing an adjustable device, called a bellows, between the lens and the camera body, it is possible to take extreme close-up photographs. Because this device moves the camera lens farther from the film plane, the lens's focusing is altered so it is possible to take photographs of objects that are quite close.

☐ *Extension tubes* Extension tubes are an adaptation of the bellows system. These tubes are placed between the lens and camera body, like the bellows, in order to move the lens farther from the film plane. Because they are rigid and not adjustable, extension tubes are not as variable as are bellows systems. However, they usually are less expensive.

Before leaving the subject of copy stands, it is important to point out that the stand to hold the camera and lights is relatively unimportant. Satisfactory copy work can be done outdoors, using sunlight. Place the original on a flat surface and photograph it in the usual way, using appropriate close-up lenses and measuring the light for proper exposure, using the exposure meter.

For indoor work at home, consider using two clip-on flood lamps mounted on chairs so that the light is directed at the original at a 45° angle. Stand directly over the original (or kneel so that you can arrive at the appropriate distance) and take the picture. If you are using outdoor film, use a color-correction filter, or use lamps that provide the correct type of light for outdoor film.

Copy Stand Hints

☐ Position your lights at 45° angles from the subject being copied. This minimizes glare. Keep all lights the same distance from the object being copied. At least two lights should be used.

☐ If you are copying something from a book that will not lie flat, use a piece of glare-proof glass over it to hold it down. Hardware stores and lumber yards sell this kind of glass.

☐ If you are copying something that is mostly light in color, and that is very reflective, or that is very dark and not light-reflecting, do not trust your camera's built-in light meter. Instead use a neutral test card to set the correct exposure. Light-colored objects fool the light meter into indicating that there is more light being reflected than there really is, so an underexposure is produced. Darkly colored objects tend to fool the light meter into indicating that less light will be reflected than there actually will be, so overexposures will be produced. The neutral density card compensates for these effects. Most photography dealers sell neutral density cards.

☐ If possible, keep the camera lens aperture as small as possible (f/11 or f/16) in order to increase its depth of field.

☐ Remember to obtain permission before copying a copyrighted work.

ARTIFICIAL LIGHTING

Even with high-speed film there may be times when additional light will be needed to obtain a good-quality, correctly exposed photograph. While available light is usually preferred, in some situations flash or floodlighting may be needed.

Flash Lighting

There are two types of flash lighting: bulb and electronic. Both produce basically the same result, although light from an electronic flash is more powerful and can penetrate to greater distances than bulb flash systems.

When either type of flash lighting system is used, film is exposed in a slightly different way than when available light is used. Flash lighting works because the camera's shutter opens before the flash goes off. When the flash unit trips, light from it bounces off of the subject of the picture taking and is reflected back through the open shutter. Then the shutter closes. Synchronization of the flash unit with the shutter must be exact for this type of lighting system to work correctly. If pictures are unevenly exposed after a flash is used (they are partly light and partly dark), it is likely that the shutter either opened after the flash went off, or closed before all the reflected light entered the camera.

To minimize the problem of synchronizing camera to flash unit, many units now have semiautomatic control devices that keep both camera and flash working together. All units also have manual calculators that are used before picture taking to set the camera's aperture and shutter in order to produce a correctly exposed photograph.

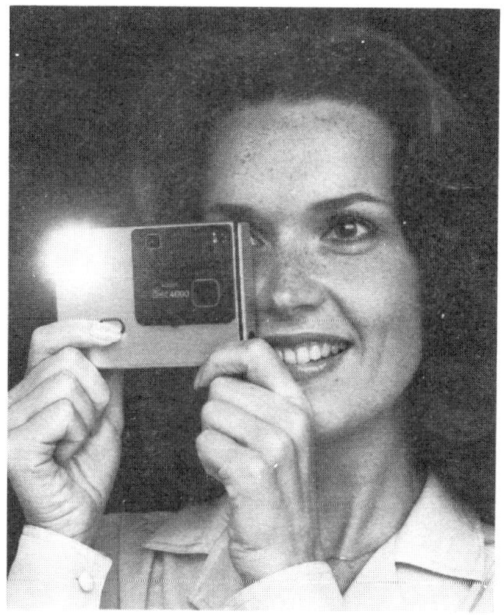

Many consumer model cameras have built-in flash units that are powered using the same battery that operates the camera (compliments Eastman Kodak Co.).

Many models of electronic flash units are available. Costs range from as little as $20 to as much as several hundred dollars (courtesy Pentax Corp.).

Subject illuminated by key light only.

Key with fill light added.

Key, fill, and background lights.

Key, fill, background, and hair accent lights.

Floodlighting

In order to avoid the harsh effect produced by flash lighting you may want to consider using floodlights to illuminate your subject. Floodlighting is recommended for studio settings, or for portraiture work. Remember to match the color temperature of the film to the lights.

There are four lights normally used in floodlighting systems—key, fill, background, and accent. The *key* or main light is the most important and brightest of the four. It should be placed about 45° to the side of the camera, slightly closer to the subject than the camera, and higher than either the subject or the camera.

The second light, the *fill* light, should be placed beside the camera, on the side opposite the key light. It should be at about the same height as the camera. The fill light should be about one-half the intensity of the key light. The fill is used to reduce shadows produced by the key light.

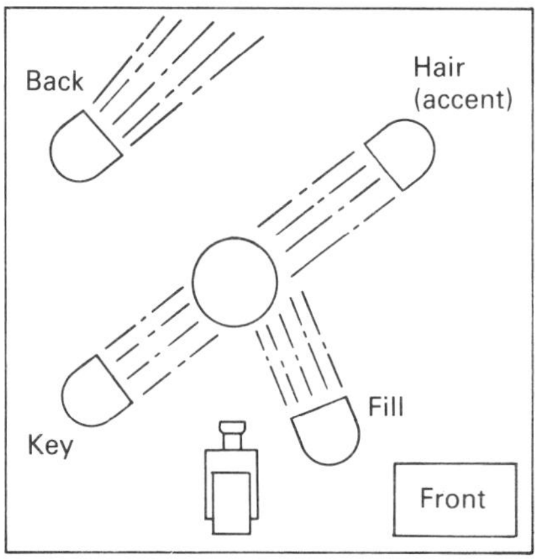

Standard location of lights for studio recording.

The third light in the system should be used to illuminate the background, and is called the *background* light. It should cover the background evenly. The background light serves to give depth to the photograph by separating the subject from whatever is behind it. Keep the subject at least three feet from the background. Usually this light should be of less intensity than the combined light falling on the subject from the other light sources.

The *accent* light, often placed high above the subject, pointing downward, is used to highlight the central point of interest of the photograph. A hair accent is popular for portraiture photography.

CAMERA LENSES

Inexpensive, amateur cameras come with a fixed lens, but single-lens reflex cameras can accommodate interchangeable lenses that greatly expand the camera's capabilities. The most important characteristics of camera lenses are their focal length and their speed.

Lens Focal Length

Lens focal length is the distance from the center of the lens to the film when the lens is focused to infinity. Lenses with different focal lengths form different-sized images on the film. If a camera lens is replaced with one having a shorter focal length, a smaller image will strike the film and a greater area is photographically recorded. A longer focal-length lens gives a larger image on the film, so a smaller area of the subject would be photographed.

Focal lengths of lenses are measured in millimeters. Each lens has its focal length printed on its front. Each camera format, such as 35mm, has a "normal" focal-length lens that produces a photograph of a subject close to what would be seen with the eye. For the 35mm format, the 50mm lens is standard (a 13mm lens is standard for a super 8mm camera). A lens with a smaller, or shorter, focal length is called a *wide-angle lens.* A lens with a focal length longer than 50 mm is called a *telephoto lens.*

Zoom lenses allow the photographer to vary the lens's focal length by moving its internal glass elements. Zoom lenses have adjustable focal lengths. The 85mm to 200mm zoom lens is a popular example.

Lens Speed

Lens speed is determined by the size of the lens's aperture. The larger the maximum aperture size, the faster the lens. This maximum size is usually printed on the lens next to its focal length. Many 35mm SLR cameras come with a f/2.0 lens as standard. F/1.4 or f/1.2 lenses are common replacement lenses for 35mm cameras that are considered to be quite fast. They are also expensive. Fast lenses make it possible to use the camera in situations where less light is available.

Most lenses will have their focal length and speed stamped on the end. This lens has a maximum aperture size of f/1.2 and a 50mm focal length.

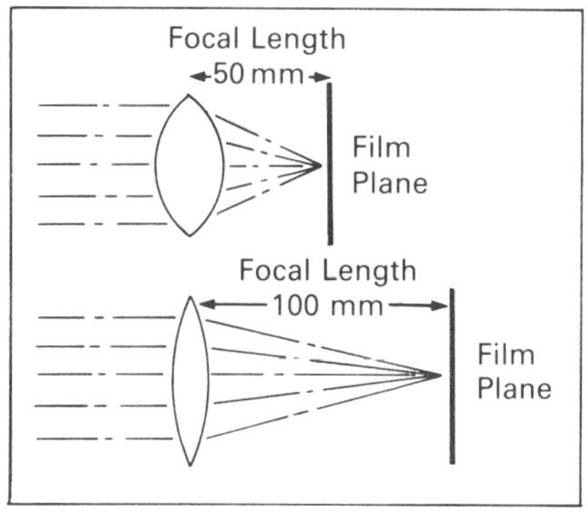

Lens focal length.

By changing the focal length of lenses it is possible to change the scene being photographed without moving the camera. This photograph was taken with a 20mm wide-angle lens.

This photograph was taken with the normal 50mm lens.

And this photograph was taken with a 200mm telephoto lens. All pictures were taken without moving the camera.

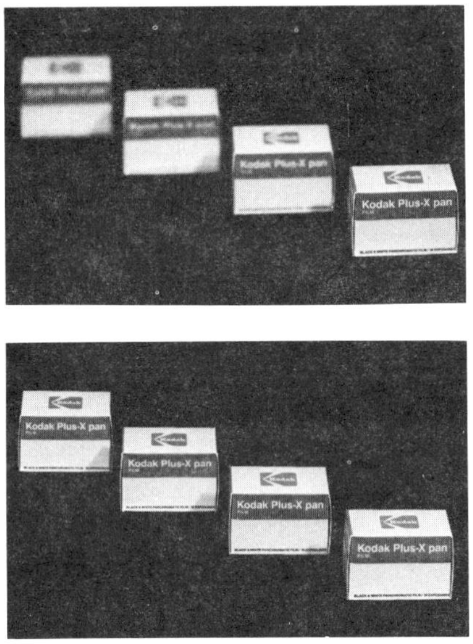

The photo on the bottom has a larger depth of field than the one on the top.

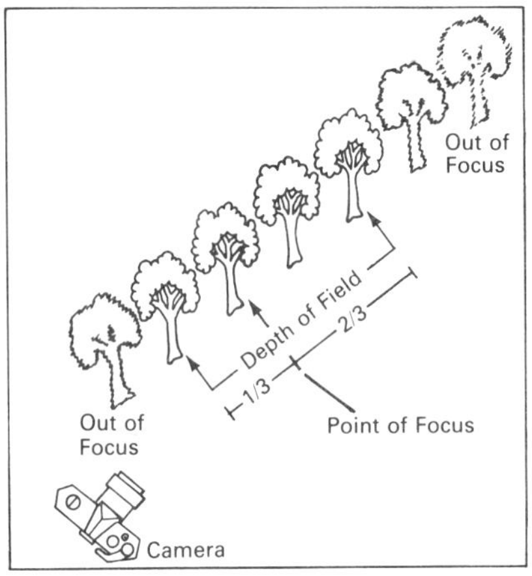

To maximize depth of field, focus on an object about 1/3 of the way into your scene.

Depth of Field

Depth of field is determined by the lens's focal length, by the aperture selected, and by the camera-to-subject distance. Depth of field is defined as the distance within a photograph that appears to be in focus.

From one set location, depth of field can be varied in two ways. The shorter the focal length of a lens, the larger the depth of field of pictures taken with it will be, as compared to lenses with longer focal lengths. A wide-angle lens will tend to have a large depth of field. Telephoto lenses have narrow ones, all other things being equal.

Aperture size also affects depth of field. The larger the aperture, the smaller the depth of field. If a shallow depth of field for a photograph is desired, select an aperture setting such as f/2.0 or f/1.4. Conversely, small aperture sizes such as f/16 and f/22 give a larger depth of field.

To maximize depth of field, use a small aperture, use a lens with as short a focal length as possible, and focus on a point about one-third of the way into your scene. Generally, the depth of field of a photograph should be proportioned so that about one-third of it is in front of the point of focus and two-thirds behind it.

PROCESSING AND PRINTING BLACK AND WHITE PHOTOGRAPHS

Once you have exposed the film, the task of picture making is only half done. You must next process your film into negatives (for negative-type films), and then make positive prints. Most often photographers have this done for them by commercial film processors. Photographic processing services are widely available, either locally or by mail. However, one of the most interesting and exciting photographic activities is processing negatives and printing photographs.

Color processing tends to be a little expensive and complicated. However, black and white work is fairly simple and costs less than having this work done by others. In the next section processing, printing, and enlarging will be discussed.

Processing Black and White Film into Negatives

Processing or film developing refers to the chemical reactions that film undergoes when the exposed silver particles are made permanent and visible and the unexposed silver is washed away. Processing is done in absolute darkness, in a light-proof developing tank. Chemicals and supplies needed for processing can be purchased at most photography stores.

EKTA- versus KODA-film

Since EKTA-films can be processed locally while KODA-films must be sent to a Kodak lab to be processed, you may want to buy EKTA. Local competition between EKTA processors may result in rate reductions and rapid turnaround times.

Small, tabletop processing and drying systems for disc film can process up to 70 discs per hour (compliments Eastman Kodak Co.).

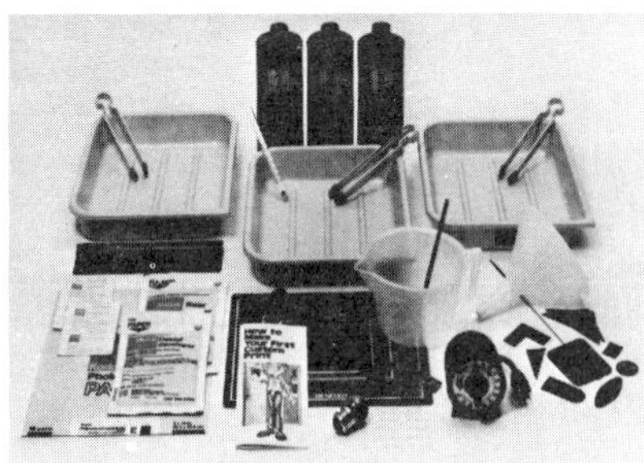

Darkroom supplies can be purchased at most photography stores (courtesy Beseler Corp.).

PRODUCE

Processing Exposed Black and White Film into Negatives

You need: a roll of exposed black and white film, a developing tank, chemicals (developer such as D-76, stop bath, fix), water, scissors, pliers, developing instructions that come with film, a totally dark room or a processing dark bag

Procedure

1. Remove the film from the camera.
2. In total darkness, remove the exposed film from its cassette. Use the pliers to pry open the cassette, and the scissors to cut the film from the take-up spool and to square its ends. Load the film into the developer tank reel. Be sure the film is wound evenly and that the film does not touch itself as it sits in the tank's spool. Cap the tank. Now you can turn the lights on, because the developing tank is light-tight.
3. Follow the directions on the instruction sheet that comes with the film. Mix the developer. Check its temperature. Pour it into the developing tank for the appropriate time indicated on the instruction sheet. Agitate the tank every 30 seconds by swirling or shaking the tank. While the film is developing, mix your stop and fix chemicals so they are ready to use.
4. About 15 seconds before the developing time is up, pour the developer out. Fill the tank with water of the same temperature, agitate, and pour the water out. A chemical stop bath can be used instead of water. Just pour it into the developing tank and leave it for about 10 seconds before pouring it out.
5. Pour the fix chemical into the developing tank for the time specified on its directions. The fix should be at approximately the same temperature as the developer. When the recommended time for fixing has elapsed, pour the chemical solution back into its bottle for reuse.
6. Open the developing tank and run water over the film to wash the chemicals and excess silver off. Wash for 20 to 30 minutes in running water.
7. Near the end of the washing period, use a small amount of a wetting agent such as Kodak Photo-Flo mixed with water. This will eliminate water spots on the film as it dries. Leave in Photo-Flo for about 60 seconds.
8. Hang the film to dry. You can squeegee it dry if you wish. Be careful not to scratch the emulsion.

TRY THIS

Expose a series of photographs using black and white film. Take some out-of-doors, some indoors, ond some on a copy stand. Be sure to keep your log. Process your film into negatives. Keep the negatives so you can make a contact print from them later.

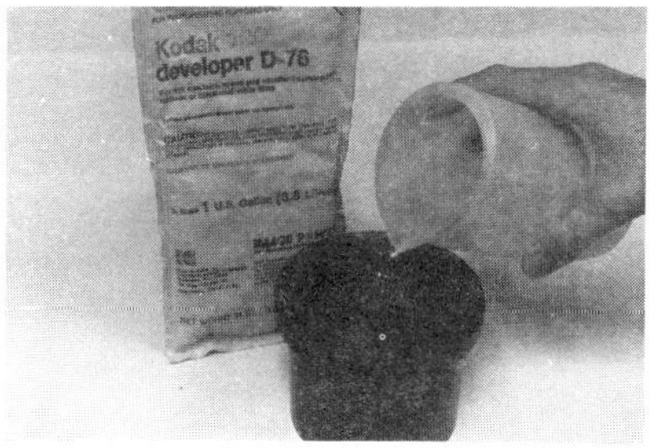

Pour the developer into the developing tank. Agitate every 30 seconds or so.

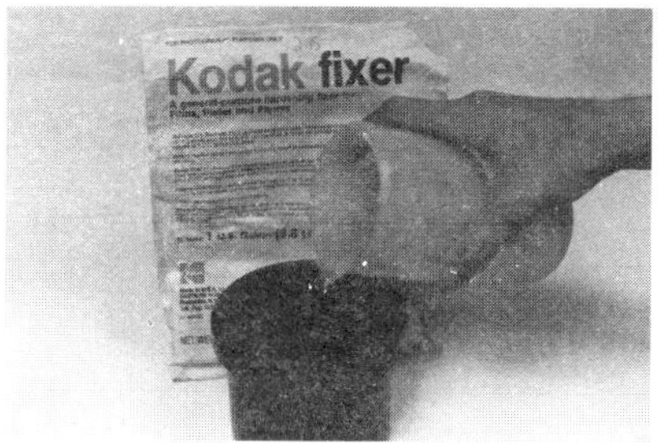

After the developer has been poured out, pour the fix solution into the tank.

The negatives can be washed in the developing tank. Be sure to use a wetting agent before the wash cycle is completed.

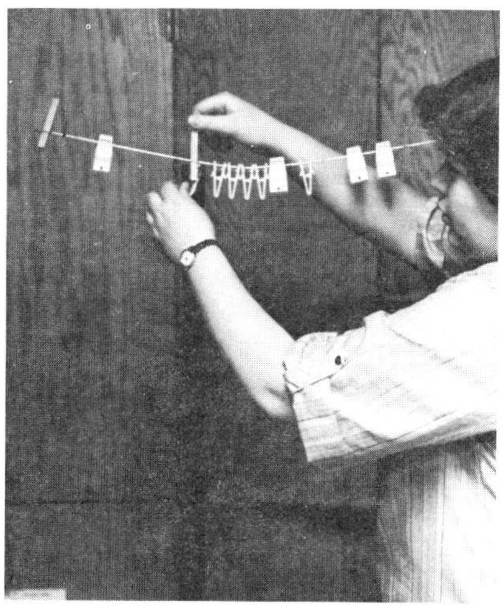

After the negatives have washed, hang them in a dust-free area to dry.

Hints about Processing

☐ Detail should be obvious, even in very dark or very light portions of the negative.
☐ The negative should be transparent enough to see through even in the dark areas.
☐ The borders of the film should be clearer than any part of the negative image.
☐ No scratches or specks should be visible on the negative.

Processing Time and Temperature Charts

Panatomic-X (ASA 32)

KODAK Packaged Developers	Developing Time in Minutes*									
	Small Tank (agitation at 30-second intervals)					Large Tank (agitation at 1-minute intervals)				
	65° F 18° C	68° F 20° C	70° F 21° C	72° F 22° C	75° F 24° C	65° F 18° C	68° F 20° C	70° F 21° C	72° F 22° C	75° F 24° C
D-76	6	5	4½	4¼	3¾	6½	5½	5	4¾	4
D-76 (1:1)	8	7	6½	6	5	9	7½	7	6½	5½
MICRODOL-X	8	7	6½	6	5	9	7½	7	6½	5½
MICRODOL-X (1:3)†	–	–	11	10	8½	–	–	12	11	9½
POLYDOL	6½	5½	5	4½	3½	7	6	5½	5	4
HC-110 (Dilution B)	4¾	4¼	4	3¾	3¼	5½	4¾	4¼	4	3½

†For greatest sharpness.
Note: Do not use developers containing silver halide solvents.
*Unsatisfactory uniformity may result with development times shorter than 5 minutes.

Plus-X (ASA 125)

KODAK Packaged Developers	Developing Time in Minutes									
	Small Tank (agitation at 30-second intervals)					Large Tank (agitation at 1-minute intervals)				
	65° F 18° C	68° F 20° C	70° F 21° C	72° F 22° C	75° F 24° C	65° F 18° C	68° F 20° C	70° F 21° C	72° F 22° C	75° F 24° C
HC-110 (Dilution B)	6	5	4½	4	3½	6½	5½	5	4¾	4
POLYDOL	6½	5½	4¾	4¼	3¼	7½	6	5½	4¾	3¾
D-76	6½	5½	5	4½	3¾	7½	6½	6	5½	4½
D-76 (1:1)	8	7	6½	6	5	10	9	8	7½	7
MICRODOL-X	8	7	6½	6	5½	10	9	8	7½	7
MICRODOL-X (1:3)	–	–	11	10	9½	–	–	14	13	11

Note: The development times have been changed as a result of a recent comprehensive study
of all the factors affecting development.

Tri-X (ASA 400)

KODAK Packaged Developers	Developing Time in Minutes*									
	Small Tank (agitation at 30-second intervals)					Large Tank (agitation at 1-minute intervals)				
	65° F 18° C	68° F 20° C	70° F 21° C	72° F 22° C	75° F 24° C	65° F 18° C	68° F 20° C	70° F 21° C	72° F 22° C	75° F 24° C
HC-110 (Dilution B)	8½	7½	6½	6	5	9½	8½	8	7½	6½
D-76	9	8	7½	6½	5½	10	9	8	7	6
D-76 (1:1)	11	10	9½	9	8	13	12	11	10	9
MICRODOL-X	11	10	9½	9	8	13	12	11	10	9
MICRODOL-X (1:3)	–	–	15	14	13	–	–	17	16	15
POLYDOL	8	7	6½	6	5	9	8	7½	7	6
DK-50 (1:1)	7	6	5½	5	4½	7½	6½	6	5½	5
HC-110 (Dilution A)	4¼	3¾	3¼	3	2½	4¾	4¼	4	3¾	3¼

Note: Do not use developers containing silver halide solvents.
*Unsatisfactory uniformity may result with development times shorter than 5 minutes.

Each step in the processing sequence is important. Be sure to follow the directions concerning developing time and chemical temperature closely. The film *developer* acts on the silver that was exposed to light during picture taking. These silver particles will turn black after contact with the developer. Developer temperature influences the speed at which this reaction occurs. Warmer developers make the reaction happen more rapidly. Temperatures between 65°F and 70°F work the best.

The *stop bath* or the *rinse* removes the excess developer from the film and stops the developing process. *Fixing* is important because this chemical sets the developed image and hardens the emulsion so that it will not crack. Fixing prepares the silver that was not exposed to light to be removed during washing. *Washing* gets rid of all the chemicals on the film and further sets and hardens the image.

Process and Print Equipment

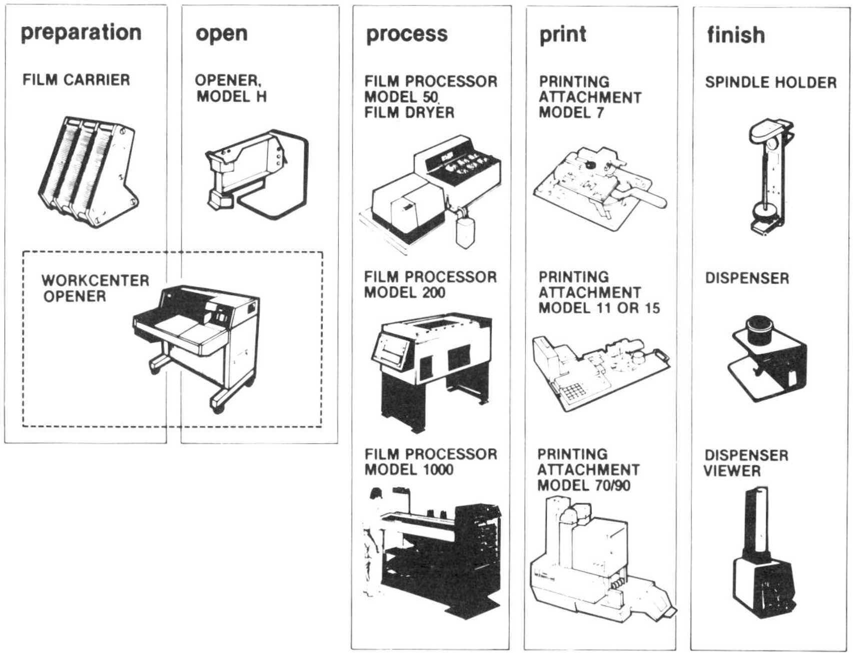

The procedure followed by commercial laboratories for processing and printing is basically the same as that followed by individuals. This five-step process is used by Kodak labs for the disc-format film (compliments Eastman Kodak Co.).

Printing Positives from Negatives

The next step in black and white darkroom work is to make positive prints from the processed negatives. *Contact printing* is the production of positive prints that are the same size as the negatives. *Enlarging* is when positive prints are made larger than the negative.

Printing Paper In order to make positive prints from negatives, another light-sensitive substance is needed. Photographic printing paper is similar to film, except that the light-sensitive emulsion is placed on opaque paper rather than transparent plastic. There are a variety of paper types, just as there is a wide selection of films. Some characteristics of photographic paper are size, contrast, image tone, surface texture, gloss, and base.

Typically, paper is available in packages of 10, 25, 50, 100, and 500, and in sizes such as 4 × 5 inches, 5 × 7 inches, 8 × 10 inches, and 11 × 14 inches. Paper can be cut in a darkroom to whatever size is needed.

Paper contrast is measured on a scale of 1 to 6 with 1 being low-contrast paper and 6 the highest-contrast paper. A low-contrast paper will create a photograph with many gray tones. A high-contrast paper will produce very few grays when it is printed. It will have white whites, and black blacks. Often photographers like to purchase polycontrast print paper and contrast filters. Polycontrast paper has variable contrast capabilities. Contrast can be varied by changing the filter that is used in the printing process. Polycontrast filters are rated on a scale of 1 to 4 with 1 being the lowest contrast and 4 the highest.

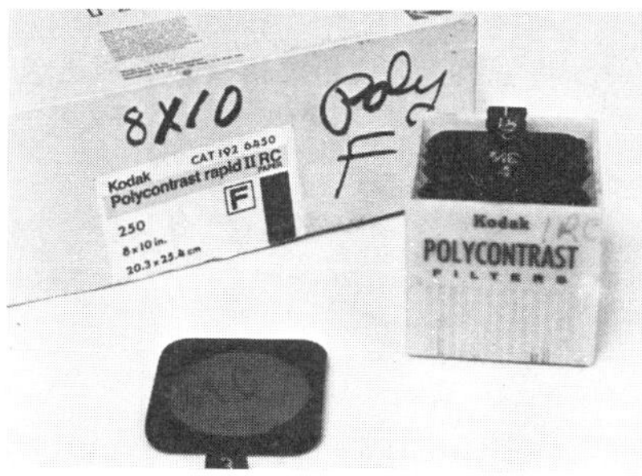

Polycontrast printing paper and filters make it possible to print photographs with varying contrasts by merely switching the filter.

Image tone for print paper refers to the coldness or warmness of the blacks produced. Tones range from warm to cold. Warm-tone papers have a brownish-black image. Cold-tone papers have a crisp, coal-black image. *Textures* of printing paper are smooth, which shows no texture; fine-grained, which has a pebble-like surface; rough; and silk, which has a surface like cloth.

Printing paper can also be purchased with glossy, lustre, or matte finishes. Glossy paper tends to show more image detail on its shiny surface. Lustre paper is a subdued gloss, and matte paper has a dull surface that shows little detail.

One of the most popular paper types now in use is called RC (for resin-coated) paper. RC paper has a water-resistant coating, and tends to dry quickly without the need of a print dryer.

Contact Printing Before you make enlargements of your negatives you should produce a contact print. A contact print gets its name because the negatives are placed in "contact" with the print paper when it is exposed to light. Contact prints of negatives are used to analyze the photographs in order to select the ones for enlarging, and so image cropping can be planned.

Mix all chemicals before printing.

Place the negatives on top of the printing paper with the emulsion side down. Put a piece of clean glass on top of the negatives to hold them flat. Expose the print paper for a few seconds.

PRODUCE

Contact Printing from Black and White Negatives

You need: black and white negatives, print paper, chemicals (paper developer such as Dektol, stop bath, fix), water, three printing trays, tongs, a darkroom equipped with a safelight

Procedure

1. Mix your chemicals according to the directions that come with them. Pour each chemical solution into a separate, clearly marked developing tray. While chemical temperature is not as critical in printing as it is in processing, try to keep all chemicals between 65°F and 75°F.
2. Turn off all lights except the safelight. Remove one sheet of 8 × 10-inch print paper from its box. Tightly reseal the box. Place the paper, emulsion side (the shiny side) up on a flat surface. Place your negatives on top of the print paper, emulsion side (the dull side) down. Place a sheet of clean glass on top of the negatives to hold them flat.
3. Turn on the room lights for two or three seconds. If you are using an enlarger light, turn it on for 5 to 7 seconds. Be sure it covers the entire surface of the print paper.
4. Remove the exposed print paper and place it in the developer solution for 90 seconds. You should see the positive image begin to form after 10 to 15 seconds.
5. With the tongs, lift the paper out of the developer, let it drain into the developer for a few seconds, and drop it into the stop bath solution for about 10 seconds.
6. Using a second set of tongs, remove the print from the stop bath and place it in the fix solution. Fix the print for the time recommended for the paper type being used. Ten minutes is normally adequate.
7. After fixing, place the print in the wash tank for 10 to 30 minutes or so. Water should be circulating in the wash tank.

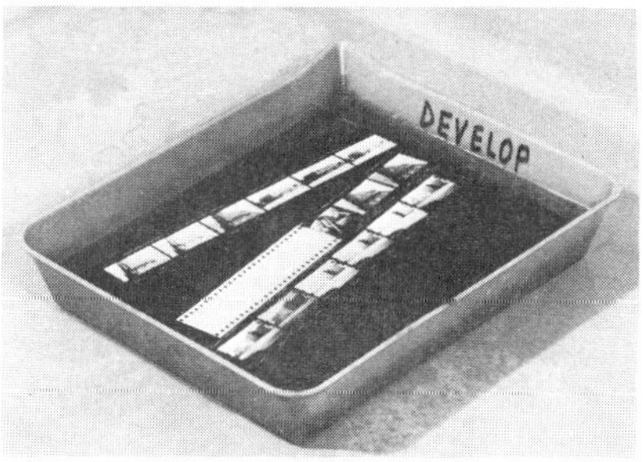

Develop the exposed paper for 90 seconds.

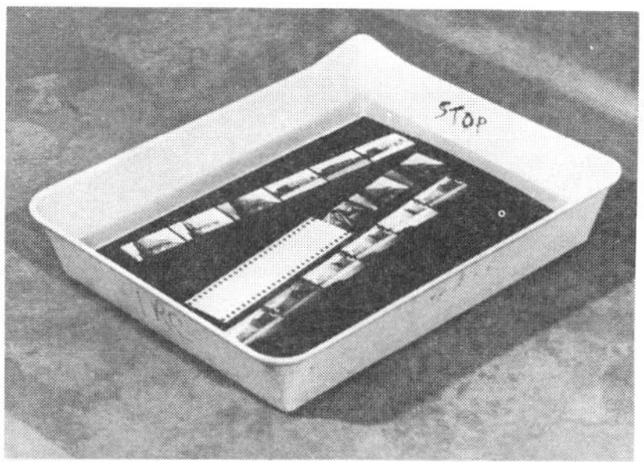

After developing, place the contact print in the stop solution for about 10 seconds.

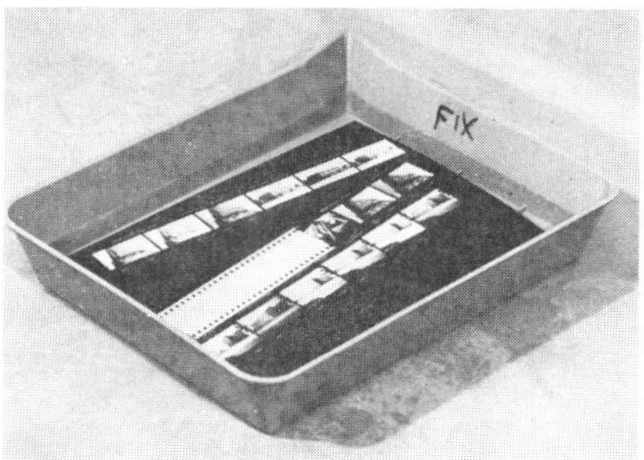

Fix for 10 to 20 minutes.

Wash the contact print to remove all chemicals. Before removing the contact print from the wash, use a solution to prevent the forming of streaks and spots on the paper. Kodak Photo-Flo works well.

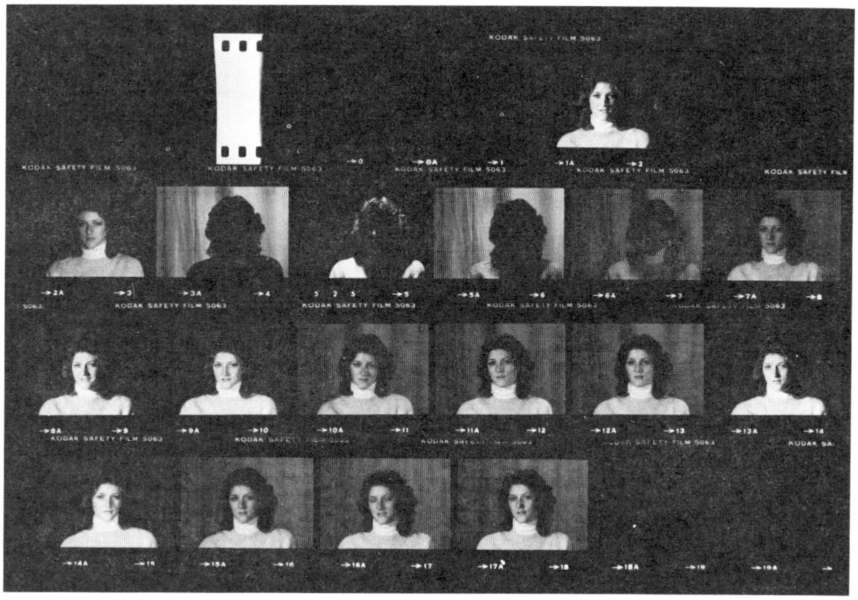

A contact print is made by placing the negatives "in contact" with the printing paper and exposing both to light.

TRY THIS

Make a contact print from the negatives you processed earlier. Analyze the positives using your log. Mark the ones you want to enlarge, and indicate how they should be composed using a felt marker.

Photo enlargers can be purchased for as little as $100 (courtesy Beseler Corp.).

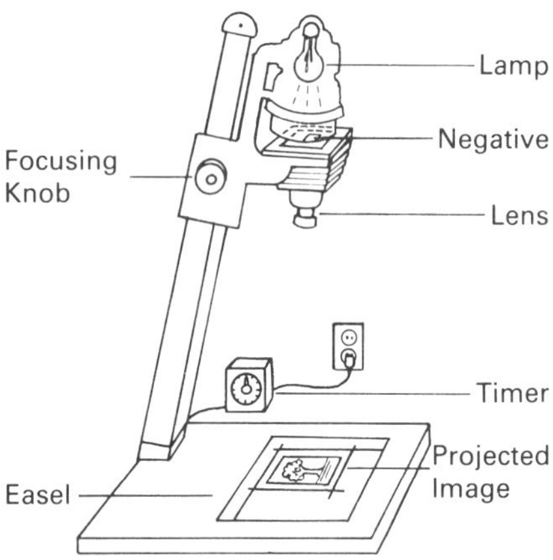

Darkroom enlarger.

A magnifier makes focusing easier.

Enlarging Once the contact print has been analyzed, the negatives can be used to make enlargements. A device called a darkroom enlarger is needed for making positive prints larger than negative size. Enlargers have three component parts. The first is a light source. The light source is controlled by an on/off switch. The next component is an adjustable aperture. Last, there is a stage where pictures can be composed before printing.

OPERATE

How a Darkroom Enlarger Works

1. Place a negative in the negative carrier, emulsion down. (There are different-sized carriers for different film formats.) Insert the carrier in the enlarger between the lamp and lens.
2. Open the enlarger's lens aperture as wide as possible. Place the photo paper easel under the lens on the stage of the enlarger.
3. Turn out the lights in the darkroom, and turn on the enlarger lamp. Darkroom safelights can be left on. Focus and compose the image projected through the negative that falls on the stage of the enlarger. Compose by raising or lowering the entire enlarger head and by moving the easel. Focus by turning the knob that moves the lens only.
4. Use the magnifier to focus the image that falls on the easel. When focusing is completed, stop the lens down about two clicks (2 f/stops) to increase the depth of field, and insert the contrast filter you desire, if you are using polycontrast print paper.
5. Turn the enlarger lamp off and place a sheet of print paper in the easel.

PRODUCE

Printing an Enlargement

You need: black and white negatives, chemicals (paper developer, stop bath, fix), a wash tank, printing paper, an enlarger, tongs

Procedure

1. On the easel, compose the image projected through the negative by the enlarger. Stop down your enlarger lens and turn the enlarger lamp off. Insert the appropriate contrast filter if you are using polycontrast print paper.
2. Make a test print. Place a sheet of print paper in the easel. Cover four-fifths of it with a piece of cardboard. Set the enlarger timer for a 5-second exposure.
3. Expose the print paper for 5 seconds. Move the cardboard to uncover the next one-fifth of the print paper, and make another 5-second exposure. Repeat this process three more times so there is a series of five exposures in 5-second intervals.
4. Remove the print paper from the easel and develop, stop, and fix.
5. Analyze the test print to determine the correct exposure time. Repeat the process, if necessary, using a shorter or longer exposure time.
6. Set the enlarger timer to the correct setting and make a print. Develop, stop, fix, and wash.
7. Repeat steps 1 through 6 for each additional print.

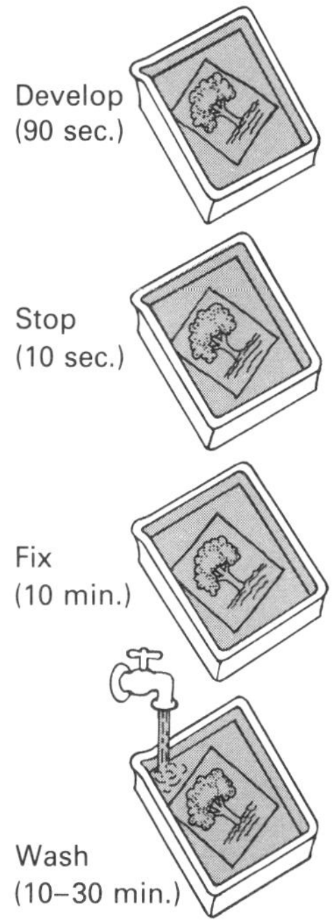

Steps for printing an enlargement.

A sheet of cardboard can be used to make the test print.

TRY THIS

Make an enlargement from one of the negatives you processed earlier. Try several contrast filters without changing anything else, and compare the results.

Processing and Printing Hints

☐ Use RC polycontrast papers at first. They are the most versatile, and easiest to use.
☐ Temperature control is critical for processing. Keep chemical temperatures between 65°F and 75°F for the best results. Remember, as chemical temperature goes up, developing time goes down.
☐ Do not try to salvage a dark print by removing it from the developer before 90 seconds. Results will not be satisfactory. Make another print using a shorter exposure time.
☐ A quart of developer, stop, or fix is usually good for about 15 to 20 8 × 10-inch prints.
☐ Using two fixing trays can give better results. Put your print in the second tray after about one-half the fixing time is up.
☐ Agitate the fixing tray several times during fixing. If several prints are fixing at the same time, rotate them from the bottom to the top of the tray.
☐ A hypoclearing-agent neutralizer can be used before washing prints. This reduces washing time to 5 to 10 minutes.

A FINAL WORD

In this chapter we have attempted a lot. Photography can be a very complex subject. Obviously, this discussion has only brushed the surface of the topic. However, the skills and techniques presented here are the foundation for expanded uses of photography. The reference section lists sources for further, in-depth study.

A good book on photography can help a great deal, but trial and error is how most photographers develop and sharpen their skills. Remember, photography is not only a valuable skill for the teacher or trainer to possess, it can also become a rewarding and relaxing hobby.

Based on your study of this chapter, you should now have an understanding of how film and cameras work, and how photography is possible when they are combined. Remember the Ping-Pong story. Photography is really that simple. Light (you can think of light as Ping-Pong balls, if you wish) enters a light-proof container called the camera. If too much light enters, the light-sensitive film in the container is overexposed. If too little light enters, the film is underexposed. The size of the hole (aperture) and the length of time the hole is open (shutter speed) determine how much light will enter the camera. The film's ASA rating determines what amount of light is correct. The camera's light meter measures the light available in a photographic situation and tells you how to set your camera's aperture and shutter correctly.

Photographic printing is a similar process. Light from the darkroom enlarger lamp passes through your negative. Where the negative is dark, light is blocked. Where it is clearer, more light is allowed through. This light strikes light-sensitive printing paper and a latent positive image is formed. Chemicals are used to make the image visible and permanent.

The enlarger lens opening and lamp timer are used to control the amount of light that strikes the print paper. That is all there is to it. All the other techniques discussed in this chapter, and bantered about by your photographer friends, are just embellishments on this simple process.

LOOKING AHEAD

In this chapter still photography has been presented. The transition to Chapter 12 will be easy if you remember what has been discussed here and add one other concept—continuous picture taking. Motion pictures are merely a series of still pictures. A motion camera will normally expose either 18 or 24 individual photographs each second. A movie projector shows these still pictures back at the same rate. Because our eyes cannot distinguish between each individual image, they seem to melt together, and we perceive motion. Actually, this is a very simple idea. Add a motor to the camera, and to the projector, and you are all set.

REVIEW QUESTIONS

1. Which photographic product does not require a machine in order to be used during teaching?
2. How large are standard-sized transparent slides, in inches?
3. What is the active ingredient of film emulsion?
4. What are the three components of the basic camera?
5. As f numbers increase in size from f/2.0 to f/16, what happens to aperture size?
6. How do focal-plane shutters get their name?
7. When you focus a camera by finding the exact camera-to-subject distance, what kind of focusing technique are you using?
8. Explain why single-lens reflex cameras are the type most widely used by teachers.
9. What are the three categories of photographic products?
10. A 35mm SLR camera would use what format film?
11. As ASA rating increases, what happens to film sensitivity?
12. Define *color temperature.*
13. An incident-light meter measures light reflected off of an object. TRUE/FALSE
14. What is the purpose of a light-meter indicator needle found in the viewfinder of many SLR cameras?
15. Why are instant-film cameras different from roll-film cameras?
16. What is a bellows?
17. The light used to eliminate the shadows produced by the key light is called ________.
18. Which view is one produced with a telephoto lens?

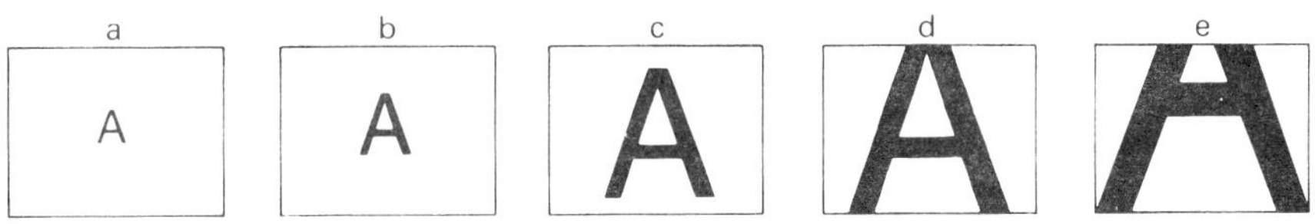

19. Describe the steps for processing exposed black and white film into negatives.
20. What is polycontrast printing paper?

ANSWERS TO REVIEW QUESTIONS

1. The study print does not require a projector to be used.
2. Two inches square is the size of the standard slide.
3. Silver halide crystals are the light-sensitive substance in film emulsion.
4. Light-proof box, aperture, and shutter are the three components of all cameras.
5. As f number increases, aperture size decreases.
6. Focal-plane shutters are placed behind the camera's lens in the focal plane of the film. That is how they get their name.
7. Rangefinding is the focusing process of finding the exact subject-to-camera distance.

8. Single-lens reflex cameras have through-the-lens focusing so close-up photography is possible. It is possible to change lenses on SLRs. The price of SLR cameras has come down substantially in recent years. Recent models of SLRs have built-in light meters. Some will even set the aperture and shutter speed automatically.
9. Color transparent slides, color prints, and black and white prints are the three photographic products obtained from amateur films.
10. 135 film is used by 35mm cameras.
11. Sensitivity to light increases as ASA increases. Every time ASA doubles, sensitivity also doubles.
12. Color temperature indicates the whiteness of a light source and is measured on the Kelvin scale.
13. False; an incident-light meter measures the light that falls on an object.
14. The light-meter indicator needle in the viewfinder of many cameras is used to calibrate the camera's shutter speed and aperture so the correct amount of light will enter the camera to expose the film. Centering this needle by changing the aperture, the shutter speed, or both, is a commonly used technique for setting camera exposure.
15. Instant-film cameras use sheet film that is made up of a sandwich of light-sensitive print paper and chemicals. When exposed instant film is taken from the camera, the chemicals mix together to develop the latent image into the visible print. With an instant-film camera it is possible to take one picture at a time and to see the results almost immediately. Unfortunately, it is not easy to make copies or enlargements of instant-film prints, nor are transparent slides usually available.
16. A bellows is an adjustable device that fits between the camera body and camera lens. Since the lens is moved away from the film plane, extreme close-up photography is possible when a bellows is used.
17. The fill light is the light used to eliminate shadows produced by the key light.
18. "e" is the telephoto image, if any are.
19. Load the film into a developing tank in total darkness, develop with film developer, stop with a water rinse, or with stop bath, fix, wash, and dry.
20. Polycontrast paper is paper that can be used with polycontrast filters to produce prints of any contrast type. Changing the filter for different prints produces positives of different contrasts.

REFERENCES

Adams, A. *Camera and Lens* (Hastings-on-Hudson, N.Y.: Morgan and Morgan, 1970).

Advanced Camera Techniques for 126 and 35mm Cameras, Pamphlet AC-56 (Rochester, N.Y.: Eastman Kodak Company).

Adventures in Existing Light Photography, Pamphlet AC-44 (Rochester, N.Y.: Eastman Kodak Company).

Atkinson, S., and Locatis, C. *Media and Technology for Education and Training* (Columbus, Ohio.: Charles E. Merrill, 1984).

Basic Developing, Printing, and Enlarging, Pamphlet AJ-2 (Rochester, N.Y.: Eastman Kodak Company).

Benedict, J. A., and Shenenberger, T. W. *Creative Photography: Camera and Darkroom Manual,* 2d ed. (Tempe.: Arizona State University, 1976).

Bergin, David P. *Photojournalism Manual: How to Plan, Shoot, Edit, Sell* (Dobbs Ferry, N.Y.: Morgan and Morgan, 1967).

Brown, J. W.; Lewis, R. B.; and Harcleroad, F. F. *AV Instruction: Technology, Media and Methods,* 5th ed. (New York: McGraw-Hill, 1977).

Bullough, R. V. *Creating Instructional Materials,* 2d ed. (Columbus, Oh.: Charles E. Merrill, 1978).

Close-up Pictures with a 35mm Camera, Pamphlet AB-10 (Rochester, N.Y.: Eastman Kodak Company).

Coffin, C. *Photography as a Fine Art* (Hastings-on-Hudson, N.Y.: Morgan and Morgan, 1971).

Color Photography Outdoors, Pamphlet E-75 (Rochester, N.Y.: Eastman Kodak Company).

Composition, Pamphlet AC-11 (Rochester, N.Y.: Eastman Kodak Company).

Data: Kodak Black and White Films—Roll, 135 and 126, Pamphlet AF-16 (Rochester, N.Y.: Eastman Kodak Company).

Davis, P. *Photography,* 2d ed. (Dubuque, Iowa.: W. C. Brown, 1975).

Hedgecoe, J. *Pocket Guide to Practical Photography* (New York: Simon and Schuster, 1979).

Hurley, Gerald D. *Visual Impact in Print: How to Make Pictures Communicate* (Chicago: Visual Impact, 1977).

Kemp, J. E. *Planning and Producing Audiovisual Materials,* 4th ed. (New York: Harper and Row, 1980).

Kodak Color Films for Still Cameras, Roll, 135, and 126, Pamphlet AE-41 (Rochester, N.Y.: Eastman Kodak Company).

Life Library of Photography (New York: Time-Life Books, 1970).

Photographer's Handbook (New York: Time-Life Books, 1971).

The Sourcebook: Kodak Ektagraphic Slide Projectors, Pamphlet S-74 (Rochester, N.Y.: Eastman Kodak Company, 1977).

Speechmaking: More Than Words Alone, Pamphlet S-25 (Rochester, N.Y.: Eastman Kodak Company, 1979).

Upton, B., and Upton, J. *Photography* (Boston: Education Associates, 1975).

Visual Marketing: A Program for the Design Profession, Pamphlet VI-36 (Rochester, N.Y.: Eastman Kodak Company, 1978).

Volker, R. P., and Simonson, M. R. *Media For Teachers,* 3d ed. (Dubuque, Iowa.: Kendall/Hunt, 1981).

Wittich, W. A., and Schuller, C. F. *Instructional Technology: Its Nature and Use,* 5th ed. (New York: Harper and Row, 1973).

GOALS

☐ Gain knowledge about the uses of film in teaching.
☐ Develop the ability to design and produce film for teaching.
☐ Enable students to produce films in your subject area as alternative methods of preparing projects.

OBJECTIVES

☐ Describe film formats and their advantages and weaknesses.
☐ Define terms and describe steps in film production.
☐ Operate selected cameras and projectors.
☐ Calculate appropriate times governing time-lapse and slow-motion films.
☐ Splice and edit film.

Thomas Edison's seventy-seventh birthday party was given in New York by members of the motion picture industry on February 15, 1924. He had these remarks to make about his invention of the motion picture: "...Whatever part I have played in its development was mainly along mechanical lines. The far more important development of the motion picture as a medium for artistic effort and as an educational factor is in your hands." From the beginning, Edison thought highly of the educational possibilities of motion pictures, for both popular audiences and the classroom. He claimed, in 1923, that "... in twenty years children will be taught with pictures and not with books . . ." and that ". . . motion pictures are 100 percent perfect for disseminating knowledge."

In the early days there was little more for a teacher to learn—relative to film—than how to operate a projector to show a film, set up the screen, and prepare the class for the topic. Recently, however, with the availability of inexpensive cameras and film it has become possible for teachers and their students to make their own films. This chapter is about the use of films in teaching.

We are assuming that many teachers would like to produce their own films and, in addition, learn enough about the process to help students produce films. The film-making process involves ideas, skills, techniques, and characteristics from a number of other areas of educational media. For example, the "language, structure, and expression" of film use techniques that are similar to those in computer progamming, design of a hook and loop story, or the process described in scripting techniques. In addition, the mechanics of motion photography are similar to those needed for still photography. Movie cameras have shutters and apertures and require certain types of film and photographic techniques that are very similar to those needed when using still cameras. Thus, you can see that film making relies on fundamental techniques such as:

☐ Developing a scenario, scripting, storyboarding
☐ Basic photographic techniques; camera operation
☐ Audio recording
☐ Editing, rearranging

12 MOTION PHOTOGRAPHY

To get an idea of the types of film that you and your students might want to produce, we begin the chapter with a look at the ways film can be used in teaching. Then, the making of films is described. Finally, in the third section of the chapter, we will consider special projects using film to expand or contract time and space for certain special effects.

The first part of the chapter, then, will concentrate on the ways prepared film can be used in teaching, where you find it, how you prepare for its use, the equipment needed, and certain special applications.

USING PREPARED MOTION PICTURES

Although many teachers enjoy making their own films, not everyone feels qualified to do so. For these people, the answer is to use prepared films.

Finding Films

One of the easiest tasks, related to film, for a teacher is locating a film for a specific teaching task. Over 30,000 educational films are currently available. Several sources are available—as well as several levels of educational service agencies—to help teachers locate films. Some of them are:

☐ *Local collections* Sometimes school districts or companies have a modest collection of films. Check the local catalog to determine if there are any films available for use.

☐ *Area media collections or regional collections* A larger consortium of schools, counties, or businesses may have formed a media center that has a collection of films. Motion pictures in a collection of this sort are generally chosen by an advisory team or group of teachers who reflect the needs and interests of the institutions that are served by the center. Films usually may be obtained quickly from a regional media center, through frequent delivery systems. If you need a specific film you can recommend its purchase.

There are numerous film catalogs that list films available from many sources.

Many super 8mm projectors, such as this one, have variable projection speeds in both forward and reverse and stop-action capability so an individual frame of film can be viewed and analyzed.

This Kodak Pageant 16mm projector is an example of a manually threaded projector.

A Technicolor super 8mm film loop projector (compliments Technicolor Corp.).

TRY THIS

Locate a source of films for your subject area. Identify one or two suitable films that you could use in teaching.

☐ *State university film libraries or state departments of education* Once quite popular and well used by schools and teachers in many states, the film libraries of certain state universities continue to provide help to the teacher who needs supporting materials through film. In addition, state agencies often collect and maintain film libraries for special needs. If these sources do not have the film you are seeking, they probably have catalogs that you can use to locate films from other sources.

Preparation, Use, and Follow-up

Seeing a film is somewhat like seeing a play, going to a concert, or watching a sports event. It helps if you know something about what you are watching *before* you attend. When planning to use a film, determine the important ideas to be covered in the film, and involve the class in some questions or situations that will sensitize them to what they are going to see. Remember that film and television can evoke a range of emotional responses—audiences can laugh or cry, feel happy or sad or angry, when events are shown through these media. Point out these potential feelings to the students, so that they can monitor their feelings as they see the film.

During use it may be important to stop the projector and talk about some portion of the film—or rerun a portion—to emphasize what took place or provide more opportunity for students to learn the information.

After the film, conduct a discussion—very much as a group of spectators would after a sporting event. You have observed this at commercial movie theaters, as you walk out after the show and hear others commenting on what they saw. As a guideline, go back to the original questions used during the preparation before the film. If students have questions, or a point of controversy comes up, show that portion of the film over again.

Projecting and Projectors

This is an aspect of film use over which you may not exercise much control, but if you can help determine what sort of projector the school needs, consider the various types:

☐ *Autoload projectors* Those that automatically move the film through the threading path, and feed it out the end of the path; some projectors of this sort will even catch and thread the film onto the take-up reel.
☐ *Manual-loading projectors* Sometimes the most versatile because you can load a film in its middle rather than having to load it from the beginning.
☐ *Film loop projectors* Designed to handle film that has been put into cartridges. There is no "head" or "tail" but rather the ends of the film have been spliced together to make a continuous loop; 8-track audiotape is packaged on the same principle.

Another method for projecting film, similar to the single-concept film loop idea, is available in a variety of models from several manufacturers. In one case, the film is mounted in a container that fits into the projector in such a way that the film may be threaded into the projector automatically. In this method, however, the film does not need to be in a loop, but may be mounted into the container so that it is open-ended. This makes it possible to reverse the film and show a portion over again—something that is not possible with a single-concept film loop system.

Single-Concept Films

When films were first developed for use in the classroom, they were used to portray a series of events, a field trip, an extended situation, or a story. Even though the film took less time than the event itself, it was used to represent the entire activity. Naturally, the editorial point of view in portraying an event, the nature of the audience, its age and background, and the social conditions at the time had an impact on the "story" the film was representing.

Some types of information did not particularly lend themselves to this type of treatment, however. Certain physical or biological events, such as the fertilization of an egg, the hatching of a tadpole, or the explosion of a bullet from a gun, were not long and did not involve stories. Clearly, what was needed was the capability of a teacher to show ideas of this sort in which the event was introduced as part of a larger educational setting. Thus, the use of single ideas, or *single concepts,* was introduced into education some time ago. The films used to show these ideas, usually three or four minutes long, became known as single-concept films, and were designed to be inserted into classroom instruction without any particular preparation by the teacher. The projector, often a film-loop projector, was easy to use and inexpensive enough that it could be kept in the classroom. The films, selling for $20 to $30 each, were also inexpensive enough to be kept in the classrooms where they would be used.

Single-concept films can be inserted into regular classroom instruction and are used in the same way as a blackboard drawing, a chart, a transparency, or any other media. There is no need to consider the use of a single-concept film as in any way different from that of other media. As you consider the types of content you will be teaching, you may wish to identify potential ideas for producing single-concept films.

Film Production as Student Projects

You can probably imagine the value of certain types of film for use in your subject area, either to show a longer, larger, more extended view of an idea, or to provide a single-concept view. You also may realize that it is not always possible to obtain a film that meets your needs exactly. Because of drawbacks in the use of commercially available films, such as availability, cost, or suitability, it is important to consider production of your own motion pictures. One of the best ways to do this is by involving students. As students respond to what they have learned in a course, they ordinarily "read-a-book," "write-a-paper," and "take-a-test." There may be class discussion, interaction with other students, some question-and-answer activity in class, and a few additional opportunities for students to put their new knowledge to work, but the scope of these activities is somewhat limited.

This super 8mm projector loads automatically. Just insert the film into the film channel and turn the projector on.

OPERATE

A 16mm Film Projector

The following steps are generally applicable to 16mm projectors of various types. For specific instructions refer to the diagrams and operating procedure that accompany the projector.

1. Load the film on the front forward spindle located on an arm extended near the lens. Film should wind off of the feed reel in a clockwise direction as you face it, and the sprocket holes should be on the margin nearest you.
2. Remove about three feet of leader if the projector is a manual-threading type. For automatically loading projectors, a few inches of leader is sufficient.
3. Thread the film. For automatic projectors, turn the projector on, lock autoload switch, and insert the leading edge of the film. When film is threaded through the entire path, turn off projector and insert leading edge into the take-up reel. For projectors that must be threaded manually, do not turn the projector on yet, but follow the sequence below:
   a. Open all sprocket covers, film gate, and other devices so that the film can be inserted.
   b. Insert film, making sure that it passes over or through the upper sprocket (leave a loop), the film gate, the lower sprocket (leave a loop), and the sound head.
   c. Close sprocket covers and film gate.
   d. Insert free end of the film into the take-up reel.
4. Turn projector on, noting the possibility that the switch may have a "motor only" position, followed by a "motor and lamp" position.
5. With the motor and lamp on, the film should be visible on the screen. Adjust height of the projector so that image is centered on the screen, and then focus the image by rotating the front lens or manipulating the appropriate focus control. The image can be framed by turning the framing control.
6. Adjust volume level and tone so that the sound is satisfactory. Be sure to turn the amplifier switch on, if the projector has a separate one.

To rewind: Procedures vary a great deal from one projector to another. Consult the instructions for the machine you are using.

We suggest throughout this text that your students become involved in media production for two important reasons:

- □ They will acquire skill in media production, communication, and the use of contemporary technology.
- □ They will learn a great deal about the content you are teaching if they can present it to others with a variety of media.

There is another important benefit: Well-done media produced by students can be used in teaching other students. Under your guidance and leadership, students might script a single-concept film on a particular topic in your subject area. As you help the students develop the script for the film, they must face questions such as "What should the viewer be able to *do* after seeing the film?" "What is the most appropriate sequence for portraying the ideas?" and "What techniques should be used to tie the visual segments together?" Thus,

OPERATE

A Super-8 Projector

The operation of a super-8 film projector is similar to that of a 16mm projector. Various designs are available, ranging from automatic-threading to manual machines. Since many super-8 films have no soundtrack, the projector you are likely to encounter may not have a sound head. This simplifies the threading procedure.

1. Load the film on the front forward spindle, located near the lens. Film should wind off the feed reel in a clockwise direction as you face it, and the sprocket holes should be on the margin nearest you.
2. If the projector is the automatically threading type, insert film according to instructions for the machine. Turn on the projector. When leading edge emerges after threading, insert it into take-up reel. For manual threading, follow the sequence below.
   a. Open all sprocket covers and film gate.
   b. Insert film so that it passes over or through the appropriate sprockets and film gate. Refer to the threading diagram for the projector. Close sprocket covers. Insert leading edge into take-up reel.
   c. Turn on the projector and allow the film to be projected. Adjust height of the image on screen, and focus.

To rewind: Follow instructions for the particular machine in use.

Super-8 projector.

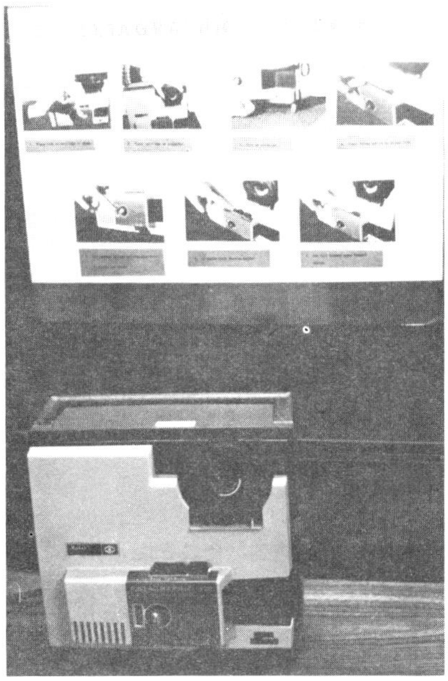

While most super 8mm projectors work in basically the same way, it is always a good idea to follow operating instructions that come with the projector.

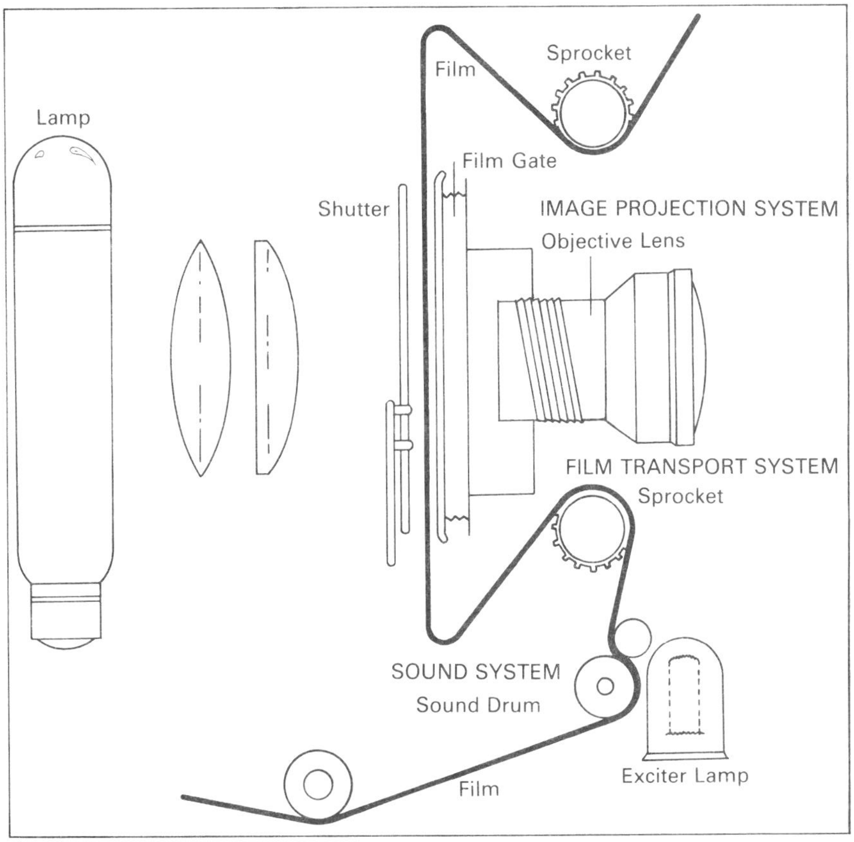

Components of a motion projector.

Rear-screen projectors can be used in a room with the lights left on (compliments Technicolor Corp.).

student-produced media are a means by which they can learn content, and learn to communicate ideas as well. Your role as teacher, then, becomes one in which you are an advisor—a sort of "executive producer" of the material that students are using as they make the media.

If you are to use some of the ideas we have mentioned in this first section—especially those in the latter portion where student production was described—you naturally will become interested in the method for producing film. The next section of the chapter provides an overview of that process.

MAKING FILMS

This section of the chapter is particularly important because it is concerned with techniques that apply to both film and television. If you plan to make a film—single-concept or a longer one, or a videotape—the methods used to communicate through the medium of motion are nearly identical.

Choosing a Format

With the current availability of several film formats, you can decide whether you wish to work with super-8 film or 16mm. The older regular-8 format is no longer used widely, although regular-8 film is still available.

Both 8mm and 16mm film function in the same way when it comes to recording images. Light enters the camera lens and is focused on the film, where it activates various layers of silver and dye. When the film is developed, these layers each take on the color to which they are sensitive; when the layers are viewed together the entire range of color is seen.

There are some differences between brands of color film that determine whether the developing process will be simple or more complex. The Ektachrome process, for example, is simple enough to be done at home. On the other hand, the Kodachrome process requires specialized equipment and temperature controls that are not readily available in the typical training center or school.

Scripting Your Own Films

The process for scripting a film is outlined in a general way in Chapter 9. Of course it is necessary to determine the scope of the film, the topic, the objectives to be reached by the viewers, and other important parameters. However, while these characteristics are of general interest and use for any medium, it is important to apply some additional criteria to the script if it is for a film.

Types of Super-8 Film

Brand	ASA	Description
Ektachrome 160	ASA 160	Color film available as silent film with a magnetic stripe for sound recording
Plus-X Reversal (No. 7276)	ASA 50	A medium-speed panchromatic film used for exterior photography
Tri-X Reversal (No. 7278)	ASA 200	A high-speed panchromatic film for interior photography
Kodachrome 40	ASA 40	Color film available as silent film and with a magnetic stripe for sound recording
Ektachrome EF (No. 7242)	ASA 125	High-speed color reversal film used for low-light conditions indoors or outdoors
4-X Reversal (No. 7277)	ASA 400	A very-high-speed panchromatic film for use under very limited lighting conditions

Comparison of regular-8 and super-8 film

	Regular-8	Super-8
Length	25 feet	50 feet
Normal film speed	18 frames/second	18 frames/second
Packaging	Reel	Cartridge
Magnetic soundtrack	Not available on raw film stock	Can be purchased with or without magnetic track
Projection time	3 minutes, 42 seconds, for 50 feet	3 minutes, 20 seconds for 50 feet
Frames per foot	80	72

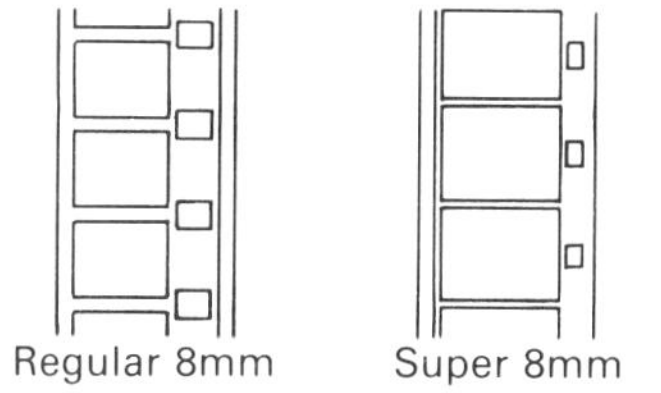

Frame configuration for regular-8 and super-8

Because film (and videotaped programs) realistically portrays events as we see them in the world around us, a sort of rhythm, flow, and liquid motion can be developed. A medium that is capable of showing motion uses its own "syntax" and "grammar" in the same manner as a language. In fact, it is possible to think of the visual portrayal of events as a kind of language; the techniques of grammar, syntax, connectives, and other elements of language are associated with motion.

The comparison to written and spoken language will be helpful, because we can then describe the *movie sentence,* the *movie phrase,* the *paragraph,* and so forth. Scenes that contribute to these concepts are illustrated visually here and on the next page, and you will recognize some of the language we have used in previous chapters. It is helpful to study comic strips, too, because they are in a sense "frozen" sequences of action that look very much like a script or storyboard for a film or videotape.

A long-shot, medium-shot, close-up sequence is often called a visual sentence.

Visual paragraph.

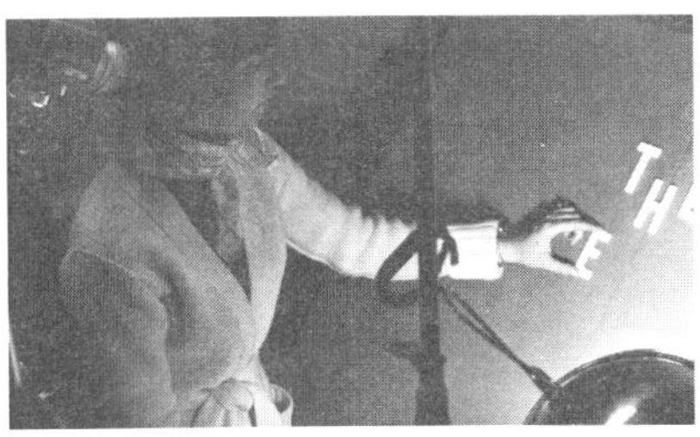

Proper lighting is an important element of successful film making, just as it is for any photography.

In addition to the language of film, there are other parallels to written or spoken communication. A novel or story, for example, has a plot, a theme, and various characters. The story is set in a certain set of circumstances. People and events are introduced in a sequence that will most appropriately contribute to the development of the story. This development of a point of view, a setting, and a cluster of circumstances that will establish what the message is and why the message is important is called the *treatment.* A film treatment is one of the most difficult aspects of film making to develop—and, of course, one of the most important. For films that are intended to express an attitude, a moral, or a point of view, the treatment is very important. But even for those films that are how-to-do-it instructions, a treatment is necessary. For example, if you intend the camera to be used as the observer—to become the eyes of the person doing the work, or serving as the center of the action—you must write such instructions into the script. Such use of a camera is called *subjective.* On the other hand, if the camera sees the scene from the other side—the side opposite the actor—the use in this case is called *objective.*

Some possible ways to organize a treatment that might be used for a teaching film include:

☐ *The personal testimony* Portray the subject through the eyes of a person who has had personal experience with the problem, process, or idea.
☐ *The historical approach* Portray the subject over time, beginning in the past and gradually bringing the audience up to date.
☐ *The problem/solution approach* State or portray a problem, and then show how several solutions may be applicable.

Once the script is developed and laid out as a storyboard, production of the film can be planned. However, before determining such things as lighting, type of film, and other production information, it is necessary to examine the script one more time. This time you have to rearrange the scenes in the script, changing it from a storyboard into a *shooting* script. The shooting script permits you to take advantage of people, props, and places as they would occur naturally.

Shooting the Pictures

Now that the script has been developed, and you are ready to begin production, you must consider the equipment you will need.

Lights As outlined in Chapter 11, where lighting was discussed in relation to still photography, there are four fundamental places for lights. Since the subjects may be moving, however, there is the possibility that shadows may develop during the action in certain scenes. One method for reducing the shadows, or at least softening them, is to use diffused light. Placing a diffuser over the light source, using available light, or reflecting the light off a ceiling or wall reduces the chance that distracting shadows will occur.

With the wide range of film types now available, it is possible to use film with a relatively high ASA/ISO rating. Thus, it might be tempting to make films in available light, such as that found in an ordinary classroom, in an auditorium, or in an office. While there is a sufficient *quantity* under these circumstances to provide for the proper exposure of the film, the *quality* is not satisfactory.

Light *quality* refers to the amount and intensity of the colors that comprise white light. Perhaps you recall the colors by thinking of the mnemonic ROY G. BIV, or by remembering the colors in the rainbow: red, orange, yellow, etc. Photographic film is designed to respond to these colors in somewhat equal intensity. That is, the red end of the spectrum becomes as powerful or as strong as blue, indigo, or violet. In ordinary sunlight the colors differ in intensity, with those in the middle of the rainbow being somewhat stronger. On a cloudy day certain colors are muted or reduced by the clouds. Certainly at sunrise and sunset, when natural light is quite dim, some colors are almost absent from the spectrum. (See Chapter 11 for a discussion of color temperature.)

The quality of light can be measured precisely, and is referred to as color temperature. Think of color temperature as the color given off by the end of a metal rod, heated "red hot." If you increase the temperature, it becomes "white hot." The temperature, measured on the Kelvin temperature scale, provides a standard way of designating the quality of light. Thus, light may have a low color temperature but might still be of sufficient intensity for you to take pictures.

Types of Cameras As you review Chapter 11, you will recall that two of the fundamental types of cameras are the *viewfinder camera* and the *single-lens reflex camera.* These designations are used for both still and motion cameras. While fundamental camera features and operation are similar for both still and motion cameras, there are some important distinctions. They include:

□ *Type of film and ASA setting* The most popular motion picture cameras used by the general public are called super-8 cameras, because they require a film that is 8mm wide. The "super" portion of the name refers to the size of the picture, as compared to "regular 8." Some twenty years ago the regular-8 format was changed to super-8, providing a picture that was 50 percent larger.

The super-8 format is almost universally used by those who produce films for home use, or in connection with local educational projects, since the cost is far less than for 16mm film and equipment. However, the types of film generally rented for use in classrooms are 16mm, and are produced with 16mm cameras and equipment. With the development of super-8 film in the early 1960s several other significant changes took place. One of them related to the packaging of the film. To avoid bothering (or confusing) the consumer with unnecessary instructions for use, a new film cartridge was designed to be loaded into the camera easily and quickly. Called the "instamatic" system, the super-8 cartridge was designed for "home movies of average length." Fifty feet of film were put into a plastic cartridge containing a full-feed chamber and an empty take-up chamber. Various notches were placed on the margins of the film cartridge. These notches, sensed by the camera, set the ASA/ISO number automatically.

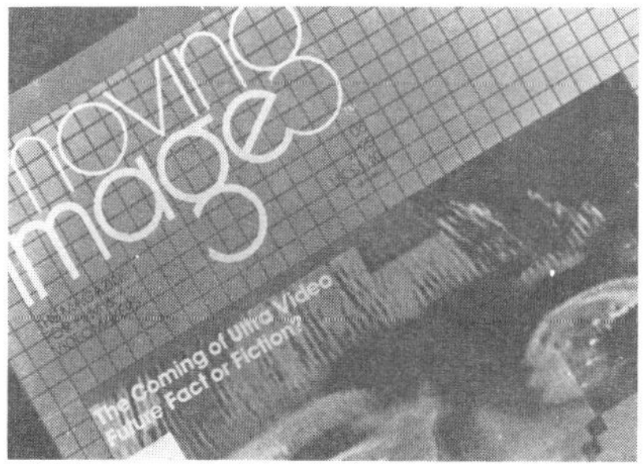

Photography and film-making magazines are an excellent source of up-to-date information about techniques and equipment.

When the entire 50 feet of super 8mm film are exposed, the cartridge can be removed from the camera and sent to the processor. The word *exposed* will be visible on the film.

To "critically focus" a camera with a zoom lens, first zoom in tight on your subject and focus carefully, . . .

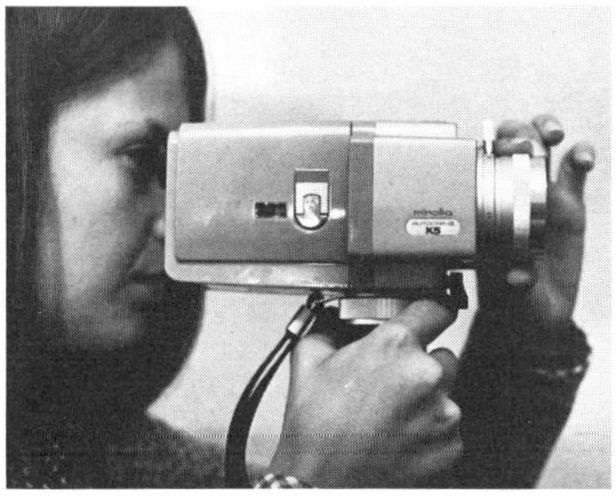

then zoom back to frame the scene correctly.

☐ *Lenses* While film was being redesigned from regular to super 8, other changes were engineered into the equipment to make it convenient and easy for amateurs to produce films. One of these changes had to do with the lenses. Formerly, to change from a wide-angle lens to a telephoto lens, it was necessary to rotate a turret on the front of the camera, bringing another lens into place. Perfection of an optical system that permitted the focal length of a single lens to change with the movement of a lever made it possible to cease the turret-lens arrangement, and use a single "zoom" lens. It is tempting for amateurs to use a zoom lens when filming a scene. It is equally tempting to hold the small, lightweight camera up to the eye and begin filming. These two improper techniques of zooming and hand-holding the camera lead to a finished film that is unpleasing and ineffective.

Fundamental Camera Technique

A fundamental general rule for camera use is: Treat the camera as if it were your eyes. To apply this rule, we will consider some characteristics that you probably take for granted. They include:

Length of the "Take" As you view a scene or converse with another person, you use your eyes in a series of short glances. Your eye contact with another person or an object is usually 5 to 10 seconds. Because of this natural method of using the eyes, many viewers of films and television seem to accept the average length of the scenes as 5 to 10 seconds.

Avoid Panning or Tilting When you look at a scene you tend to hold a steady gaze on the subject; few people attempt to get a panoramic view unless they are watching a tennis game (horizontal pan) or are observing a flag being raised or lowered (vertical pan or tilt). Panning movements with a camera yield scenes that appear to be unnatural because such movements are rarely used in everyday circumstances.

Avoid Zooming Similarly to the pan, the zoom is also unnatural. If you want to see the subject more closely you will walk over to it and get a closer view. But you rarely fix your gaze on the subject as you walk toward it. Use the long-shot/medium-shot/close-up sequence in place of a zoom. Remember, the utility of the zoom lens lies in its ability to provide lens settings of different focal lengths in the same manner as the older-style turret lenses provided. If you want to photograph a scene with a telephoto lens and then change to a close-up lens, you can change the lenses with a single zoom lens, but the change in lens setting should be done while the camera is turned off.

Avoid Hand-Holding the Camera Always mount the camera on a tripod to hold it steady. A hand-held camera can never be held perfectly steady; a small amount of movement will always be the result, no matter how careful you are. The resulting film will appear as if you were operating the camera from a rowboat that was gently bobbing on the water.

OPERATE

A Super-8 Camera

Since operating procedures vary from one camera to another, it is important to use the instructions furnished with the camera you use. Some operation considerations include:

1. *Load film* Insert the cartridge into the camera, noting that the camera's footage indicator should reset to zero. (If you remove a cartridge before it is fully used and insert it later, the indicator will again reset to zero. Mark the amount of unused film on the cartridge if you remove it from the camera.)
2. *Make settings* Determine if the ASA/ISO rating on the film must be set manually or if the camera adjusts automatically. (The latter case probably applies since the film cartridge is manufactured with notches that register the proper ASA/ISO upon insertion into the camera.) Set operating speed if necessary. The standard is 18 f/s (frames per second). The camera may be capable of more rapid speed, resulting in a slow-motion film. A setting for single frames may be available, too, for use in animation.
3. *Focus lens, and the eyepiece* View a distant object through the camera; set lens at infinity. Then focus the eyepiece so that the image becomes clear as viewed through the camera. Once focused, lock the eyepiece in place.
4. *Adjust zoom* Use the zoom lens to make the viewed object larger or smaller. Note that when zoom is adjusted to provide a close-up view, sharp focus becomes more noticeable. To maintain sharp focus for a given scene, remember to focus at the zoomed-in stage. This is called critically focusing the camera.

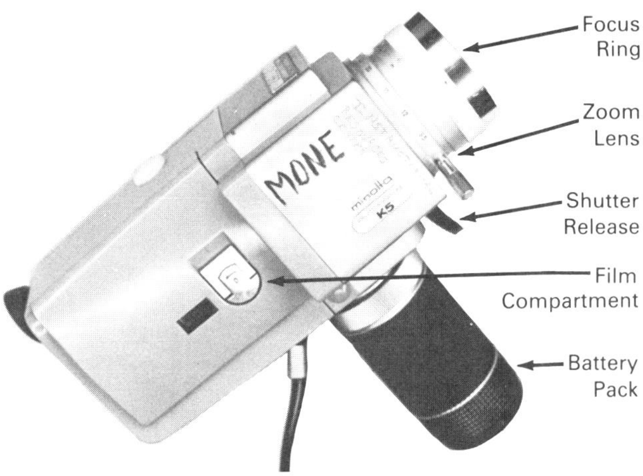

A Minolta super 8mm camera.

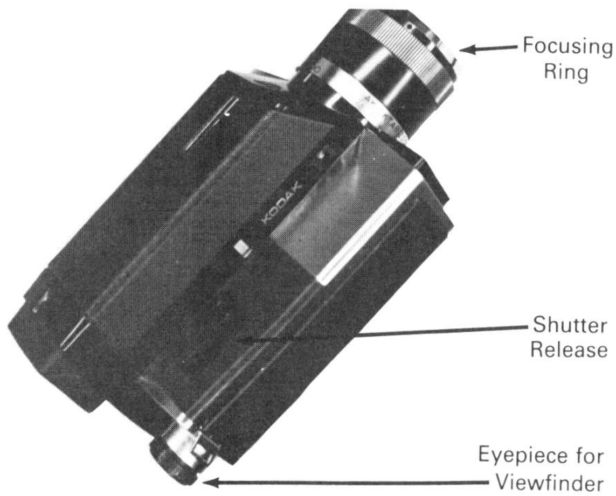

A Kodak XL-55 super 8mm camera.

This super 8mm camera will record sound on a magnetic stripe on the film at the same time as the picture is photographed.

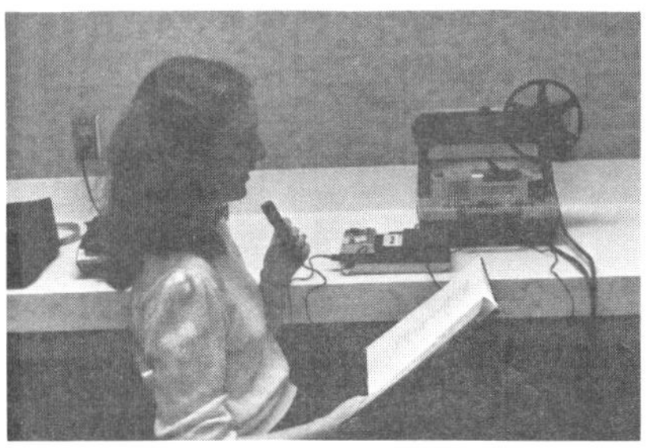

After the film has been processed and edited, an audio narration can be made. Since super 8mm films are usually short, it is possible to record the narration on an audio casette and play both the film and the tape simultaneously.

The Set

Arrange the props, background, and other elements that will be seen by the camera so that unnecessary objects are not in view. Amateur film makers frequently overlook the need to do their picture taking in a place that has been cleared of distracting materials. Bright floral patterns of wallpaper, magazines or newspapers on tables, miscellaneous objects in view of the camera, and an overall lack of attention to the visual details that make up the set will mark the film with a home-movie look. Study the set by viewing each scene through the camera viewfinder to check for unwanted shadows and other improper elements.

You will find it helpful to use large backdrops such as curtains or blankets that can be hung on the wall. Try to confine your scenes to medium or close-up views, once the establishing shots are taken. Cover the table or other working surface with a large piece of fabric if you need to provide a clear and neutral area. Choose a background color that contrasts with the subjects being photographed so that objects can be seen easily.

Recording Sound

If you are planning to produce super-8 films, it is possible to record sound and photograph pictures with the same camera at the same time. This technique is especially important if you are producing a film of someone talking, or are photographing an event in which the sound must be synchronized with the action. Films of this sort are called "lip-sync" films; the term applies to simultaneous sound/picture production. To produce films of this type the camera must be capable of recording sound, and the film must have a narrow stripe of magnetic material along one margin.

If you do not have a camera that is capable of making lip-sync films, it is possible to add sounds on the film after it is processed. The film, of course, must have the magnetic stripe in place. If you have purchased special super-8 sound film in the first place, the sound can be recorded after the film is processed, as well as recorded during the filming. If the film does not have a magnetic stripe it is possible to have one added, either by sending the film to a special laboratory that specializes in that process, or by adding the magnetic stripe yourself with a machine designed for that purpose. In either case, adding sound after shooting the film (a process called post-syncing) must be done with a super-8 projector that is properly equipped to handle the job.

Two additional approaches are possible if you desire to add sound to your film. The first uses an audio cassette that is intended for playback while the film is projected. While lip-sync effects cannot be produced in this manner, there remain several applications in which the technique is useful. A narrative, sound effects, and background music can be incorporated into an audiotape that makes an effective companion to a super-8 film. Keep in mind, however, that there is a chance that the film and tape may operate at slightly different speeds, which will cause one to fall behind the other. When working with short films, this problem is relatively minor. The standard length of a super-8 film (50 feet) requires 3 minutes and 20 seconds for projection at 18 frames per second. In this relatively short time, the difference between film and tape speed will not cause extreme problems during playback.

A second possibility for adding sound involves the use of videotape. After your super-8 film has been processed you might consider videotaping it, and then adding sound to the videotape. While the tape/film speed problem discussed above is eliminated here, a disadvantage of another type is introduced. Unless a special projector, designed for transfer of film to television, is used, the videotape version of the film may contain a distracting light/dark pulsating effect. This is caused by the difference between the film projection rate (18 f/s) and the television scanning rate (30 f/s). While the problem is a minor one, you may wish to experiment with your equipment to determine if the results are acceptable.

Editing

The two major types of editing are described below. You can choose one or the other, or combine the two, depending on the needs of the production.

In-Camera Editing While this is the simpler of the two methods, it does not yield the most professional result. In-camera editing is more often used by amateur film makers who want to save time.

The purpose of editing is to select the best take and to obtain sharp connections between scenes. As you can imagine, the beginning of a scene in which there is action usually will not *start* with the action. The traditional "lights, camera, action" order from the film director serves to introduce those elements one at a time. First the lights are turned on, then the camera is started, and then the talent begins the action. If your intent is to film this scene:

Person picking up telephone (LS) → Putting phone to ear (MS) → Saying hello (CU)

it will be very difficult to edit the sequence in the camera because of the need to preserve smooth continuity of the action. Each scene must begin at precisely the correct frame or a jump in the action will be the result, as the sequence at the right shows.

On the other hand, if detailed and connected action is not called for, the camera can be turned off at the end of one scene, reset, refocused, and turned on to photograph the next scene. This is in-camera editing.

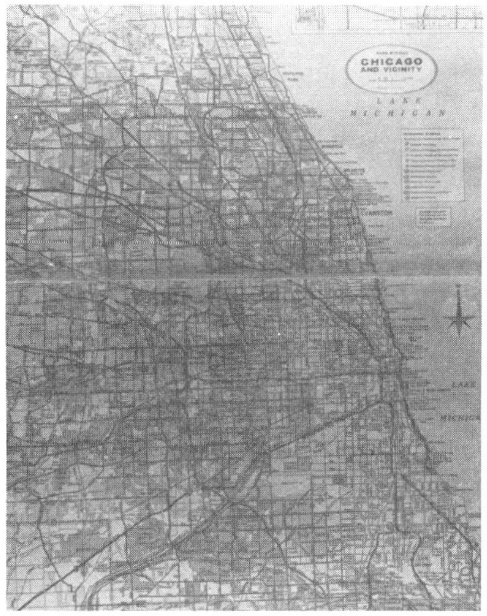

Long shot.

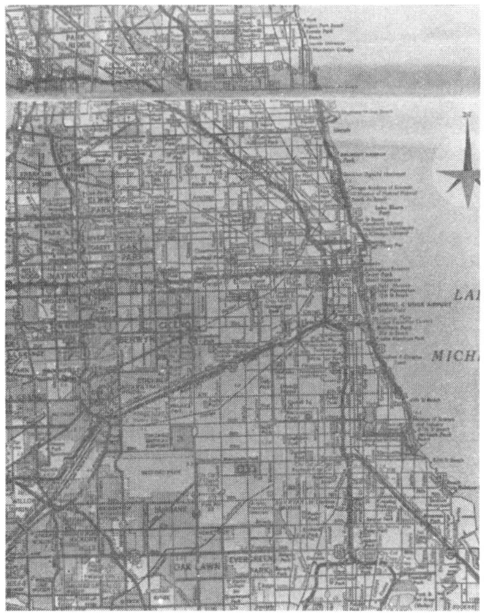

Medium shot.

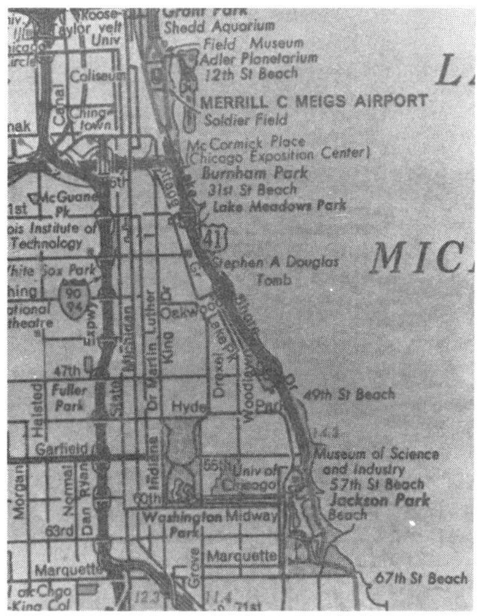

Close-up.

A viewer, with rewinds, makes it possible to preview a film one frame at a time.

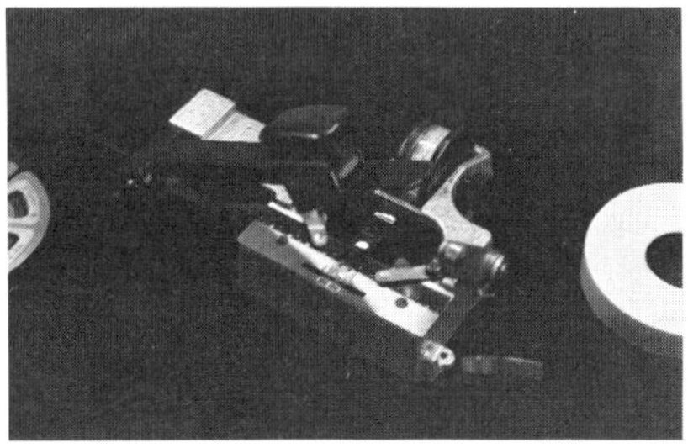

A splicer holds scenes together with a thin, transparent plastic that has an adhesive on one side.

Out-of-Camera Editing The problems described above can be avoided if you plan to edit the film after it is processed. Working from your script, select the best takes by viewing the film on an editing machine or viewer. Cut each scene from the film and hang it on a numbered pin in a trim bin. When you cut the film into scenes you can decide precisely where to begin and end each scene, although it is generally easier to save that decision until later.

Once all the scenes have been selected you can splice them together, producing a rough cut of the entire film. As you view the rough cut, using the viewer, you will develop a feeling for the rhythm and flow of the visual elements. This is the appropriate point at which to decide where to shorten certain scenes—and as you make this second series of cuts you are proceeding from the rough-cut stage to the fine-cut, or final, stage. The rough-cut/final-cut process is time-consuming and somewhat tedious, but the resulting production will be more satisfying and pleasing to the viewer.

Production: A More Professional Approach

Professional film makers (cinematographers) carry out many more steps in their approach to film production. Even though the study of film production can require several years of apprenticeship and experience, an overview of the process is provided here. Several important differences between amateur and professional methods are highlighted.

A/B Roll Printing This technique refers to the use of two rolls of original film: roll A and roll B. When a final print is to be made, the original scenes are arranged alternately on two rolls, with the even-numbered scenes on one roll and the odd-numbered scenes on the other, with black opaque leader between the scenes, checkerboard fashion. The purpose of this technique is to make the splice lines invisible in the final print. Another benefit is that it enables the laboratory to make fades or dissolves between scenes.

PRODUCE

Super 8mm Film Editing

You need: a film splicer and splicing tape, a filmviewer, film rewinds, a trim bin

Procedure

1. Shoot footage. Use your shooting script and allow for several "takes" of each scene.
2. Process exposed film.
3. Cut processed film into scenes. Label each scene with masking tape and hang strips in trim bin.
4. Sequence scenes according to script.
5. Select best "take" of each scene. Edit each scene to desired length.
6. Splice scenes together. For super 8mm splicing, the Kodak Presstape "dry" splice is easy to use and very effective.
7. Preview and evaluate.

Note: Sometimes interesting and fun films can be made from the "outs" (the scenes you do not use), so do not throw anything away. Save your "outs" on a spare take-up reel until you have a quantity and then try making a special-effects film.

Workprints Workprints are copies of the original footage. To avoid scratches and unnecessary handling of the original film, it is never projected for viewing. Instead, it is used to print a copy that can be viewed, edited, and worked with as the finished motion picture is put together. Once the workprint has been edited in the proper sequence of scenes, and the scenes have been cut to the desired length, the entire film can be viewed to determine if it should be the final version, or if additional revisions are necessary.

The workprint cannot be used to create fresh copies of the final version because it usually becomes scratched and soiled due to much handling. It is necessary to go back to the original film—the one from which the workprint was made—and cut that footage to match the edited workprint. An exact, frame-for-frame matching can be accomplished because film carries a photographic imprint of numbers on its margin. These frame numbers are printed just as the photographic elements of the pictures are printed when copies are made from the original.

When the original film is cut to match the edge numbers of the workprint, alternate scenes are spooled together on roll A, and scenes between those occurring on roll A become the scenes on roll B. Thus the A/B roll system is created from the original footage.

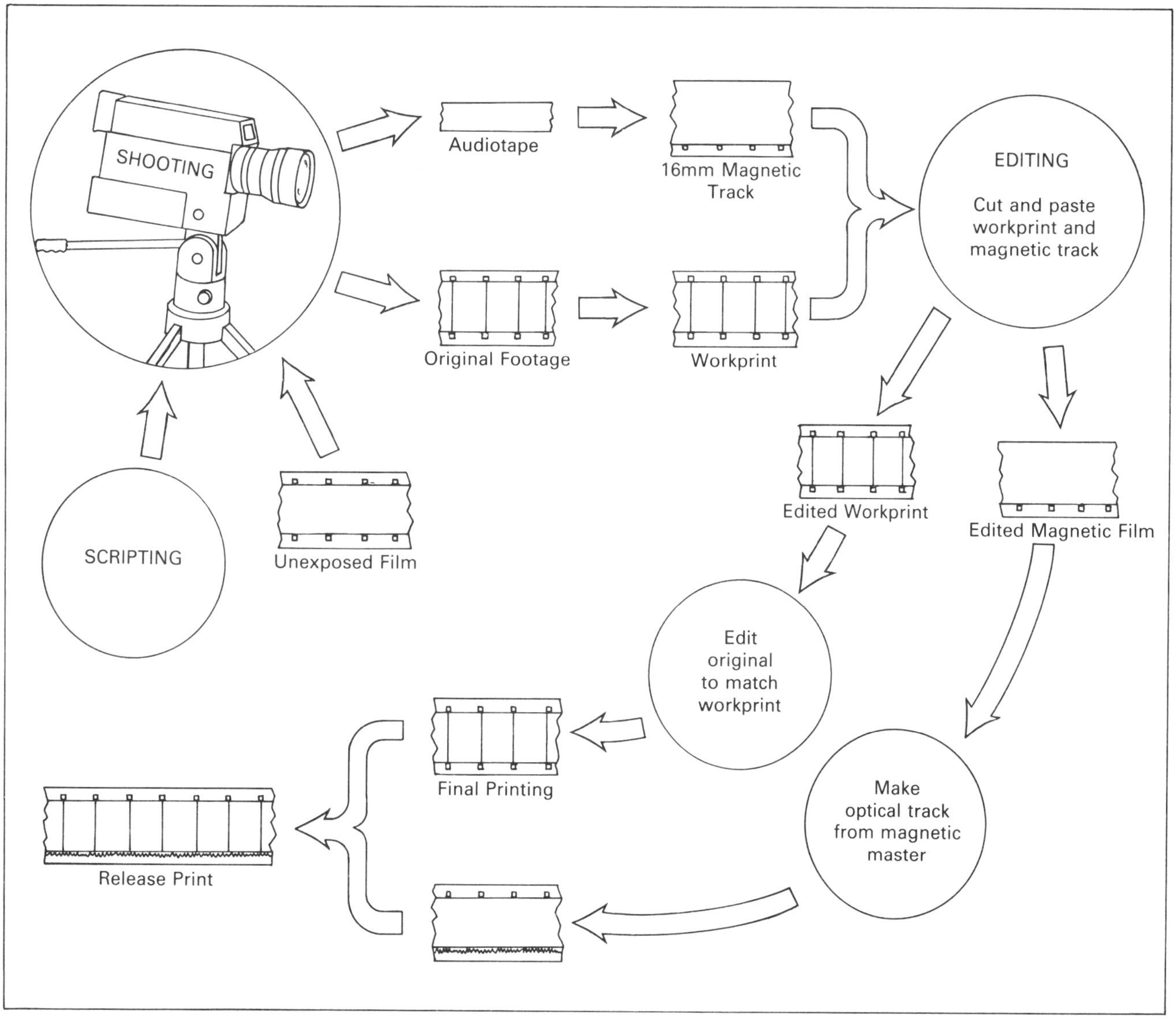

The film-making process.

PRODUCE

Film Production Review

1. *Getting started* The script is the basis for production of a film. Block out all the necessary action and specify the number of seconds each scene should run. It is very important to mount the camera on a tripod to prevent movement that would call attention to the camera. Light the set to prevent harsh shadows.
2. *Shooting film* Study each scene as you look through the camera to make sure you will photograph exactly what you want. Watch for clutter in the background. Take pictures over the shoulder of the actor, or position the camera in front of the actor, depending on what is to be shown. Keep the footage counter on the camera in mind, and remove and process the film when finished. One cartridge of super-8 film is good for 200 seconds of filming.
3. *Editing* You probably won't be able to shoot the entire film in sequence, but the finished production can be arranged in the proper sequence by cutting and rearranging the developed footage. Splices are stick-on Mylar tabs, making the editing task easier. Choose the proper projector based on *what* you want to show, and *how* you want to show it.

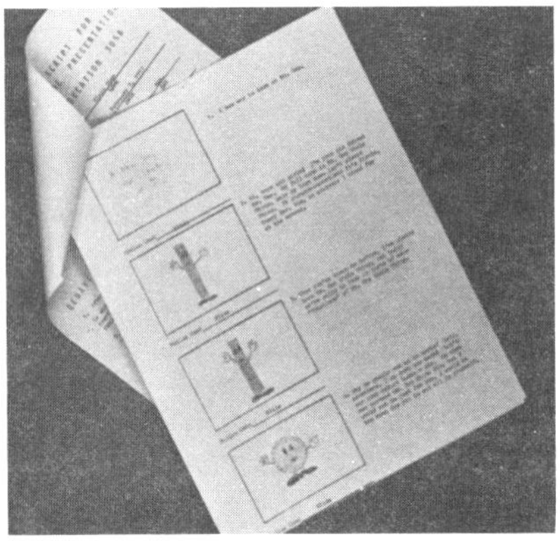

1. Start with a script.

2. Mount the camera on a tripod.

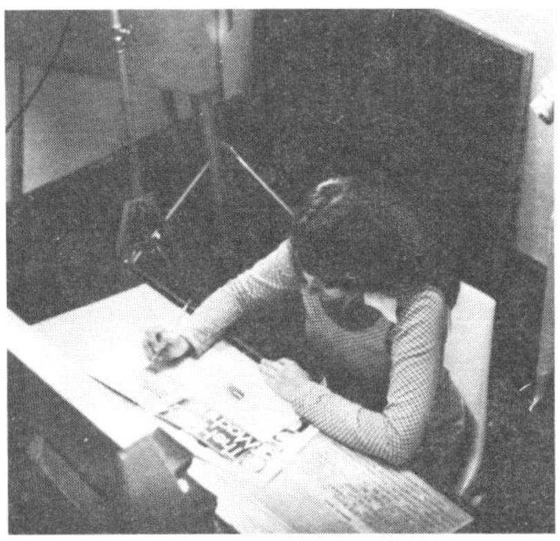

3. The set may look like this.

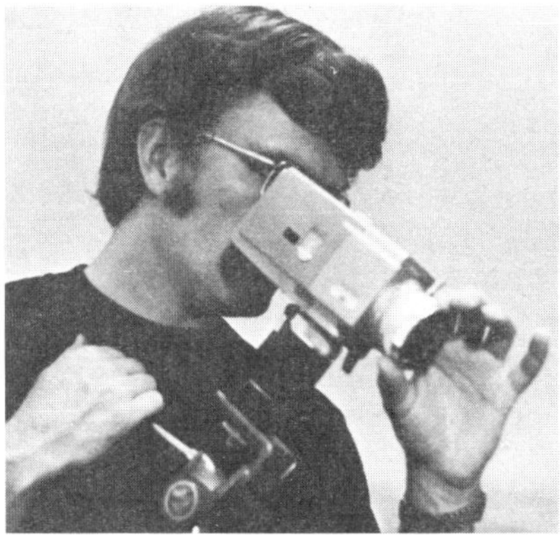

4. Carefully frame the shot you want.

5. This is the scene the photographer decided to use. It is called a take.

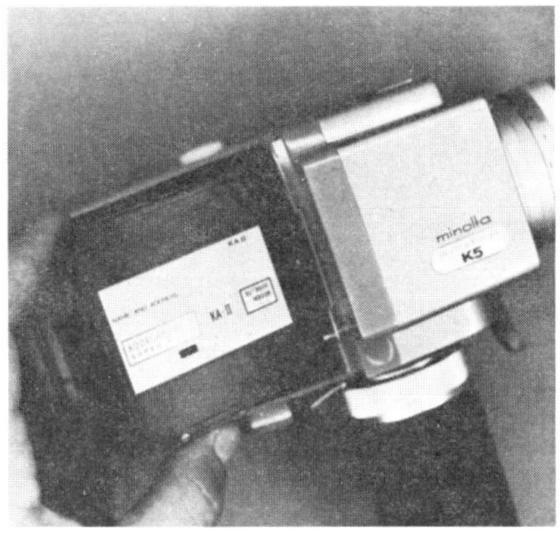

6. When you are finished filming, remove the film cartridge from the camera and have it processed.

7. Edit the film after viewing it on a viewer.

Adding Sound Unlike super-8 film production, in which sound can be recorded while the film is made, production of 16mm film requires a tape recorder, a microphone, and usually a crew of workers who have primary responsibility for production of the sound. To guarantee a match between the picture and sound, for lip-sync production, a clapboard or slate is used at the beginning of each take. When the top gate of the device closes, it makes a sharp noise which can be precisely located on the soundtrack. It also makes possible the precise location of the click by visual means, as the gate at the top of the clapboard is closed.

Since the picture and the sound must be edited together, the tape recording made during shooting of the scenes must be converted to the same sort of film as the picture. Then both the picture and the sound films can be edited with the same equipment.

8. You may have to splice takes together, using a splicer.

9. Load the edited film and view it using the appropriate projector.

SPECIAL PROJECTS WITH FILM

Now that you have learned basic film techniques, it's fun to attempt some specialized approaches.

Videotape Your Film, Adding Sound

As mentioned previously in this chapter, it is possible to add sound to your film after the editing process, if you transfer the film to videotape first. If you have videotape editing equipment available, it might be equally effective to transfer the entire film to tape—producing a sort of "workprint" on tape—and then edit the videotape. The more widely accepted method for editing videotape, where postsynchronization or a match to the soundtrack is concerned, is to prepare the edited soundtrack first, and then cut the visuals to match the sound. If you try this technique with super-8 film you will have to produce the desired soundtrack, record it on videotape, and then edit the visual elements to correspond to the sound.

Animation

Animation is the use of drawings, cartoons, and other graphic materials to create motion pictures. Most of us are familiar with cartoons used on television and at motion picture theaters. The process of creating animated productions will be presented next.

The important consideration in developing both animation and pixilation films is calculating the screen time required for certain actions. If the film is super-8, there are 18 frames projected per second because they are usually silent; 24 frames per second in the case of 16mm film because they generally have sound. In either case you must decide how long the desired action is to appear on the screen and then calculate how many frames are needed to produce that action.

By using a ruler as a boom to support a microphone, an interview can be recorded during filming.

A clapboard is used in film making when pictures and sound are to be synchronized during editing.

One simple, and fairly effective, method for transferring a film to videotape is to use a video camera and record the projected image off of the screen. Be sure to keep the room as dark as possible.

Pixilation: A Special Case of Animation

Rather than use a series of drawings to represent motion, 3-dimensional objects are used in pixilation. Generally, pixilation refers to animation using people. Also, cardboard cutouts, stick figures, small plastic dolls, or other materials can be photographed frame-by-frame to create motion on the finished film.

Time Lapse and Slow Motion

Time lapse and slow motion are methods for compressing or expanding time. There is a relationship between the number of frames made during the action and whether or not the screen time will be stretched or compressed:

Slow motion

Photograph at more than 18 f/s
Project film at 18 f/s

Ordinary motion

Photograph at 18 f/s
Project at 18 f/s

Time lapse (fast motion)

Photograph at less than 18 f/s
Project at 18 f/s

Scratch-on/Write-on Films

This technique uses clear or frosted white film. To make a film, scratch figures on the frosted film or, if you are using clear leader, "paint" or draw shapes and figures on the film with felt-tip pens. Remember that each frame line is designated by a sprocket hole, so you may wish to make a guide or marked template to use when you draw on the film.

Slow Motion

A drop of water splashing off a plant leaf requires one second of real time, as it hits the leaf and shatters into droplets. Suppose you want the event to be shown on the screen for ten seconds in slow motion.

Question: At how many frames per second should the film be shot using super-8 film?

Here are the calculations (using the proportion method):

1. Set up proportion:

$$\frac{18 \text{ frames}}{1 \text{ second}} = \frac{x \text{ frames}}{10 \text{ seconds}}$$

2. Cross-multiply:

$$\frac{18 \text{ frames}}{1 \text{ second}} \times \frac{x \text{ frames}}{10 \text{ seconds}}$$

3. Solve for x:

$$x = 180 \text{ frames}$$

In other words, 180 frames need to be photographed to produce a film 10 seconds long. Those 180 frames must be shot in one second.

Time Lapse

A flower requires around 7 hours to open. Suppose you want to collapse that time so that the event could be shown on the screen as a 10-second film.

Question: How many frames per second (or per minute) should be shot using super-8 film?

Here are the calculations:

1. Set up proportion:

$$\frac{18 \text{ frames}}{1 \text{ second}} = \frac{x \text{ frames}}{10 \text{ seconds}}$$

2. Cross-multiply:

$$\frac{18 \text{ frames}}{1 \text{ second}} \times \frac{x \text{ frames}}{10 \text{ seconds}}$$

3. Solve for x:

$$x = 180$$

In other words, 180 frames must be photographed over a period of 7 hours so that the resulting film of 180 frames, when projected at the rate of 18 frames per second, runs 10 seconds on the screen.

1. Set up proportion:

$$\frac{180 \text{ frames}}{7 \text{ hours (or 420 minutes)}} = \frac{1 \text{ frame}}{x \text{ minutes}}$$

2. Cross-multiply:

$$\frac{180 \text{ frames}}{420 \text{ minutes}} \times \frac{1 \text{ frame}}{x \text{ minutes}}$$

3. Solve for x:

$$180x = 420$$
$$x = \frac{420}{180}$$
$$x = 2.3 \text{ minutes}$$

In other words, to photograph 180 frames in 7 hours (420 minutes) the camera will have to shoot a frame every 2.3 minutes.

TRY THIS

Suppose you wish to produce a 20-second film that shows, in slow motion, a person shooting an arrow from a bow. The bow-and-arrow shot happens in about three seconds, from the time the arrow is inserted into the bow string and pulled back, until the arrow is released and leaves the bow. How many frames per second should be photographed in order to produce a super-8 film twenty seconds long?

TRY THIS

Suppose you want to show the build-up of traffic at the stop lights of an intersection during several cycles of red/green light changes. One cycle requires forty-five seconds. Plan to photograph five cycles to make a time-lapse film that runs ten seconds on the screen. What should be the frame exposure rate?

Time Line for Visual Developments

Early sketches on cave walls	5000 B.C.
Renaissance paintings	1500–1800 A.D.
Still photography	1840s
Motion photography	1890s
Television	1920s
Videotape	1960s
Videodisk	1970s

A FINAL WORD

As you think back over this chapter, you may get the impression that it is important to become a better-than-amateur film maker in order to produce films as a classroom teacher. While considerable skill is certainly helpful, it is not absolutely necessary. In this case, the process may be more important than the product. The process of planning the scenario, determining the types of shots, scenes, and camera angles, and scripting the film so that it communicates an idea may be the significant learning events, not the film itself. Perhaps one would learn nearly as much about film making—and about the subject area—if *only* the planning were done.

As a teacher, it is important to have available a variety of options that students can use in learning. Certainly reading and writing are important, as well as discussions, tests, and laboratory work. This chapter suggests an additional option—scripting and producing a film. If students must communicate ideas by means of a film they design and produce, there is ample opportunity to learn the subject matter. There is the added benefit of learning about film production.

LOOKING AHEAD

There was a time when it was impossible to record sights without drawing or sketching. Eventually the still camera came along. Then movies. What now?

The next chapter is important because it discusses—and provides techniques for—utilizing the contemporary methods of recording life situations. The use of a camera that imprints information magnetically on tape for later playback is the state-of-the-art method for preserving the "sights and sounds of the subject area."

Answers to "Try This" problems on page 243:

1. 126 f/s
2. One frame each 1¼ seconds

REVIEW QUESTIONS

1. List some sources of prepared films that are of use in teaching.
2. What is the importance of preparing a class to use a film, and holding a follow-up discussion?
3. List the types of film projectors.
4. What is a single-concept film?
5. What is the value in having students produce films?
6. What is the difference between subjective and objective camera use?
7. What is the difference between a storyboard and a shooting script?
8. List the elements of good camera technique.
9. What is meant by *lip-sync*?
10. What is the difference between in-camera and out-of-camera editing?
11. Of what use is A/B roll printing?
12. What are workprints and how are they used?
13. Describe: animation
pixilation
slow motion
time lapse

ANSWERS TO REVIEW QUESTIONS

1. Local school district collections, area media centers, state film libraries.
2. Preparation sensitizes the class about what to watch for, and follow-up emphasizes important points.
3. Autoload, manual load, film loop.
4. A short film that portrays a single idea.
5. There are two benefits: Students learn about film production and film content.
6. Subjective: a case in which the camera is in the position of the subject, as in over-the-shoulder shots.
   Objective: the camera is viewing the action from across the table or on the edge of a scene.
7. A storyboard is a sequence of scenes arranged to tell a story, but a shooting script is a rearrangement of the scenes in which all similar locations or actors are clustered so that a given series of shots can be made at one filming session.
8. Take short scenes; avoid panning, zooming, and hand-holding the camera.
9. The recording of sound while the film is being made so that the movements of the lips can later be synchronized with the words.
10. In-camera editing requires that the film be shot in sequence, so that the resulting scenes match the storyboard. Out-of-camera editing implies that scenes are shot out of sequence, to be spliced together later in the appropriate order, so that the resulting film matches the storyboard.
11. Used to accomplish dissolves in the final film, and to hide splice lines.
12. Workprints are exact copies of the original and they are used in editing. Fingerprints and dust, which may find their way onto the film while handled during the editing process, make no difference in the workprint stage. Such imperfections would damage the original.
13. Animation—the use of drawings, photographed frame-by-frame to create a film. Each drawing differs from the next in a minor way, but when projected as a film the small changes add up to movements such as walking, talking.
   Pixilation—similar to animation, except that real objects (people, machines, clay models) are photographed frame-by-frame to create movement.
   Slow motion—used to slow down an event that happens very rapidly. Slow motion is accomplished by photographing at a faster rate than the frame rate at which the film will be projected.
   Time lapse—used to speed up an event that happens very slowly. Time lapse is accomplished by photographing an event at a slower rate than the frame rate at which the film will be projected.

REFERENCES

Cinematographer's Field Guide (Rochester, N.Y.: Eastman Kodak Company ., 1980).

Filming Sports (Rochester, N.Y.: Eastman Kodak Company., 1981).

Gaskill, A. L. *How to Shoot a Movie Story* (Dobbs Ferry, N.Y.: Morgan and Morgan, 1970).

Johnson, Lincoln F. *Film: Space, Time, Light, and Sound* (New York: Holt, Rinehart and Winston, 1974).

Laybourne, Kit. *The Animation Book* (New York: Crown, 1979).

Lipton, L. *Independent Filmmaking* (New York: Simon and Schuster, 1972).

Maynard, Richard A. *The Celluloid Curriculum* (Rochelle Park, N.J.: Hayden, 1971).

Rehrauer, George. *The Film User's Handbook* (New York: R. R. Bowker, 1975).

Williams, Catherine M. *Learning from Pictures* (Washington, D.C.: Association for Educational Communication and Technology, 1973).

GOAL

☐ Understand how to produce video programs for instruction.

OBJECTIVES

☐ Set up video equipment to show a live program, to show a closed-circuit program, to record "off-the-air," to record using a one-camera system, and to record using a two-camera system in a studio.
☐ Explain how videotape stores a message.
☐ Describe the physical make-up of a videotape, videotape formats, and videotape tracking patterns.
☐ Describe the four components of a video equipment system.
☐ Describe and carry out the process of in- and out-of-camera editing.
☐ Describe what is accomplished during preproduction activities, production activities, and post-production activities.

In many elementary schools, students participate in a simple science experiment using magnets. For this experiment each student is given a bar magnet, a ten-penny iron nail, and a steel paper clip. First, the students are asked to touch the nail to the paper clip. Naturally, nothing happens. Next, the students are directed to pick up the bar magnet and rub it up and down against the nail, twenty times. Then the nail is again brought in contact with the paper clip. This time the paper clip sticks to the nail. Obviously, the teacher is introducing magnetism to the students. Any appropriate object that comes in contact with a magnet will become magnetized.

While this experiment may be fun for the students, in many classrooms it is never fully explained. Actually, this simple process is the underlying principle behind magnetic recording. All magnetic recorders—audio, video, and computer—are based on this concept.

The magnetic tape or magnetic disk used in recording is similar to the ten-penny nail. The oxide on the tape is magnetized by the recording head (which is actually an electromagnet) in the recorder, forming a pattern that represents a message. During playback the magnetized tape comes in contact with the playback head (another electromagnet) in the recorder/player, and the information is read out. This very weak signal is amplified by the machine and transduced, or converted, from magnetic impulses to electrical signals by the playback head. The electrical signals may be converted into sound by a loudspeaker, or into a picture by a television tube. Of course, whether the result is a picture or sound, or both, depends on the nature of the original recording.

In other words, unrecorded videotape is similar to the experimental nail. Tape can be magnetized by a recording head, just as a nail can be magnetized by a bar magnet.

In this chapter, the process of video recording will be explained. Techniques for the recording of television programs will be presented first. Then, more elaborate one- and two-camera video-recording procedures will be outlined. Finally, complex topics such as video editing will be described. This chapter will include information on how all components of video-recording systems work, and how they can be used to produce video programs for teaching.

13
INSTRUCTIONAL TELEVISION AND VIDEO RECORDING

TEACHING WITH TELEVISION

Two trends of the last decade have made instructional uses of television and videotape recording increasingly attractive. Both trace their roots to the technological innovations of the space program and the computer industry. These trends are:

1. The steady reduction in cost of video equipment at all levels of complexity, so that in constant dollars the prices of fairly sophisticated video equipment systems are now one-fourth to one-tenth of what they were ten years ago.
2. The increased simplification of video-system components so that operation requires little or no formal training.

Video equipment systems have nearly reached the stage of instamatic, or disk, one-step photography.

As a matter of fact, video recorders have become as much of a consumer item as TV sets were twenty-five years ago. *Time, Newsweek,* and daily newspapers contain ads for home videotape recorders at prices as low as several hundred dollars.

Video is here to stay. Its popularity and use seem likely to increase in the future.

Use of Broadcast Television

Television used in classroom instruction comes from one of two sources: broadcast or closed circuit. *Broadcast* television refers to programming that is available from commercial and public television stations and that is sent through the air. The primary sources of broadcast television are the four major networks: ABC, CBS, NBC, and PBS.

Closed-circuit television usually refers to programming that travels through a wire rather than through the air. Many schools are equipped with closed-circuit systems that allow programs to be sent from one room in the building to another, or from one building to another. When a teacher plays back a videotape using a videotape player and a television set, the signal is sent to the receiver from the video player through a wire. This is a simple example of a closed-circuit television system.

Cable television is also sent through a wire, from the cable company to subscribers. While this is actually closed circuit, a new term has been coined, called *cable casting,* to more accurately describe the cable television process. For cable casting, the signal is sent to subscribers over a wire rather than through the air, but because cable programs are available only when the cable company decides, the use of cable television in the classroom is limited, just as the use of broadcast television is limited. While it is possible to record broadcast and cable programs for later classroom use (off-air recording will be described later in this chapter), additional equipment and planning are necessary.

Most public television stations publish comprehensive program guides that list all programs they will broadcast. Often, information on how the program can be used by the teacher is included.

Sometimes programs broadcast during school hours may be used by teachers. To be aware of what broadcast programs are available for classroom use, a teacher should routinely study television station program guides. Standard guides list when shows are available. Most public television stations publish comprehensive schedules of their shows. These schedules can be obtained by ordering them from the station.

In order to present a broadcast program to a class, a teacher needs only a standard television receiver. A *television receiver* is a set that can receive only radio signals of predetermined frequencies called channels. There is a VHF band (for *V*ery *H*igh *F*requency), with twelve channels (2 through 13), and a UHF band (for *U*ltra *H*igh *F*requency), with 70 channels (14 to 83).

Television signals are broadcast through the air or through wires on modulated signals called radio frequencies (RF). Standard television receivers, such as those sold for home use, are designed to receive these RF signals and to convert them into pictures and sound. This conversion process is called *demodulation.*

Showing a Live Program with a Closed-Circuit System

Often it is effective to use video equipment to bring live events that are *occurring in one area of the school to another* over a wire. This is one example of a *closed-circuit* television system. A play that is sold out could be shown in another room over TV while it is being performed. A guest speaker's speech could be carried live to several classrooms for students to see and hear, or lunch rooms could be monitored in the principal's office with a TV camera and monitor. These are all examples of closed-circuit television.

Basically, closed-circuit programming is the delivery of video and audio signals through wires. For this type of closed-circuit system, a television monitor, a video camera, a microphone, and connecting cables are needed.

TV Monitors A TV monitor looks like a standard TV receiver, but is actually quite different. Monitors cannot receive RF signals of very high frequency (VHF) or ultrahigh frequencies (UHF). Instead, they receive unmodulated, lower-frequency video and audio signals over wires. These signals are sometimes called *raw video* and *raw audio.*

Video cameras convert light into electrical signals that may be viewed with a television monitor. A TV monitor converts electrical signals from a video camera back into visible images that can be seen on its picture tube.

A closed-circuit system being used in a school assembly.

A video monitor can receive only video signals that have not been modulated (compliments Sony Corp.).

OPERATE

Setting Up a Standard Color Television Receiver

You need: a TV receiver with a built-in or rooftop antenna

Procedure

1. Plug in and turn on your receiver.
2. Extend the built-in antenna, or connect an external antenna to the VHF antenna posts. Be sure that the two antenna wires do not touch each other, or anything other than the antenna posts.
3. Set the channel selector to a station in your area. Turn down the sound.
4. Use the fine-tuning knob to obtain as clear an image as possible. Turn off the automatic fine tuning control before doing so. Turn the television's fine-tuning knob until a herringbone pattern appears on the screen, then tune back until this pattern disappears.
5. Adjust the color and hue controls. Color should be turned up until the colors on the screen look good, or begin to saturate. Hue should be adjusted so flesh tones look correct.
6. Turn the automatic fine tuning and color-lock controls back on.
7. Place the receiver where it can be seen easily by all students, generally the front corner of the classroom. If there are windows along the side of the classroom, the position near the windows is best. Be sure the light from the windows or from classroom lights does not fall on the television screen and wash out the image. The optimum viewing angle for a television image is about 90°, or 45° either side of a line extending perpendicular from the screen. It is possible to have students sit in an arc up to 120°, if necessary. No one should sit closer than about five feet or further than sixteen to twenty-four times the screen's vertical size.
8. Adjust the volume. If you are doing this when the classroom is empty, turn it a little louder than seems correct. When twenty to thirty students are seated in the room the sound will not carry as well as when the classroom is empty.

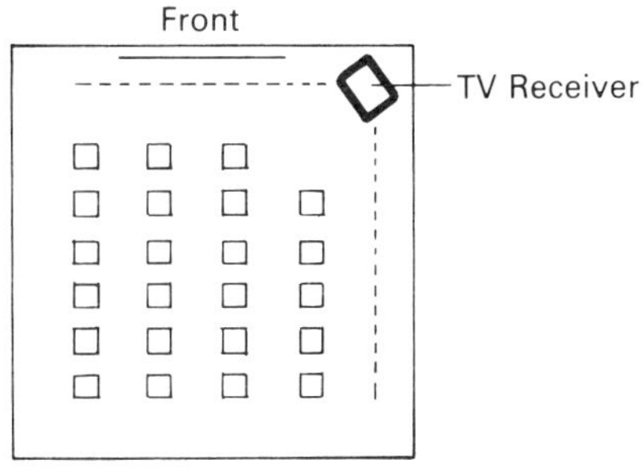

Correct location for viewing TV in a classroom.

It is possible to obtain television sets that can function as either a receiver or a monitor. A receiver/monitor can process RF signals that are broadcast, or those that are not broadcast but are received over a wire as unmodulated signals. TV receiver/monitors are often used in closed-circuit systems.

To view images and hear sound, two input devices are needed in a closed-circuit system. They are a video camera and a microphone.

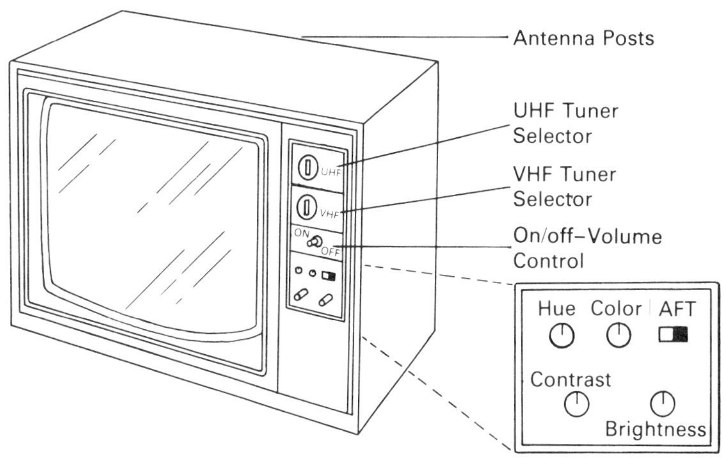

TV receiver.

This Sony Model 1640 color video camera is a portable, one-tube, high-quality camera (compliments Sony Corp.).

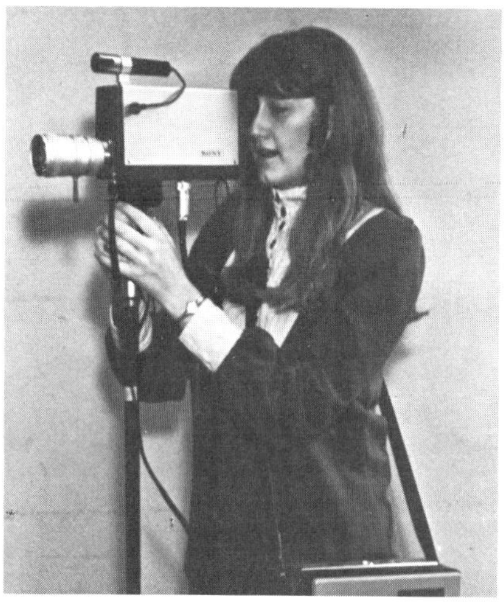

This monochrome camera was first introduced more than ten years ago, but is still being used in many educational and training applications.

An electronic viewfinder.

Video Cameras Many types of video cameras are available. Most cameras used in educational recording are the single-tube type. These cameras have a component called a vidicon tube that is light-sensitive and reacts to light that is reflected or emitted from objects at which the camera is pointed. The vidicon tube is similar to photographic film. More expensive color video cameras use three tubes, one for each of the primary colors.

The purpose of the camera tube is to change light into electronic signals that can be sent through a wire. In single-tube systems, a special color filter produces the color information for the vidicon tube to transfer. Generally, the quality of the image picked up by these simple cameras is excellent, even under low-light conditions.

Remember the term *transducer,* used in Chapter 10? Review the definition, and think about its applications in the case of a video camera.

In recent years the cost of all video equipment, including cameras, has declined significantly and, until recently, there was a significant difference in cost between color and monochrome cameras. When they were first introduced, color cameras cost at least $10,000. Gradually, prices have declined until it is now possible to purchase a color video camera for as little as 15 percent to 20 percent more than monochrome cameras. Since almost all new video recorders now have color recording capability, it makes good sense to closely examine color video cameras when purchasing equipment.

While there are several slight disadvantages to color cameras besides their cost, such as slightly poorer operation in very low light situations, and the need to match camera color temperature setting and light source, in the future most video cameras will probably be color units.

Video Camera Features: Viewing Systems Generally there are two types of video camera viewing systems—electronic and viewfinder. Electronic viewing systems are by far the most preferable and most widely used. Viewfinder, external view systems merely provide the camera operator with a framing mechanism and, while greatly reducing the cost of a camera, they also severely limit its usefulness.

Electronic view systems, which actually are small 1.5-inch TV monitors, show the operator exactly what is being recorded.

Electronic viewfinders make video cameras comparable to single-lens reflex cameras used in still and motion photography because they provide the camera operator with an exact view of what the camera is recording. Some electronic viewfinders also function as playback monitors. This capability is very convenient for locating starting points on tapes, or for verifying that a signal has been recorded when a regular television monitor is not available. Many viewfinders also contain light-level and recording indicators.

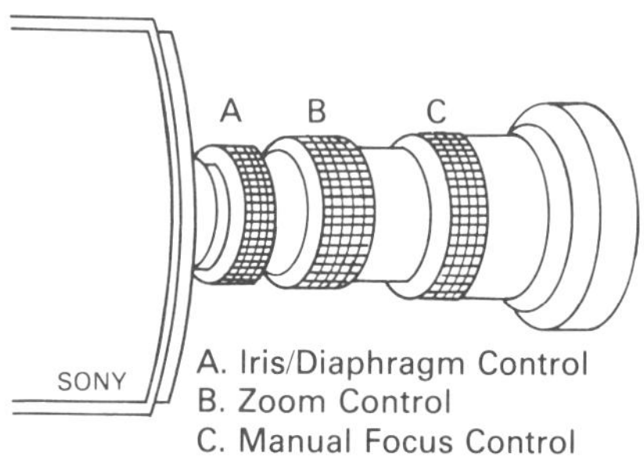

Typical video camera lens.

Zoom Lenses A zoom lens is actually many lenses in one. Three-to-one (3:1), four-to-one (4:1), and six-to-one (6:1) zoom lens ranges are common for video cameras. This means that an object can be made three times (3x), four times (4x), or six times (6x) larger merely by adjusting the zoom lens. Zoom lenses are convenience devices provided so that the operator will not have to change lenses or move the camera between scenes. Remember, good motion-recording techniques require using the camera to capture events as the eye would and, since eyes do not zoom, it is not a good idea to overuse the zoom control during recording. The zoom control is most appropriately used between, rather than during, scenes. When using a camera with a zoom lens, the camera operator should remember to *critically focus* the camera, just as is done with a movie camera with a zoom lens. To focus the camera critically, the operator should point the camera at the subject, zoom in to an extreme close-up of the subject, and then focus. Then when the operator zooms back to correctly frame the subject, the camera will stay in focus.

Color-Temperature and White-Balance Controls Color cameras must be adjusted for specific lighting circumstances, just as photographic film must be matched to a light source. Otherwise, the recorded images will have an unrealistic tint. Different light sources have different color temperatures, so the video camera must be adjusted to match the lighting. Usually, this is accomplished by selecting the correct setting on the camera's control unit (CCU). CCUs are usually small boxes to which the camera is connected before the box is connected to the recorder. Sometimes CCU controls are found on the camera body itself.

For accurate color, an additional adjustment is required on many cameras. The *white-balance* control consists of two knobs, one labeled "red" and one "blue". These knobs permit calibration of the camera to what white should look like under the prevailing lighting conditions. Once white balance is properly set, other colors will also appear correctly. To manually set for white balance, the camera should be pointed at a large white object. Then the red and blue controls should be adjusted until the

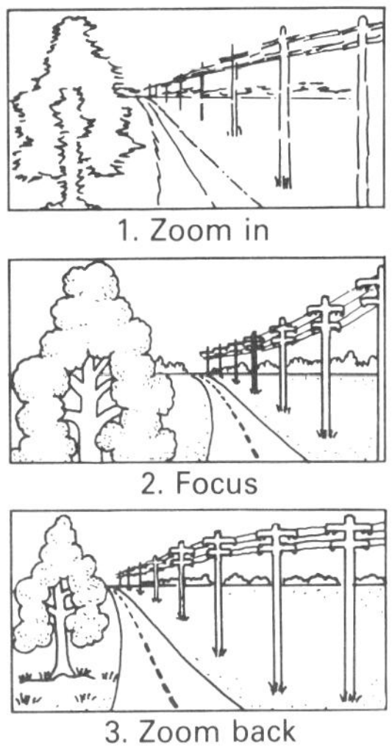

To focus critically.

Color Temperature Chart		
Designation	**Situation**	**Approx. Temp.**
	Cloudy Day	7000 K
	Sunny Day	6000 K
	Fluorescent Light	3200 K
	Incandescent Light	3000 K

Color temperature chart.

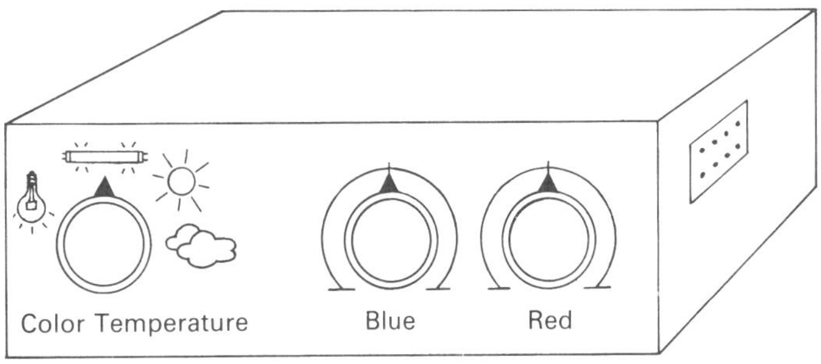

Color-temperature and white-balance controls on a camera control unit.

camera's white-balance meter registers correctly, according to the camera's operating manual. Remember, white balance and color temperature must be reset every time the camera is moved from one lighting situation to another.

Microphones Nearly any standard microphone can be used to capture sound for a video program. Naturally, better microphones produce better sound. Many video cameras have built-in microphones, so that when the camera is turned on to record pictures, the microphone is also activated. Quality of sound is usually poor from these built-in microphones, however. Despite the unidirectional pick-up pattern of built-in microphones, they do not do a good job of capturing sounds generated more than a few feet from the camera. A hand-held or stand-mounted microphone should be used whenever possible (see Chapter 10 for a detailed discussion of microphones).

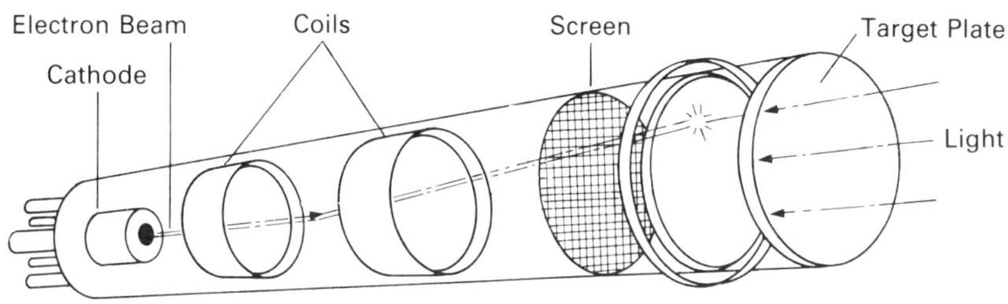

A camera pick-up tube.

Use of an external microphone.

TEACHING WITH PRERECORDED VIDEOTAPES

One advantage of the video medium is that it is possible to record live programs for later use with a video recorder and videotape. These programs can then be saved and played back when their content best fits the needs of the curriculum. Video recording greatly increases the flexibility of the video medium.

Both commercially and locally produced videotapes can be used in teaching. To play back a prerecorded videotape, two additional components must be added to the video system. They are the videotape and the videotape player. This section explains how videotape works, the physical make-up of videotape, and videotape formats. Then, the videotape player will be examined.

OPERATE

Using a Video Camera

You need: a video camera, a tripod, a monitor

Procedure

1. Mount your camera on the tripod.
2. Connect the camera to the camera control unit (if needed), and connect into the proper power source. Turn the power on.
3. Focus the camera and adjust the aperture so a clear image can be seen in the viewfinder. Do not point the camera at windows or bright lights.
4. For color cameras, adjust the color-temperature control for the correct type of light that is illuminating the scene. Set the white balance to a maximum reading, if the camera has a white-balance control.
5. For closed-circuit viewing, connect the camera to the TV monitor. Use a video cable. Connect from the *video out* jack on the camera or the camera control unit to the *video in* jack on the monitor. If a receiver/monitor is being used, be sure it is in the monitor mode. There should be a switch marked either *LINE* or *MONITOR* that should be on.
6. Turn the TV monitor power switch on. Adjust image on the monitor, or in the camera viewfinder, or both, for proper contrast, brightness, color, and hue.
7. If the camera has a built-in microphone, connect the *audio out* jack on the camera (or the camera control unit) to the *audio in* jack on the TV monitor. Adjust the sound level using the TV monitor's volume control. An external microphone can be used, also. Connect it to the *audio in* jack on the TV monitor.

Videotape

As is true with most forms of media, the software (videotape) is the determining factor used to define various categories or types of video equipment. How videotape works in and fits a particular machine determines how that piece of video equipment is described. For example, one of the most popular video formats is the *¾-inch U-matic cassette.* This format, or category, of video equipment derives its name from the fact that the videotape used in it is ¾ of an inch wide, is stored in a cassette, and is automatically threaded in the machine in a U-shaped pattern. Other systems derive their names similarly.

Videotape Make-up Videotape can be magnetized because its surface is covered with the oxides of magnetizable metals (usually iron or chromium). Actually, videotape consists of several layers of material. The most important are the magnetic oxide, the plastic backing, and the adhesive that binds the two together.

Three popular formats of videotape used in training and education are the ½-inch open reel, the ½-inch cassette, and the ¾-inch U-matic cassette.

Videotape Tracking Patterns In the early 1960s, a major development in videotape recording was introduced. Until that time video equipment was heavy and cumbersome, and used large-sized videotape. *Helical scan* video recording changed all of that. Also called "slant-track" recording, helical scan (from the word *helix*) involves the placing of magnetic messages across the oxide surface of the videotape diagonally, thus allowing for the placing of longer tracks of message information on narrower widths of magnetic tape.

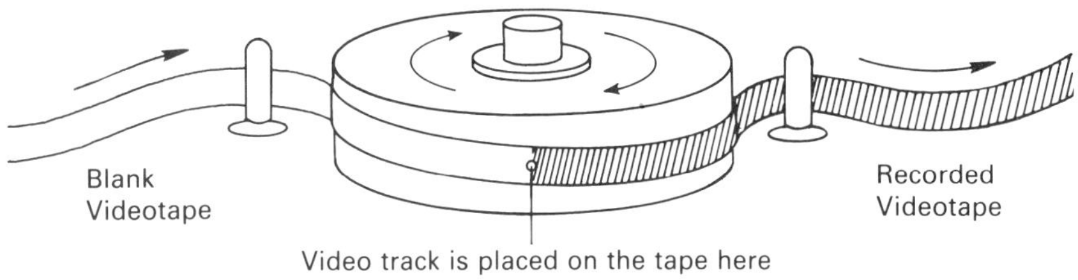

Helical scan recording.

Smaller tape reels, simpler motor drive systems, and fewer machine components resulted in more portability and lower cost, necessary requirements for educational and consumer use of video. Today, nearly all noncommercial, nonbroadcast video systems have helical scan tracking patterns.

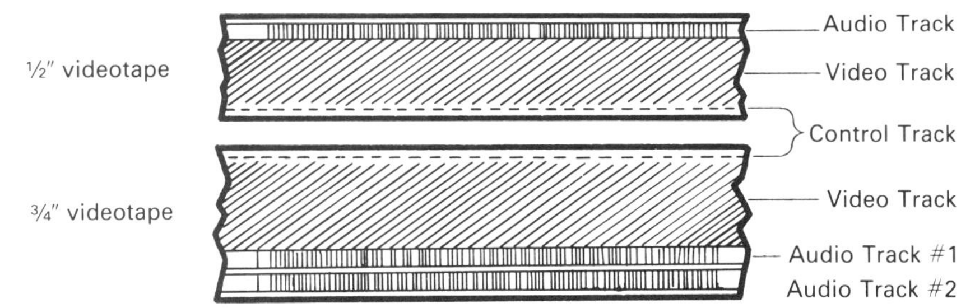

Tracking pattern refers to how the magnetic signals are placed on the tape. There are three tracks on most recorded videotapes—the video track, the soundtrack, and the control track.

As stated above, videotape is a storage medium. It is like movie film in that respect, except that information is stored magnetically on videotape rather than chemically as it is on film.

Film also has three components, and they are similar to the components of videotape. Video recorders place thirty "frames" of visual information on the videotape every second during recording. While not visible to the naked eye, these "frames" are similar to the twenty-four frames recorded on film each second by a 16mm movie camera.

Actually, on a videotape each frame is a single magnetically recorded image. During recording, the videotape recorder's recording head places magnetic tracks on the videotape. During playback each of these tracks, or frames, is read from the tape by the playback head, and transferred to the television receiver. The TV's picture tube, or cathode ray tube (CRT), shows sixty separate partial images, or sweeps, per second. Because more of these sweeps occur per second than the eye can distinguish, they all blend together and motion is perceived by the viewer.

The videotape also has an area reserved for audio information. The *audio track* is separate from the video track, and is placed on the videotape by an audio recording head. Many newer recorders have two of these soundtracks so that two sound sources can be placed on the tape (e.g., English and Spanish versions of a program). Actually, the audio portion of the videotape is very similar to a simple audiotape recording.

The third track placed on a videotape is the *control track.* The control track is a series of magnetic dots placed on the videotape during recording. The control tracks provide for the correct alignment of tape and machine during playback. The control track has a function similar to that of the sprocket holes in film. If there is an unsteady, bouncing image seen on the screen during projection of a film, it is often because the projector's sprockets and the film's sprocket holes have become misaligned. The same is true for video players. If the image viewed on the television receiver is bouncing and unsteady, it is often because the video player has not correctly aligned itself to the control track on the tape. Tracking problems can often be manually corrected by adjusting the player's tracking knob.

Video Formats The format of a medium is usually determined by its physical size and method of packaging. There are three video formats in general educational use and one system used almost exclusively by commercial TV stations.

The ¾-inch U-matic cassette, the ½-inch cassette, and the ½-inch open reel are the formats in widest use in education. Additionally, there are two types of ½-inch cassettes, the Beta format and the VHS format. The 2-inch Quadraplex system is limited almost exclusively to commercial video-recording operations.

A general rule, that the size of the software used to carry messages will determine the technical quality of that message, all other things being equal, partially determines why commercial producers of videotaped programs use tape that is 2 inches wide, and why educational and home video systems use narrower formats. Videotape recordings made commercially for broadcast must be of the highest quality. For educational and local production, however, recording quality is not nearly as important as are cost, portability, and ease of use. In teaching and training, small-format video systems are most suitable.

The video format comparable to 16mm film is probably the ¾-inch U-matic cassette. Introduced by the Sony Corporation in the early 1970s, and incorporated as a standard by other manufacturers, the ¾-inch cassette became the first automatically threading educational video system in widespread use. It was the standard until the late 1970s. Advanced component technology made the ¾-inch cassette system capable of recording nearly as high a quality message as some 2-inch systems. As a matter of fact, the ¾-inch cassette system began to be used by many commercial TV stations for their electronic news gathering (ENG), or minicamera systems. It is also the format used by many video libraries which distribute programs for rent or loan.

While the ¾-inch cassette system revolutionized the local production of videotapes, it was the development of the ½-inch cassette that made local production video recording truly widespread. One-half-inch videotapes are inexpensive, equipment seems to become less expensive almost daily, and technical quality is nearly as good as that found in the original ¾-inch systems.

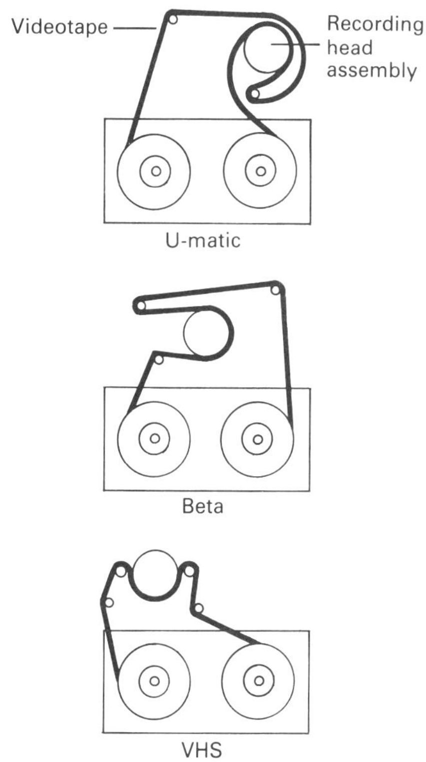

Threading tracks for video recorders.

The first ½-inch systems were introduced in the late 1960s as an open-reel format. Videotapes used in these systems had to be manually threaded, and images were recorded in black and white. While open-reel systems are no longer being purchased in large quantities, it is interesting to note that many are still in use. The positive results many early video users had with their ½-inch open-reel machines helped to popularize video recording in many schools and training centers.

Today, the most popular ½-inch video formats are cassette systems. There are two primary, noncompatible ½-inch cassette formats, the Beta (basically manufactured by Sony) and the VHS (produced and marketed by several manufacturers). Both use a videotape that is ½ inch wide and packaged into a cassette, but because the physical sizes of the cassette cases are different, and because the recorders function differently, the two formats are not compatible. It is impossible to record on a Beta cassette and play it back on a VHS machine and vice versa.

There are other differences between U-matic, Beta, and VHS formats. The speed at which videotapes move through each type of video recorder varies. All ¾-inch cassettes have a fixed tape speed. This contributes to the compatibility of videotapes recorded on ¾-inch machines, regardless of the manufacturer. However, the standard tape speed of ¾-inch systems limits the maximum recording time of a single cassette to sixty minutes. Beta and VHS machines have the capability to vary tape speed, which greatly increases the maximum recording time for a single videotape. The longest VHS cassette recorded in the super long play (SLP) mode can store up to nine hours of programming.

Presently, because of its ease of use and low cost, the ½-inch cassette is the prevalent local production format, comparable to the super 8mm format used for local production film making.

Care of Videotapes

- ☐ Store in a cool, dry place.
- ☐ Avoid leaving tapes inside cars on hot days.
- ☐ Avoid any kind of heating unit.
- ☐ Avoid direct sunlight and high humidity.
- ☐ Avoid dust, dirt, and moisture.
- ☐ Never touch the tape.
- ☐ Avoid magnetic fields (motors, transformers, magnets, etc.).
- ☐ Avoid physical shock.
- ☐ Never store tapes only partially wound. It's best to "tight wind" your tapes by allowing the VCR to continue in PLAY mode until the end of the tape is reached. The tape will then be uniformly wound on the reel.
- ☐ Store cassettes in their cases, standing upright.
- ☐ Remove erase prevention tabs on recorded cassettes.
- ☐ Do not smoke in rooms with videotape or video equipment.
- ☐ Allow tapes to reach room temperature before playing.

OPERATE

VHS Cassette Repair Procedure

If the tape becomes damaged or you have to cut off a piece which has become jammed in the machine, repair the cassette tape in the following way:

1. Use a narrow-tip Phillips screwdriver to remove the screws that hold the cassette together.
2. Carefully grasp the cassette and turn it over so the label faces you.
3. Depress the front panel release catch on the left side of the cassette and open the tape protector panel.
4. While holding the spring-loaded tape protector panel open, carefully lift the top cover of the cassette off the body.
5. Unjam the tape and, if necessary, cut out damaged tape and splice the remaining good tape together or onto the leader.
   a. Cut off the bad tape and overlap the ends of the good tape.
   b. Cut the tape at a right angle.
   c. Join the ends together.
   d. Carefully apply the video splicing tape.
   Note: Use video splicing tape only. Never use audio splicing tape!
6. Rewind the tape carefully—by hand—back into the cassette, making sure the tape is threaded through the proper tape guides.
7. Reassemble the cassette and replace the top cover and all the screws.

Summary Videotape recording uses tape that is covered with a metallic oxide that can be magnetized. During recording, the tape comes in contact with an electromagnet (recording head) that places a magnetic signal on the tape. After recording, the videotape stores this information, or signal, until playback, when the magnetic message on the tape comes in contact with another electromagnet (a playback head). The signal on the tape is sensed by the playback head. The process by which magnetic information is placed on a videotape, the size of the tape, and the way the tape is packaged determine the tape's format.

Video Players

Video players are less expensive than video recorders. A good way to save money is to purchase players when only playback is needed. A video recorder can also serve as a player.

Video Players: How They Work To view a prerecorded videotape, a device called a video player is needed. Video players, simplified versions of video recorders, have a set of playback heads that can read video and audio messages from magnetically recorded videotapes. The videotape's control track aligns the tape properly so that the video head can interpret the video signal on the tape, and the audio head can receive the audio signal from the audio track. Video players have audio and video output jacks so they can be connected to an output device such as a television monitor or receiver.

Video Player Formats There are video player formats to match the various videotape configurations. The ¾-inch U-matic cassette player plays ¾-inch U-matic cassettes. The Beta machine plays ½-inch Beta cassettes. The VHS-format machine plays VHS cassettes, and open-reel machines play back

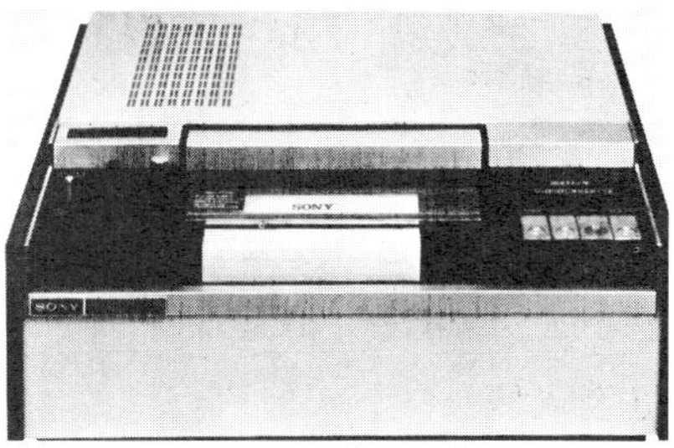

Video players can play back prerecorded videotape only. They are generally less expensive . . .

than video recorders because they have no recording head (compliments Sony Corp.).

manually threaded ½-inch tapes on open reels. Since these formats are not interchangeable, the correct format player must be used to allow viewing of a recorded videotape.

VIDEO RECORDING SYSTEMS: PRODUCING A VIDEOTAPE

The real value of the video medium is that it is possible to produce programs locally either by recording them off the air, or by recording them live on location or in a studio. Since video recording involves the use of more complex technology, a more detailed description of the video recording system and its components is needed.

An understanding of how videotape is able to store a message provides the video user with the basics for understanding how video equipment systems work. Generally, the purpose of the video recorder is to place a magnetic signal on the videotape. The role of the video player, which is often the same machine as the recorder, is to retrieve a message from a prerecorded videotape. A video equipment system consists of all the pieces of hardware that support this basic process.

Individual pieces of video equipment can be divided into one of three categories—input devices, storage/retrieval devices, and output devices. Connectors are used to tie individual pieces of equipment to one another.

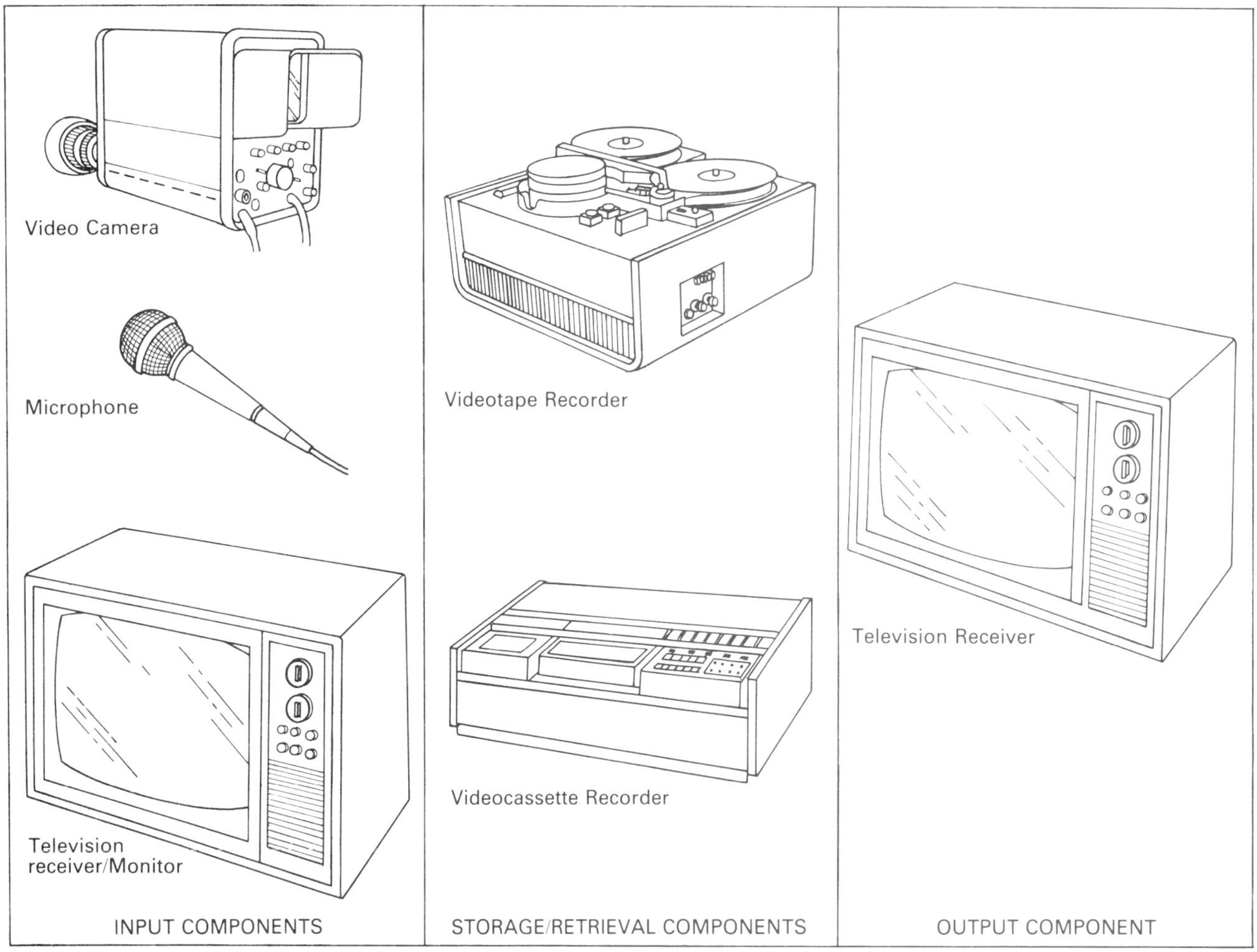

Video equipment recording system components.

OPERATE

Preparing a Video Player and Television Receiver for Viewing of a Videotape

You need: a video player, a television receiver, a videotape of the same format as the player, an RF cable with a pigtail connector

Procedure

1. Plug your video player and receiver into a power source.
2. Turn on the television and tune to a broadcast channel. Adjust the receiver for contrast, brightness, color, tint, and volume.
3. Connect the RF cable to the RF OUT on the video player and the RF IN on the receiver. Older-model receivers may require connecting the RF cable to the VHF antenna posts using a pigtail connector.
4. Tune the channel selector to channel 3 or 4, depending on the RF output of the video player.
5. Insert the video cassette or thread the videotape.
6. Turn the video player on, and depress the play, or forward, button.
7. View the program on the receiver. If the image is unsteady, or bouncy, adjust the video player's tracking control during viewing to correct this problem.

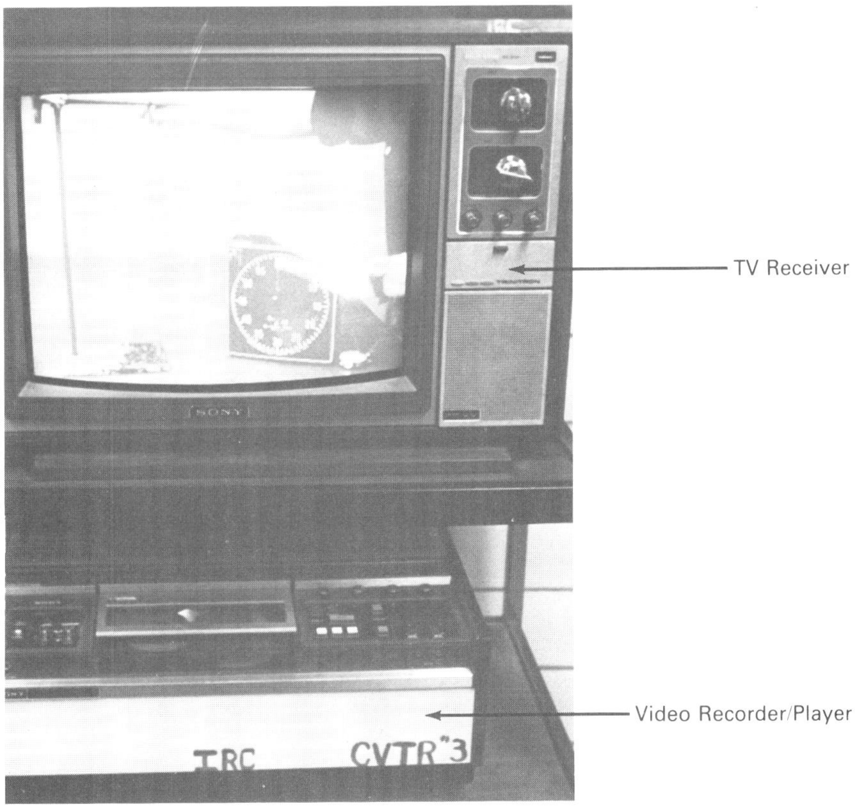

To play back a prerecorded videotape, a videotape player or a videotape player/recorder is connected to a regular television reciever with an RF cable. The RF output on the video player is connected to the VHF antenna posts on the receiver.

A video cassette player and television receiver.

If a television receiver is used, an RF cable with F connectors can be used to connect the RF output on the video player to the VHF input on the receiver.

When an RF signal splitter is used, one video player can send video and audio signals to two or more television receivers.

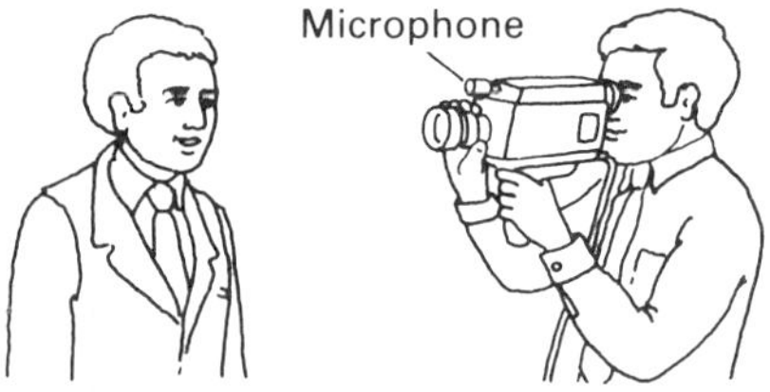

Many video cameras now come equipped with a built-in microphone.

Camera with microphone.

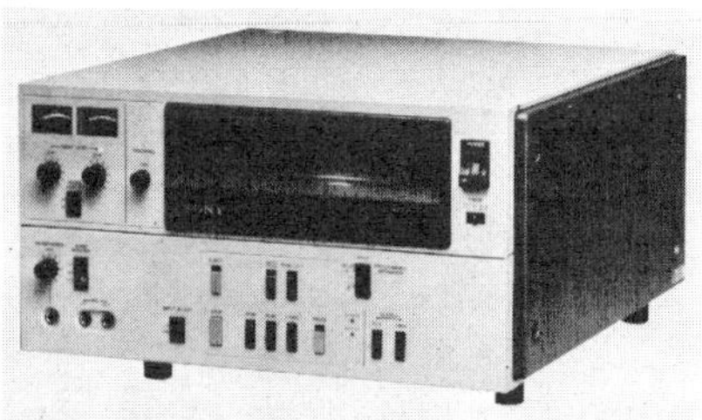

A video recorder is similar to a video player, except that a record head and record controls are added. This Sony Model VO-5600 looks nearly identical to the Sony VP-5000 video player, except for the record controls for video and audio (compliments Sony Corp.).

Video Equipment Input Components

Any piece of equipment that functions to "capture" visual or aural information for storage on videotape could be described as an input component. More specifically, inputs are devices such as video cameras, microphones, and TV receiver/monitors that are sources for information, or signals, that are to be recorded on videotape. Character generators, film chains, and video players are examples of other pieces of equipment that could be used as inputs for more elaborate video equipment systems.

Video Equipment Storage/Retrieval Components

A key component in a video recording system is the videotape recorder/player (VTR/VTP). This device receives electronic signals from input components and changes them into the magnetic information that is stored on the videotape. During playback, magnetic impulses are retrieved from the tape and converted back into electronic signals for transmission through connector cables to output components such as television receivers.

While there are various VTRs available, they all work in basically the same way. Major VTR subcomponents are:

1. *Input jacks* (where input signals transmitted through wires are connected to the recorder)
2. *Video heads* (where electronic information is changed to magnetic information for storage on the videotape and vice versa)
3. *Audio heads* (where audio signals are stored on the tape)
4. *Erase head* (where previously recorded signals are removed from the videotape)
5. *Control track heads* (where control information is placed on the videotape)
6. *Function levers or buttons* (used to operate the recorder)
7. *Output jacks* (where output signals are available for transmission to output components through patch cords)

While various manufacturers build their VTRs to work quite differently internally, the equipment receives the same type of electronic information from input components, such as cameras. These devices also deliver the same type of electronic information to output components. This makes it possible to mix equipment and to record, or dub, from one format to another. In other words, it is possible to use a Sony camera with a Panasonic recorder, or dub from a VHS player to a U-matic recorder.

Video Equipment Output Components

Output components allow recorded messages to be seen and heard. The simplest and most widely used output component is the standard TV receiver tuned to a broadcast channel so that a TV program, such as *Sesame Street,* may be seen.

A program coming to the receiver through a wire from a videotape player is another example of the closed-circuit use of a video equipment system. While either a receiver, monitor, or receiver/monitor can be used to show recorded video programs (depending on the type of signal, RF or raw, being transmitted by the player), it is more economical to use receivers to show programs. The more expensive receiver/monitors should be used primarily for off-air recording. One or two receiver/monitors may be all that are needed in a school or training center.

If a receiver is used to show recorded programs, an RF cable with a pigtail connector is used to connect the videotape player to the VHF antenna posts of the TV. If a monitor is used, both video and audio cables (or a multipin cable) are needed to connect the player to the monitor. While not often used, external speakers can be included in a video equipment system to improve sound quality during playback.

Connecting Video Components

Wires called patch cords are used to connect video components. Generally, patch cords are one of two types—those that carry either audio or video only, and those that carry both. Audio patch cords are the same as those discussed in Chapter 10. Video patch cords are similar. Usually, however, they are heavier. One of the most commonly used video-only patch cords has plugs on both ends called UHF connectors. The BNC connector is another widely used video-only patch cord. Both these cables are often called "raw-video" cables because they carry video information that has not been modulated and is considered to be "raw."

The second category of patch cords used to connect video components carries both video and audio signals. There are two types of these dual-purpose cords—*RF cables* and *multipin cables.* RF cables can be used only to connect a videotape player or recorder to a TV receiver. These cables allow for a modulated signal to be sent over a wire rather than through the air. RF cables have F connectors at both ends. Cable TV companies use this kind of cord to connect their subscribers to the central distribution point.

Multipin cables are actually several video and audio cables combined into one multipurpose cord. Most commonly used are the square 8-pin cable connector (used to connect videotape recorders to monitors) and the round 8-pin cable connector (used to connect video cameras to videotape recorders). In addition to carrying raw video and audio to and from each component, these cables often carry power and remote control operation signals.

When using video equipment, the operator must closely examine the machine's input and output receptacles, called jacks, to determine what types of patch cords are needed to connect this device to others. Generally, it does not hurt to experiment. Multipin cables fit only one way so they cannot easily be inserted incorrectly. If single-pin cables are connected improperly the system will not function. It is unlikely that patching two pieces of equipment together incorrectly will cause harm to either.

Summary

There are three components in a video equipment system:

1. *Input equipment* Video cameras, microphones, and receiver/monitors convert visual and aural information into electronic signals. This electronic information is converted to magnetic signals by the videotape recorder.
2. *Storage/retrieval* Information is stored on videotape by a videotape recorder. It converts electrical signals into a magnetic field which is stored on videotape. Information is retrieved from the videotape by the videotape player, which converts magnetic signals back into electronic information. Usually, the storage/retrieval component function is performed by the same piece of hardware.

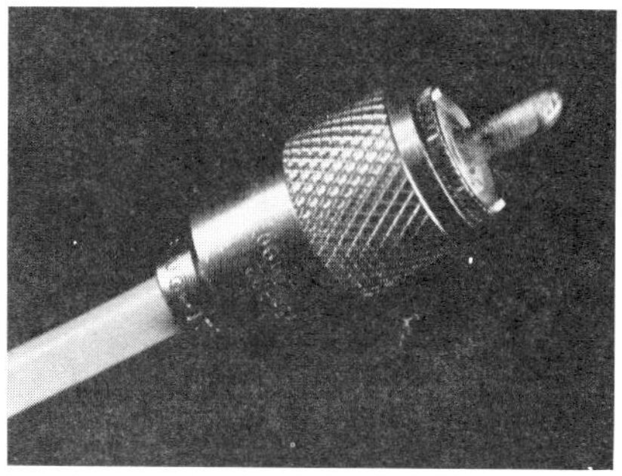

A UHF connector.

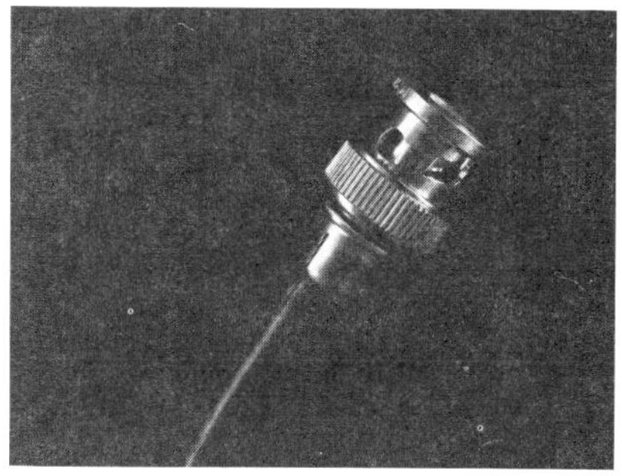

A BNC connector.

An F connector.

A square 8-pin connector.

Most home video recorders, such as the Panasonic VHS model, have built-in tuners so broadcast television programs can be recorded without the need for a receiver/monitor.

The built-in tuner makes off-air recording easy; only an antenna is needed.

3. *The output component* An example is a television receiver, which converts electronic information back into images we can see and sounds we can hear.

Cables allow all video equipment system components to be connected to one another.

RECORDING BROADCAST OR CABLECAST PROGRAMS

In order to record a broadcast or cablecast television program, a receiver/monitor is needed to convert the RF signals carrying the program into video and audio information that a videotape recorder can process.

In other words, a *receiver/monitor* is needed to receive broadcast RF signals from its antenna, or from an RF cable, and then to send this information to a video recorder in the form of nonmodulated video signals. A receiver/monitor internally changes RF signals into video. A standard television receiver can receive only modulated signals on certain predetermined channels. A monitor can process only nonmodulated video signals not on channels. A receiver/monitor can do both, and is required for most off-air recording.

Recently, most home video recorders have been outfitted with built-in tuners that act as receivers. These picture-tubeless receivers are internally connected to the recorder. With these machines it is possible to record off-air directly without using a television set, or while the connected television receiver with picture tube is tuned to a channel other than the one being recorded.

Use of Off-Air Broadcast Programs

One major problem with using broadcast programs to supplement the school's curriculum is that they are often not offered at suitable times. Most public television stations allow videotape copying of their programs by educators. However, before off-air copying is planned, station copyright regulations should be checked. Program guides will usually have detailed information about what can be copied and for how long copies can be kept. Usually, public TV stations have a very liberal copyright policy. Generally, commercial television programs, unless advertised otherwise, should not be copied without the permission of the network. Local school district media specialists should be consulted concerning the *fair-use* copying of broadcast programs by teachers and trainers.

OPERATE

Recording a Broadcast Television Program off the Air

You need: a receiver/monitor, a videotape recorder (VTR), video and audio cables, an RF cable (or a square 8-pin to square 8-pin cable that contains all video and audio connections)

Procedure

1. Plug the receiver/monitor and video recorder into the appropriate power source.
2. Tune the receiver/monitor to a broadcast station, and adjust it to obtain the best-quality picture and sound possible. The receiver/monitor must be in the TV mode to do this.
3. Connect the video and audio out jacks on the receiver/monitor to the video and audio input jacks on the video recorder using the appropriate connecting cables. A square 8-pin to square 8-pin multipin cable can be used to connect the TV receiver/monitor to the VTR instead of separate video and audio cables.
4. If you are not using an 8-pin connecting cable, connect an RF cable from the RF out on the VTR to the VHF antenna posts, or to the RF in, on the receiver/monitor.
5. If the VTR has an input switch, place it in the TV mode.
6. Turn on the VTR and thread a videotape (or insert a video cassette). Depress the record and play buttons and record the television program for a few seconds.
7. Rewind and play back the recorded program. The receiver/monitor should be in the TV mode and tuned to channel 3 (or 4) if an RF cable is being used. If an 8-pin cable connects the video recorder to the television, the VTR mode should be selected on the receiver/monitor in order to see the recorded program.
8. If the trial recording seems to be satisfactory, rewind the videotape and prepare to record the selected broadcast television program.

Note: Timers are available that can be used to automatically make a recording of a broadcast program, hours or even days in advance. Many ½-inch home recorders have these timers built into the recorder.

TRY THIS

Using a VTR and a receiver/monitor, make a recording of a broadcast television program.

Recording a Program Using One Video Camera

Single-camera recording can be accomplished with the equipment already discussed in this chapter. All that is needed is a video camera, a videotape recorder, and connecting cables. A receiver or monitor can be added so that recorded programs can be viewed during production. Both portable and studio equipment are available.

Portable video recording systems are now lightweight and easy to operate. One person can record a video program on location using one battery to power both the recorder and the camera.

Unless an electronic video editing system is to be used, a process called in-camera editing is followed during production with one-camera video systems. In-camera editing means that all scenes are recorded in the order and for the length of time that they will actually appear in the final program. All connectives are made by using the camera. For example, fades can be used between segments by turning the aperture ring on the camera to the C (for "closed") position at the end of a scene, and back to the open setting as the next scene begins. Cuts are produced by merely putting the recorder in the PAUSE mode between scenes. The slight breakup of the picture that results when this is done could be removed by electronic editing, but the imperfection may not be too noticeable.

Battery versus AC Operation

Many portable battery-operated television systems are available for videotaping where there are no 110-volt outlets. The camera, recorder/player, and television set can operate on self-contained 12-volt batteries, as well as on automobile batteries (via the cigarette lighter outlet). On their own power, these systems can be used for extended periods of time between battery charges. If standard 110-volt power is available, the system can be powered with that and the rechargeable batteries can often be charged at the same time.

PRODUCE

Steps for Planning a Video Program

1. Write a complete production script (see Chapter 9). Short programs are usually best. About 10 minutes should be a maximum.
2. Divide the script into logical subdivisions which are collections of sequential scenes. Do this by determining if some scenes can be recorded at one time without turning the recorder off. Generally, the less frequently the recorder has to be turned off and back on again, the better.
3. Once the subdivisions have been identified, time them. The length of each subdivision of the script can be determined by reading the narration, or by rehearsing the action. Write the time for each subdivision on the script.
4. Identify the connective to be used between each subdivision. Use fades between major segments and cuts between scenes that are more closely related to one another. Fades can be used as points to fall back to during production if a segment bombs out and has to be reshot. Title scenes make excellent transitions.
5. Collect all props and have actors rehearse. If sound is to be recorded, be sure everyone knows his or her lines.
6. Record the program, one subdivision at a time. Rehearse each segment just before it is to be recorded. Remember, programs produced like this have a tendency to drag, so keep things moving.
7. Before putting the props away and dismissing the actors, review the entire program. If necessary, reshoot it. Use a different videotape. The first take may still turn out to be the better one. It may be necessary to live with some minor problems in the program.

OPERATE

Recording with a One-Camera Video System

You need: a video camera, a tripod, a videotape recorder, a receiver or a monitor, a microphone on a stand, audio and video connecting cables, a videotape or cassette, a studio (classrooms work nicely), and, if possible, floodlighting

Procedure

1. Connect the camera or the camera control unit to the videotape recorder. The camera should be mounted on a tripod, preferably one with wheels.
2. Connect the video recorder to the receiver, or to the monitor, using the appropriate connecting cables.
3. Connect the microphone to the microphone input on the videotape recorder.
4. Set the video recorder tape counter to zero.
5. Turn all components on.
6. Set the input selector on the video recorder to CAMERA or LINE.
7. Depress the RECORD button until it locks into the down position. The red recording light on the video recorder should come on, and a picture should be visible on the television. Adjust the audio input level on the VTR while an actor talks. Adjust the camera's color-temperature settings and white balance. If the video recorder has a video level, adjust it to the proper setting.
8. Record a few seconds to check equipment operation. To do this, hold the RECORD button down and turn the PLAY or FORWARD control on. Picture and sound will be recorded simultaneously. Play back the sample tape to be sure everything is in working order.
9. Cue the actors and begin recording. To pause between scenes, depress the PAUSE button or move the video recorder's function lever to the PAUSE position. Do not leave the video recorder in the pause mode for extended periods.
10. To stop recording, push the STOP button, or turn the function lever to the STOP position.

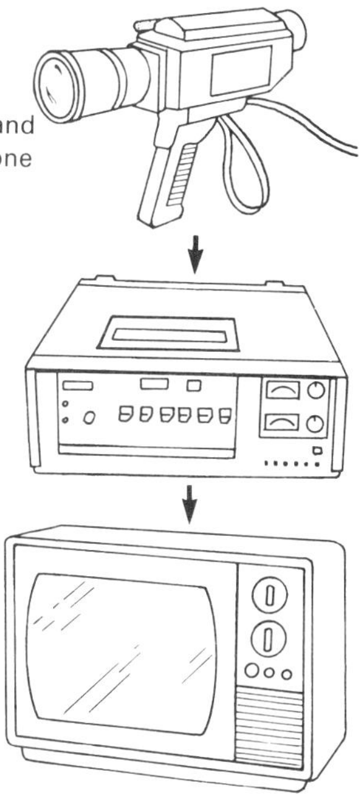

A one-camera video system.

TRY THIS

Plan a short video program on some topic in your subject area. Produce the program using a one-camera video recording system. After recording, analyze it. Be critical of the pace of the program, camera movement, and connectives between scenes. Try to identify why some portions of the program are better than others.

ADVANCED PRODUCTION: STUDIO AND LOCATION VIDEO RECORDING

Producing a program on videotape involves more than just arranging for equipment, buying videotapes, and writing a script. The production process itself is fairly complex and should be planned carefully. Generally, there are three phases to the video production process:

Phase I: Prerecording Production
Phase II: Program Recording
Phase III: Postrecording Production

Phase I: Prerecording Production

Television is distinctive when compared to other media because other media can be used as part of a TV program. Artwork and titles in the form of flat-mounted or photographic visuals can be, and often should be, used to make a video program more effective.

When graphics to be used as part of a video program are produced, three important factors should be considered: aspect ratio, image bleed area, and image resolution. If black and white video equipment is used, the contrast between different colors must also be considered before recording.

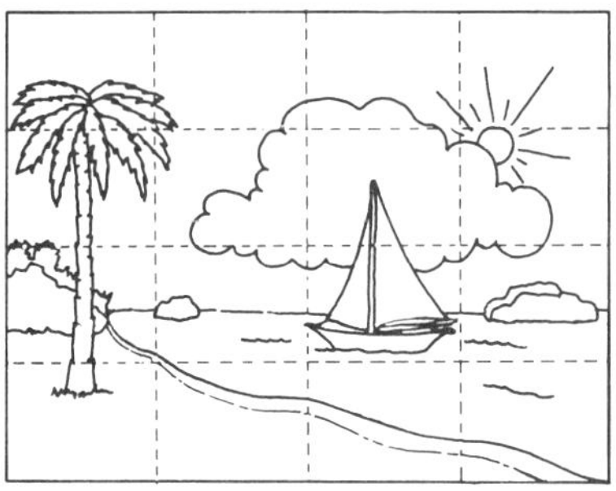

Television screen aspect ratio.

Aspect Ratio Aspect ratio refers to the height-by-width ratio of the television image. For a standard TV screen this ratio is three units high by four units wide (3 × 4), the same as for 35mm filmstrips, but different than that for 35mm slides (2 × 3). Graphic materials that are to be copied on videotape and displayed on a TV receiver should be designed to fit this ratio. If 35mm slides are used to project graphics for recording, the photographer must remember that large margins must be left around the image or information will be lost during recording.

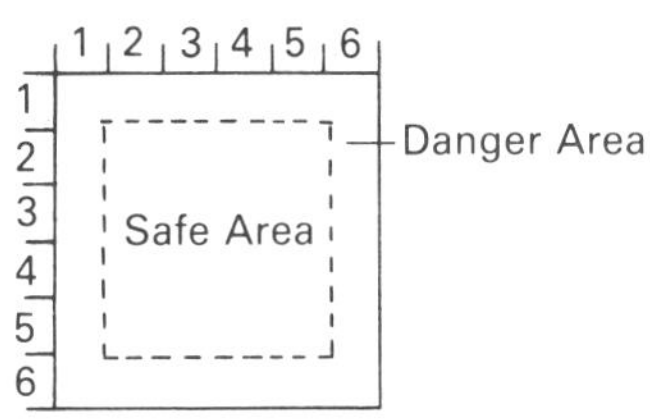

Television screen bleed area.

Bleed Area Bleed area of a TV screen image refers to that area around the margin of the image that is often lost by some TV receivers. One-sixth of the margin of each side of a graphic should not carry essential information because some televisions are misaligned slightly and it is possible that this information will not be seen during playback. To be safe, important parts of any visual should be kept in the middle two-thirds portion of the frame.

Image Resolution Image resolution, or image quality, refers to how much detail can be seen on a picture tube. The U.S. standard is stated in terms of *lines of resolution,* and 525 is the standard. This number was originally derived from how image lines were projected on the face of the picture tube (CRT). The video images we see are actually produced by a single cathode ray emission device inside the picture tube. This ray emission device sweeps horizontally sixty times per second across the rear of the tube face. A standard U.S. television should have 525 of these horizontal sweep lines to make up the image seen on the screen. (Most color images are produced by various intensities of three color dots (red, green, blue) projected on the back of the picture tube face.)

Unfortunately, not all commercial receivers have this high an image quality in actual practice. When preparing lettering and graphic information for TV copying it is much better to use several boldly lettered scenes than one with considerable detail. Graphics for video programs should:

- ☐ Use clear, bold lettering (not typing).
- ☐ Have a minimum letter height of 1/20 to 1/25 of the height of the screen.
- ☐ Be limited to lettering of three or four lines of three or four words each.

Contrast and *colors* are important concerns if monochrome (black and white) equipment is used because in monochrome sets colors are converted to various shades of gray. Some colors that appear very different when viewed in color will appear to blend together when viewed in black and white. Red lettering on a green background would appear as the same shade of gray when viewed on a monochrome receiver. Whenever there is even a remote possibility of using programs with black and white TV sets, all objects to be recorded, especially graphics, should be viewed through a black and white video system first. A chart providing a general idea of what colors look like when viewed in monochrome can be used for planning. However, it is best to actually view colors before recording.

Use of Prerecorded Materials Before recording of a video program begins, all graphic and title material should be prepared. During recording, an easel can be used to hold visuals. As display items are used during production, they can be flipped forward, out of the camera's view. This can even be done during recording.

Another easy way to use graphics is to prepare them as slides. A standard carousel slide projector can be used to project slides for video copying. The screen onto which the slides are projected should be in as dark a location as possible. Projecting into a box is one way to solve the lighting problem in a brightly lit studio. If slides are to be changed on-camera, they should be loaded into two trays in an alternating sequence, and projected using a dissolve unit. The dark period between slides projected without the use of a dissolve unit is quite distracting when viewed on the television screen.

Monochrome Shades for Colors

Color	Monochrome
Dark blue, green, brown	Dark gray or black
Light blue and pastels	White
Bright red, blue, green	Medium gray

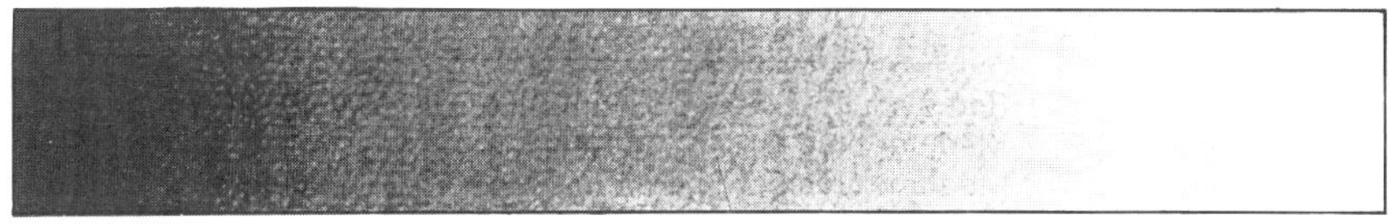

Monochrome shades of color.

Phase IIA: Program Recording in the Studio

Live video programs are recorded either on location in the field, or in a studio. Naturally, a television studio setting is preferable because it is easier to organize. However, many programs require the producer to take the portable recorder and camera right into the plant, the classroom, or the laboratory.

Easel and camera.

Video recorders, such as this open-reel recorder, have been available at relatively inexpensive prices since the mid-1960s.

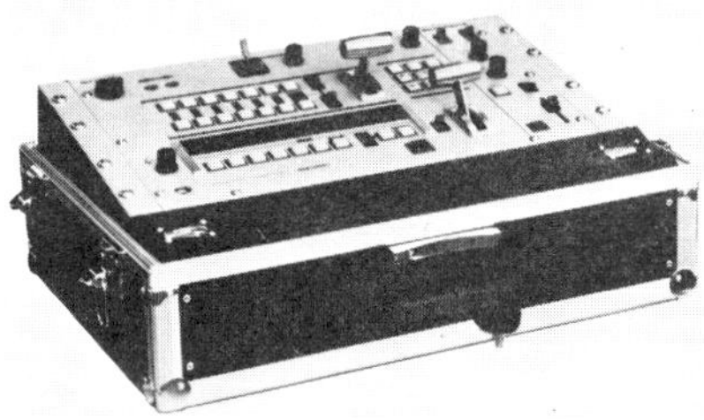

A switcher, such as this Sony Model SEG-2000, makes it possible to switch between several cameras to make a more professional-looking video program (compliments Sony Corp.).

Relatively inexpensive black and white studio switching systems have been available for over a decade.

Several cameras can be used with a switcher in a studio. Long shots, close-ups, and the title inserts can be combined in a sequence by merely pushing a button on the switcher.

Studio recording is usually easier for the producer, because the studio is a controlled environment. Whether the studio is a fancy one with a control room, boom microphone, and permanently wired equipment, or merely a converted classroom, at least the video producer is in charge. In the field this may not be the case. Several categories of preparation should be considered to ensure successful studio recording. They are the preparation of the video equipment, the lighting equipment, and the sound equipment, and the working and acting personnel.

Equipment Video equipment used in the studio should be checked out to be sure it is working properly before the production crew is assembled. While a one-camera system, with a camera and a recorder only, can be used in a studio, it is a good idea to obtain some kind of video switching capability, especially if the studio is to be used very often. A video switcher, or special-effects generator (SEG), allows two or more video cameras to be in use simultaneously. With the SEG it is possible to cut or dissolve from camera to camera while recording. This makes it possible to have different video cameras cover different portions of the action. For example, one camera could be on the entire scene (long-shot camera), one could be used for close-ups of action, and one could be used for graphics. The program producer is then able to switch from camera to camera and control which portion of the action is recorded.

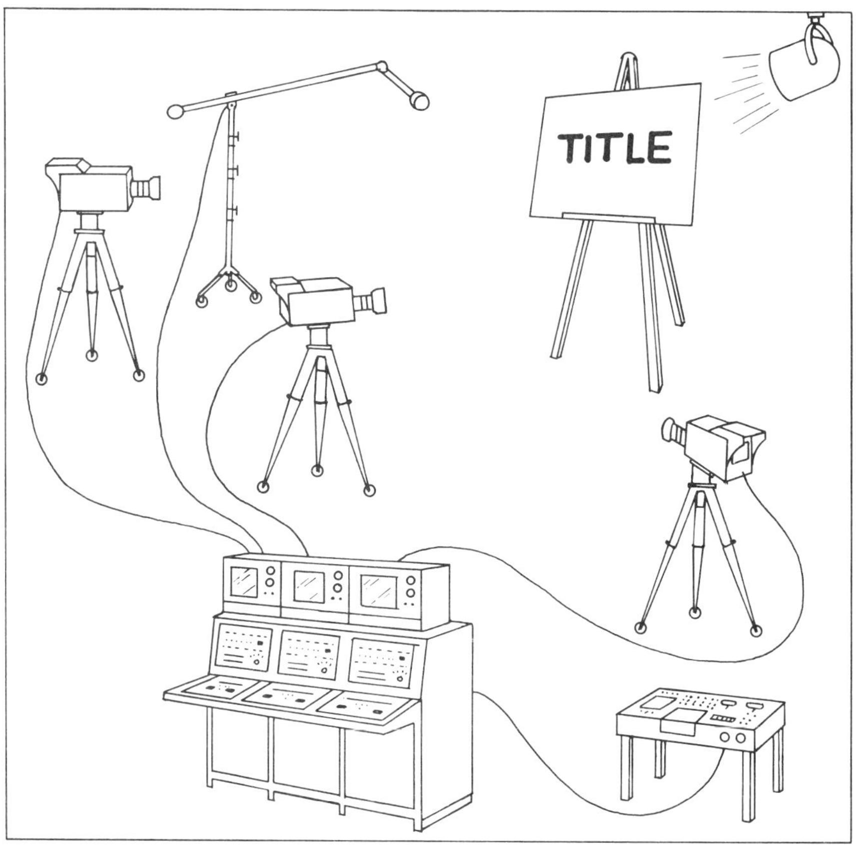

Simple television studio.

Lighting Lighting of a video studio is also an important consideration for good productions, especially those done in color. Optimally, four lights should be used (see Chapter 9, Still Photography). Key, fill, accent, and background lights should be used if possible. The color temperature of lights used is also important. Video camera white balance should be set to match the type of lighting used in the studio. Generally, fluorescent or incandescent lights normally found in rooms such as laboratories or classrooms are not going to give good-quality color or contrast to objects illuminated by them. For high quality in the final production, it will pay to use studio floodlights. Even the lights from several overhead projectors pointed at the set can vastly improve the quality of a video recording. If special light tripods are not available, lights can be clamped to microphone stands or even furniture.

Microphones Microphones can be either boom-mounted, lapel-type for each actor, or shotgun types that are pointed at actors as they speak. Remember, the closer the microphone is to the sound source, the better the quality of the recording. A boom-mounted microphone, suspended just over actors' heads, is an excellent way of recording live sound. Camera operators must be cautioned when boom microphones are used so that they do not unexpectedly zoom back or tilt their cameras upward. A microphone protruding into a scene can ruin an otherwise good recording session. Normally, the built-in microphone on the video camera should not be used to record studio sounds.

There are four lights usually used in a studio situation. The first light is the key light and is placed at about a 45° angle from the camera. This photograph shows only the key light illuminating the subject.

The fill light is placed next to the camera on the side opposite the key light. This photo shows only the fill light illuminating the subject.

The background light is placed to illuminate the background behind the subject, and serves to separate the subject from what is behind it. This gives the set some depth.

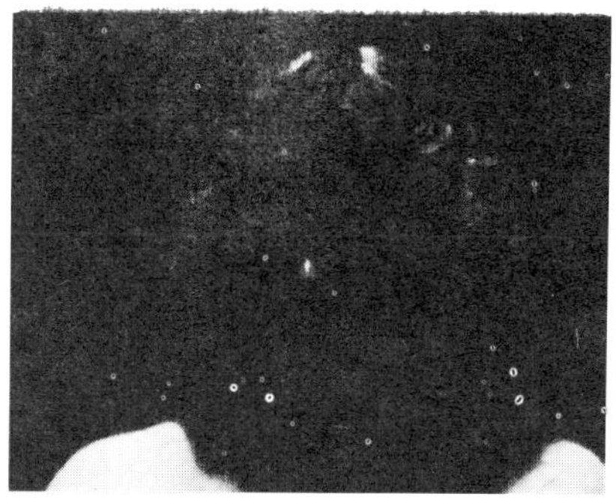

The accent light is often placed above the subject and is pointed downward to illuminate a specific portion of the scene, such as the actress's hair.

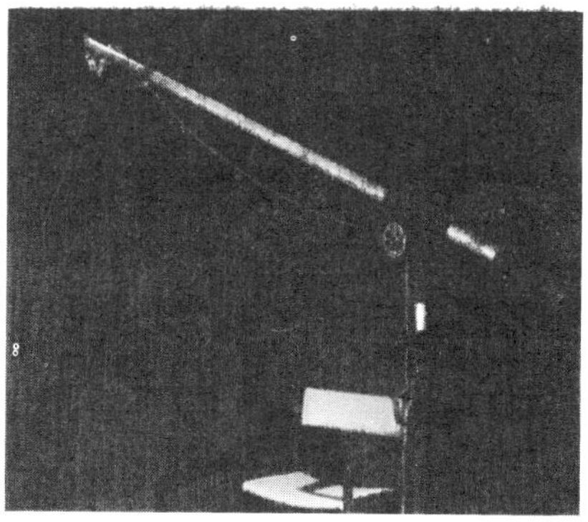

A shotgun microphone mounted on a boom.

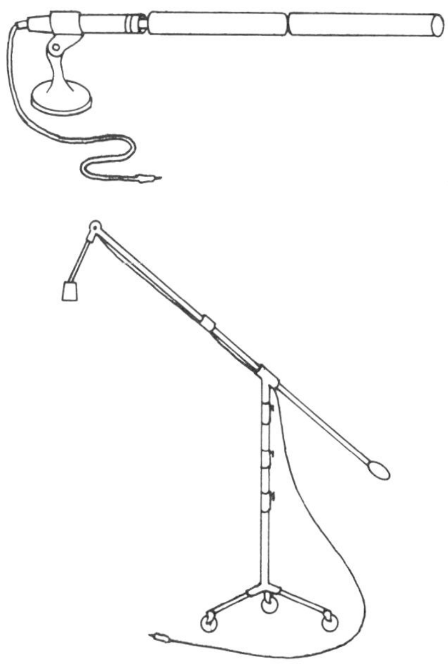

Shotgun and boom microphones in use.

Personnel Personnel involved in a studio recording session can make or break the production. The better rehearsed and trained they are, the more likely it is that recording sessions will be good ones. Generally, each piece of equipment should have its own operator. Each camera, the SEG, the microphone, the audio mixer, and possibly even the lights should have one person assigned as operator. While it is possible for one person to do everything, the likelihood of a high-quality recording being made under these conditions is considerably diminished.

Phase IIB: Program Recording on Location

One major advantage of new video systems is their portability and battery operation capabilities. New ½-inch systems have cameras no larger than super 8mm cameras, and recorders weighing ten pounds or less. When preparing to record a video program *on location,* several considerations should be remembered.

PRODUCE

Planning Steps for Location Video Recording

1. Check out all equipment before leaving the studio, and take extra equipment for backup, if possible. An equipment malfunction fifty miles from home is one of the most devastating events that can occur to the video producer. Always record a small segment and play it back in the viewfinder, both before leaving and after arriving in the field, to be sure that equipment is working and connected correctly.
2. Take extra batteries for all pieces of equipment. Battery failure is the most common problem that occurs during location recording. Also take the equipment's AC adaptor and a drop cord along. If an AC outlet is anywhere near, use it.
3. Review the script before leaving home. During the hectic process of trying to get all location scenes recorded, it is easy to forget something. Read the script over before leaving and follow it closely during the recording process.
4. Overshoot all scenes if electronic editing is planned. It is easier to take more footage of videotape than will ever be needed than to have to try to reschedule and reshoot a scene that did not turn out right. If in-camera editing techniques are being used, try to shoot the entire program more than once before returning home.
5. Take floodlights along and use them. Additional light vastly improves the quality of video recordings and should be used whenever possible. Battery-powered floodlamps that mount on the video camera can be used if tripod-mounted lights are too bothersome.
6. Use a tripod to hold the camera during recording. Do not try to hand-hold a video camera. If a tripod is out of the question, use a monopod, or a shoulder mount for the camera. It is nearly impossible to hold a camera steady during a 10- to 20-second scene.

OPERATE

Using a Portable Video System

You need: a portable battery-operated VTR, a portable video camera, a tripod or monopod, a microphone, a blank videotape

Procedure

1. *Getting ready.* Make the proper connections between camera and recorder, check battery condition, thread tape, and put microphone in place (if you use an extension cord and remote microphone). Similar to film production, it is important to mount the camera on a tripod or steady it on a monopod to minimize camera movement.
2. *Making tape.* Turn the recorder on so that you can view a picture in the camera. Practice zooming and focusing—remember to zoom all the way in first, focus, then back off with the zoom. An extension on the remote microphone is handy for interviews. Make some title cards if you wish to add graphics to your tape.
3. *Playback.* You can either play the tape on the porta-pac deck, or thread it on a studio deck. Connection can be made to household-type TV set using RF OUT (tune to channel 3). When playing tape for a class, stand near playback deck and be prepared to use pause control to freeze-frame selected portions of tape. Narrate live, or use recorded sound from tape. (See instructions elsewhere in this chapter for making audio dubs on videotape.)

1. Connect all components.

2. Thread the videotape, or insert the video cassette, depending on the format of your video system.

3. Mount the camera on a tripod, or monopod, if at all possible. Only hand-hold the camera as a last resort.

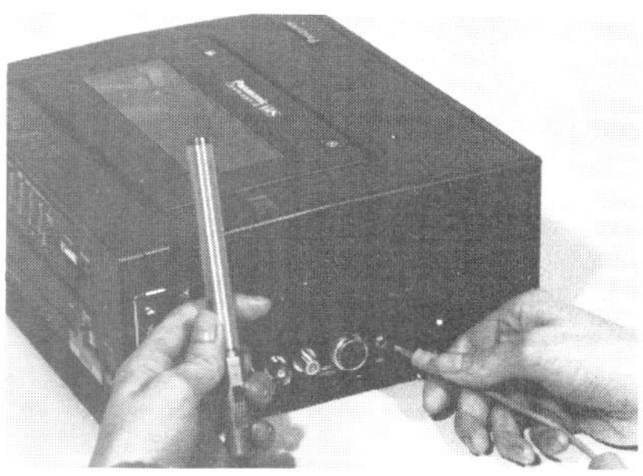

4. If an extension microphone is to be used, be sure that it is connected to the video recorder correctly.

5. Before recording, practice with the camera. Remember to critical-focus so when the zoom lens is pulled back the subject of the program will stay in focus.

6. After recording, play the tape back, either on another deck with a television or in the viewfinder of the camera.

7. Most video recorder/players now have RF outputs, so programs can be played back using a standard home television receiver.

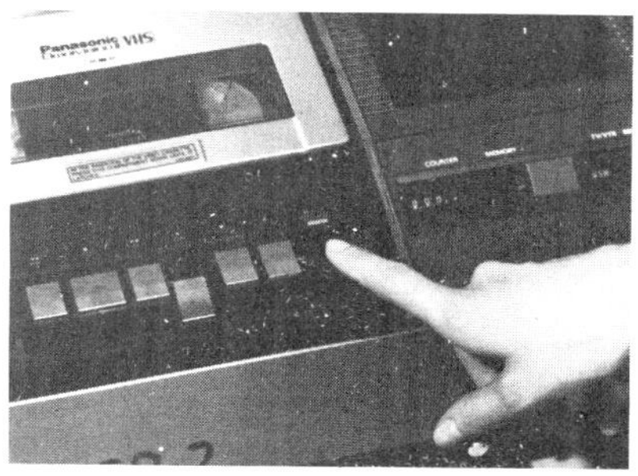

8. Both studio decks and portable decks have freeze frame, or pause, controls.

TRY THIS

Record a 3- to 5-minute program on location using a portable battery-operated video system. Play back the tape and critique the program. How might electronic editing, or video switcher and studio recording, improve your program?

On-location video recording is probably the most difficult media production technique there is. Remember, when recording is done on location it is occurring on someone else's turf. Be sure that all necessary arrangements have been made and confirmed in advance. Overplan and overschedule. It will pay off in the long run.

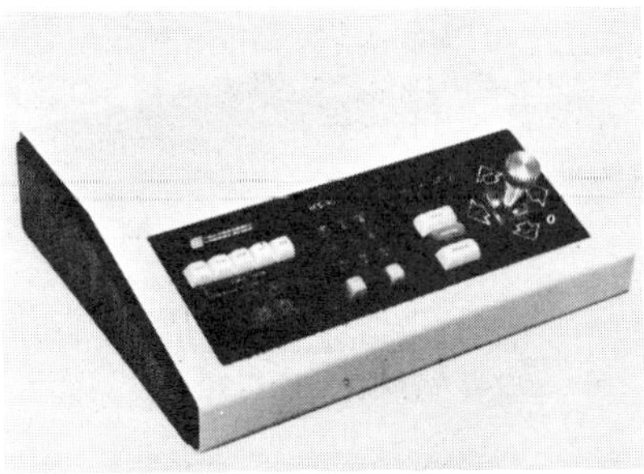

Editing controllers, such as this Convergence Corporation ECS-90, make video editing simple and quick (compliments Convergence Corp.).

Phase III: Postrecording Production

Audio dubbing, editing, and duplicating are completed once the basic program has been recorded. Audio dubs are used to add sounds, editing is done to put scenes in order, and duplicating is needed to make copies of the finished production.

Audio Dubbing One important characteristic of videotape is that there is a special portion of the tape reserved for the audio track(s). This makes it possible to record sound onto the tape independently of video recording. Whenever the record button is activated on a videotape recorder, old sound on the tape is erased and new sound is added at the same time that old picture information (video) is erased and new pictures are recorded. However, when the audio dub button on the recorder is turned on, the video signal is not affected. For this reason it is possible, and often advisable, to add audio after the visual portion of the program has been put on tape. This can be done while watching the video on a TV receiver. To audio-dub, merely connect a microphone (or other sound source) to the audio input jack of the VTR and depress the audio dub button and forward control on the VTR. As the videotape moves through the recorder, the old soundtrack is erased and the new one is recorded in its place. The soundtrack can be erased and rerecorded as many times as desired. (In ¾-inch, U-matic systems, usually only audio channel No. 1 can be audio-dubbed.)

Video Editing Video editing is the process of electronically arranging scenes to communicate a message effectively. It is an electronic process. By comparison, film editing is a manual process (i.e., the scenes are cut and cemented together manually). Two types of editing are most commonly used. In *insert editing* a scene is inserted between two existing scenes, usually replacing something that is already on the tape.

Insert editing.

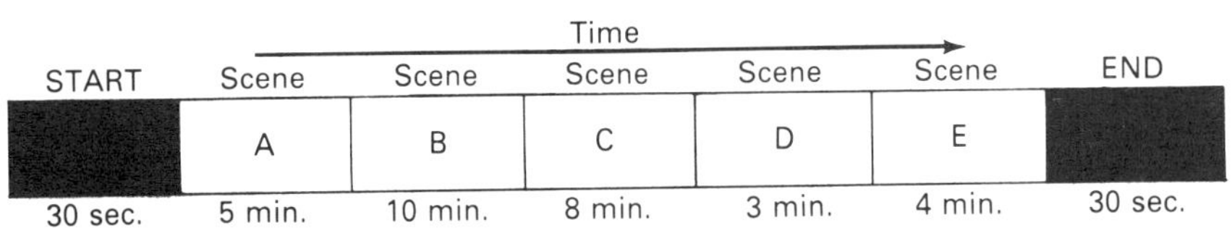

Assembly editing.

OPERATE

Audio Dubbing: Adding Sound to a Videotape

You need: a script, a video recorder, a prerecorded videotape, a TV receiver, a microphone on a stand, connecting cables, a quiet room

Procedure

1. Plug in the video recorder and the receiver. Connect them using an RF cable. Tune the TV to channel 3 (or 4). Turn the receiver's volume down.
2. Connect the microphone to the MIC IN input on the video recorder. The microphone should be mounted on a stand. The video recorder should be set up as if a tape were going to be played back.
3. View the entire videotape to be sure the visual portion matches the script. Time all scenes where audio is to be added.
4. Advance the videotape to the point where audio is to be dubbed.
5. Depress the video recorder's AUDIO DUB button and, while holding it down, depress the PLAY or FORWARD control. Immediately turn the audio volume on the video recorder to the previously selected level. The video portion of the program can be viewed on the receiver.
6. Record the new narrative. Match the new audio with the old, preceding or following it using the manual audio level control, if your video recorder has one.
7. When recording is finished, turn the video recorder's audio level down and depress the STOP button.
8. Play back the dubbed segment to be sure it is correct. If not, redub following the procedures outlined in steps 1 through 7. A videotape can be dubbed many times without affecting quality.

TRY THIS

Audio-dub a prerecorded videotape. Change the existing narration to your own.

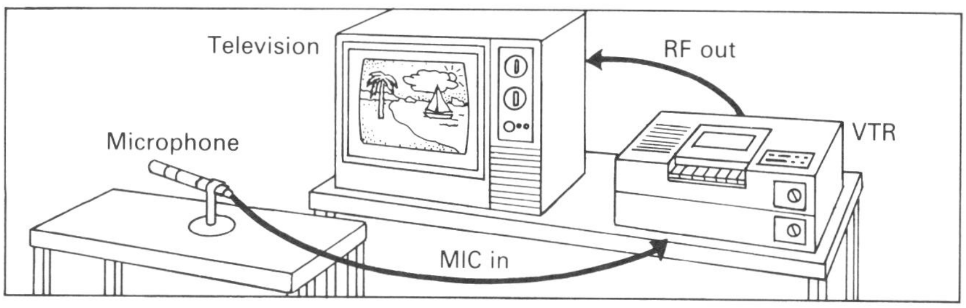

Audio dub setup.

In *assembly editing* an entire program is assembled from a series of scenes. In this process, the person doing the editing puts together, or assembles, the video program from start to finish by taking all or part of scenes shot at different times and/or in different places and putting them in order. It is even possible to assemble a program from different videotapes.

Entire editing consoles are available so sophisticated video programs can be produced fairly easily and inexpensively. This type of system costs about one-fifth of what they cost eight to ten years ago (compliments Sony Corp.).

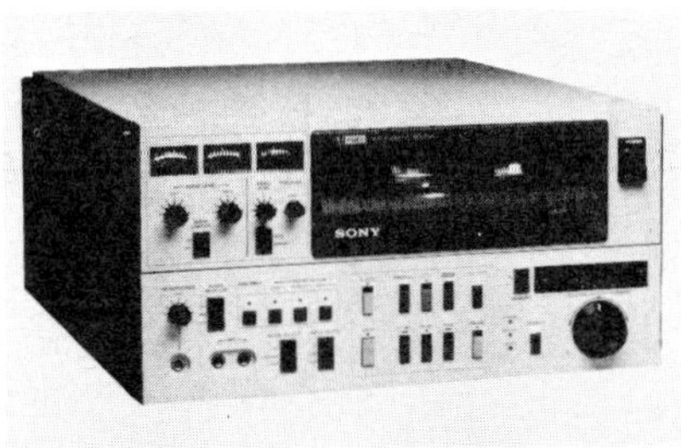

Many models of editing video recorders have fast forward and fast rewind capabilities that allow viewing of the video information on the video monitor (compliments Sony Corp.).

PRODUCE

Assemble Editing a Videotape Program

You need: *Equipment* Most editing systems consist of the following pieces of equipment.

- ☐ 1 VTR equipped to function as an editor
- ☐ 1 videotape player or another VTR
- ☐ 1 editing controller
- ☐ 2 TV monitors
- ☐ Cables to connect all pieces of equipment

As editing systems become more elaborate, the following hardware is often added to allow for more sophisticated productions.

- ☐ Time-based correctors, to equalize picture and sound recorded at different times, or by different pieces of equipment
- ☐ Audio mixers, to combine sounds from several sources
- ☐ Special-effects generators, to create dissolves, fades, wipes, and other optical effects
- ☐ Titler/character generators, for inserting letters and numbers during editing

Software One high-density editing videotape with a control track recorded for its entire length, prerecorded videotapes of program scenes, and a production script.

Since different types of video editing equipment vary greatly in their specific operating procedures, a considerable amount of time should be spent reading the manuals that come with hardware, and in practicing editing techniques. Generally, the editing process involves the following procedures.

Procedure

1. Collect all prerecorded, "rough-cut" scenes, the shooting script, and a good-quality blank videotape. This blank tape must have a control track recorded onto it. This is done using a video camera and VTR.
2. Become familiar with all scenes on tape and list their location and length on a log sheet. Scenes should be recorded with at least five seconds of unimportant action before the scene begins.
3. Locate and transfer (dub) scene No. 1 from the rough-cut tape in the videotape player to the master tape in the videotape recorder.
4. Rewind the master tape to the last frame desired in scene No. 1 and put the recorder in the pause mode.
5. Locate the first frame desired in scene No. 2 from the rough-cut videotape and place the videotape player in the pause mode.
6. Reverse both machines five to ten seconds, either automatically with the editing control device, or manually. Start both videotape machines forward, and at the appropriate point cut from the unwanted end of scene No. 1 to the beginning of scene No. 2. (This is usually done automatically by most editing control devices.) Record scene No. 2 in its entirety.
7. Review the edit. If it is acceptable, go to the next edit and repeat steps 3 through 7 for scene No. 3.

Remember, do not erase your "rough-cut" videotapes until the entire program has been edited, reviewed, and copied.

A ¾-inch U-matic cassette editing console.

Duplicating Videotapes Master videotapes of any program should not normally be used for instructional playback. Instead, copies should be made and all master or original tapes should be stored. Second-generation tapes should be used for day-to-day showing of programs.

Another important reason for duplicating copies of master video programs is to switch formats. For example, in order to use a video program recorded on a ¾-inch cassette in a ½-inch VHS player, it is necessary to make a dub from a ¾-inch player to a ½-inch VHS recorder. Fortunately, this is quite simple.

To make a videotape duplication, or dub, two videotape recorders, audio and video patch cords, a blank tape, and the original tape are needed. A TV receiver connected to the recorder may be added in order to observe the copy as it is being made.

Summary Production of video programs should be accomplished in three phases. During the *prerecording phase* all graphic and title materials should be prepared. The *recording phase* is when the videotape is actually recorded, either in a studio setting, or on location. The *postrecording phase* is when audio dubs are added to tapes, when videotapes are edited to match the script, and when copies are made for classroom use.

EQUIPMENT MAINTENANCE

Educational video equipment systems are built to be dependable and to require little maintenance. Ninety percent of all problems are due to simple human error. Most often cables are connected wrong, or tapes are incorrectly threaded. About 8 percent of the problems, such as broken plugs or wires, dirty heads, or broken switches, can be easily repaired. Only about 2 percent of all video equipment problems actually are due to electronic malfunctions requiring a service technician to repair. Before assuming equipment is "broken," it is wise to recheck the operator's manual, and try to "trouble-shoot" problems first.

Video dub Duplication of a videotape using a video recorder, a video player, and connecting cables.

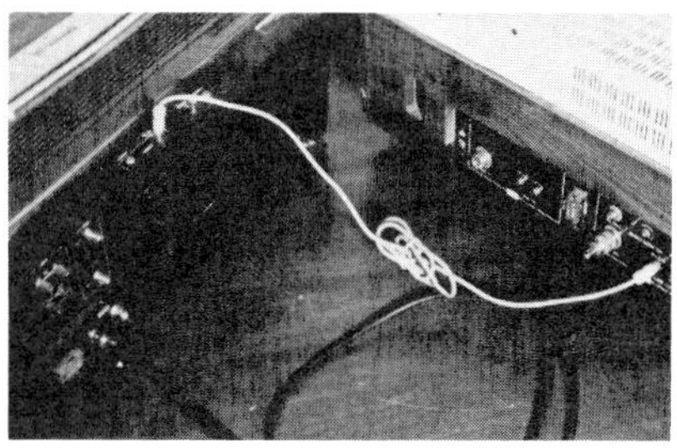

A dub is the transfer of information from one device to another. This dub requires connecting from the video and audio out jacks of a ¾-inch U-matic recorder to the video and audio input jacks of a VHS ½-inch cassette recorder.

PRODUCE

Duplicating a Videotape

You need: a recorded video program and a videotape player of the same format, a blank videotape and videotape recorder of the same format, audio and video connecting cables, a TV receiver

Procedure

1. Connect the audio and video out jacks on the videotape player to the audio and video input jacks on the video recorder using connecting cables.
2. Connect the TV receiver to the videotape recorder using an RF cable. Tune the TV to channel 3 (or 4).
3. Set the video recorder to the LINE mode, if necessary.
4. Plug in and turn on all equipment. Insert the original videotape in the video player. Insert the blank tape in the video recorder.
5. Depress the record button until it locks on the videotape recorder.
6. Play the original tape. Watch the receiver. The program on the original tape should be playing.
7. A few seconds before you want to begin duplicating, depress the videotape recorder's PLAY control while holding the RECORD button down.
8. Record a few seconds, then stop both machines. Play back the duplicated tape to be sure everything is working correctly.
9. Rewind both tapes and repeat steps 5 through 7. Duplicate the entire video program. Check out the copy when duplicating is finished.

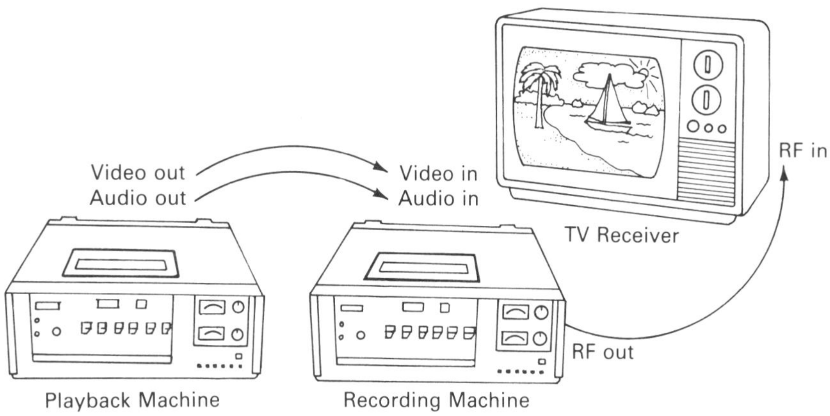

Dubbing setup.

TRY THIS

Duplicate a videotape from one format to another.

OPERATE

Vidicon Tube Cleaning

1. Carefully remove the lens by holding the camera and turning the lens counterclockwise.
2. Soak a cotton swab in lens cleaning solution, and gently wipe the face of the vidicon tube surface, which should be visible when the lens is removed. Care should be taken not to get more dirt or strands of cotton on the face of the tube. Compressed air can also be used to remove dust on the tube.
3. Replace the lens. After cleaning, it is a good idea to connect the camera to a VTR and receiver to see if the pick-up tube is indeed clean. If not, repeat the cleaning procedure.

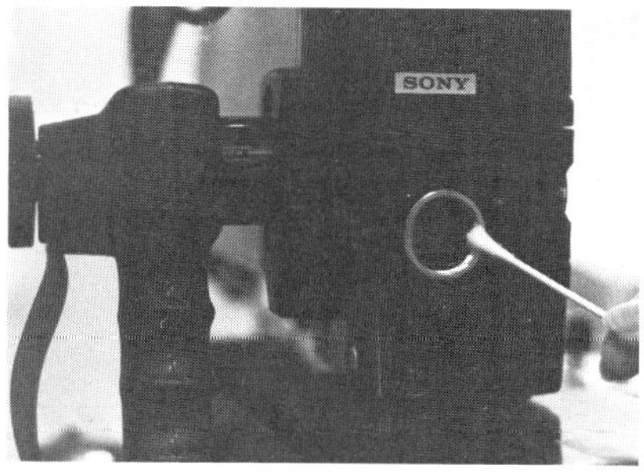

Remove the lens and use a clean cotton swab to remove lint.

Camera Maintenance

There are only two types of maintenance that should be performed locally on most video cameras—lens cleaning and vidicon tube cleaning.

Eventually, dust will settle on the camera lens, especially if the lens is not capped and if the camera is not stored in a case. Dust on the lens should be cleaned only with lens cleaning tissue or compressed air. Cleaning the camera tube is slightly more difficult.

Maintenance of the Videotape Recorder

When problems with video equipment occur, it is usually the videotape recorder that malfunctions. The VTR is the only component with moving parts. About the only type of repair that an operator can perform on a VTR is head cleaning. Video, audio, erase, and control track heads periodically become clogged with dust or oxide from videotapes. Clogging of heads will often occur when the VTR is left on PAUSE for long periods of time. This causes oxide to flake off of the tape. Oxide flaking occurs in older tape types more readily than in newer ones.

Clogged heads show up as a marked deterioration in picture and/or sound quality during playback. In extreme cases the picture may even disappear.

Inside the VTR there are two rotating video heads mounted at 180° from one another. If care is taken, these heads can be cleaned by the operator. To clean video heads, a screwdriver, head-cleaning solution, and cleaning swabs are needed. Head-cleaning kits can be purchased that include all these materials.

Except for head cleaning, it is probably not a good idea to attempt to repair a VTR. Qualified service technicians should be used.

To clean the video head, use a cotton swab soaked in cleaning solution. Clean by moving the swab horizontally.

OPERATE

Head-Cleaning Procedure

1. Make sure the VTR is turned off.
2. Remove the screws from the top cover of the machine. To remove the cover on some machines, the TRACKING knob must also be removed. It pulls directly off of the shaft.
3. Press the eject button, if it will function in the off mode, and carefully remove the machine cover.
4. Prepare to clean each of the two video heads by rotating them into position.
5. Moisten a cleaning swab with the head-cleaning solution. Very carefully, clean each head by using a back and forth, horizontal, motion only. DO NOT rub the video heads vertically. This may cause head damage.
6. Clean the video, erase, and control track heads. These are stationary heads that are more rugged than the video heads so cleaning procedures are not as critical.
7. Once the heads are dry, replace the cover and all knobs on the VTR. Turn on the machine and play a tape to confirm that the machine is working correctly. If problems persist, repeat the cleaning process.

A FINAL WORD

Video recording is one of the most complex media production techniques. This is compounded because equipment manufacturers seem to change models at just about the time a media program starts to generate enough interest in the use of television to warrant purchasing additional equipment systems. Then, instead of having two or three similar systems with similar operating instructions, several specific sets of procedures must be learned, one for each new model. Because different models and brands of equipment look different, most people assume that they work differently, and in some cases this is true.

Fortunately, if the basic principles of operation are understood, the differences between one manufacturer's models, or between components from different companies, can be easily understood. Actually, the similarities are greater than the differences. The following principles are some of the basic ones related to how to operate video equipment.

☐ A *video equipment system* has three components: inputs, storage/retrieval devices, and outputs.
☐ All components are tied together using *connecting cables.*
☐ *Inputs* serve to convert light, sound, or radio waves into electrical signals that can be sent through a connecting cable to other components.
☐ *Storage/retrieval devices* convert electrical information received from input components into magnetic information for storage on videotape, or they convert magnetic messages from videotapes into electrical signals that can be sent through connecting cables to other components.

VTR troubleshooting guide

Problem	Checkpoint
No color on TV set	PROGRAM SELECTOR, ANTENNA, or VTR/TV switch should be set to TV, even if the VTR is turned off.
	Are the antenna connections correct?
	Is the TV set to the right VTR output channel—3 or 4?
	Are the correct channels fine-tuned on both the VTR and TV set?
Badly distorted playback picture	Adjust the TRACKING CONTROL. The video heads may be clogged.
No recording or playback	Check for proper connections.
	Are the plugs broken or weak?
	Is the PAUSE switch in the normal position? Release the switch.
TV broadcasts cannot be recorded	Make sure the INPUT SELECTOR is in the correct position.
RECORD button cannot be pressed	Check to see if the cassette safety tab has been removed.
	Is the DEW indicator lamp on?
PLAY button cannot be pressed	Is the cassette properly loaded?
	Is the DEW lamp on?
AUDIO DUB button cannot be pushed	Is the cassette properly loaded?
	Check to see if the cassette safety tab has been removed.
VTR stops in REWIND before end of tape	Turn the MEMORY or SEARCH switch off.
REWIND, FWD, or FAST FORWARD mode does not function	The tape may need rewinding.
No camera or microphone recording	Are the connections correct?
	Is the camera plugged in and switched on?
	Is the microphone plugged in?
	The VTR INPUT SELECTOR switch must be in the CAMERA position.
Poor-quality camera recording	Is the lens iris open?
	Add more light.
Cassette will not eject	Turn the VTR on and push the eject button.
Whistling or screeching sound in the mic recording	Turn the TV set's volume control down or off.
	Move the mike away from the TV set.
Poor-quality sound	Improper mike or poor-quality mike.
	Microphone is too far away from the sound source.
No sound	Broken mike or cable.
	Improper connection.
Unit will not turn on	Dead battery in a portable VTR.
	VTR is not plugged in or turned on.
	Auto shut-off mechanism is activated.

Rules for Effective Troubleshooting

1. Start over. Disconnect everything, and reconnect all components. A plug may be in the wrong jack, or may not be inserted all the way.
2. Rethread the videotape.
3. Isolate equipment, cords, or videotapes that you know work correctly. Substitute a second recorder, cable, camera, or other component you know works. Replace only one item at a time.
4. Isolate the malfunction by substituting equipment. If nothing works in the entire system, the problem is probably with the power source. Check fuses or circuit breakers.

- *Outputs* serve to convert electrical signals into images we can see or sounds we can hear.
- *Videotapes,* regardless of the format, store messages magnetically for later use.
- Messages stored on videotapes are on *tracks.* There are three types of tracks: control tracks, video tracks, and audio tracks.
- Hardware and software components produced by different manufacturers can be mixed in a video system. Different formats are not compatible.

LOOKING AHEAD

Well, that's about it. We have covered most of the procedures for producing instructional media. Obviously, there are many specific techniques and types of equipment we have not mentioned. However, if the ideas presented in Chapters 1 to 13 are understood, then almost any other specific types of instructional media production will be so closely related that using, producing, or operating them will be simple.

There is one idea that we have not spent a great deal of space discussing: the act of teaching with media. In the following part a discussion of the process of preparing for, presenting, and evaluating media in instruction is presented. This section discusses ideas that can be used to get you off on the right foot when you start using media for teaching and presenting.

REVIEW QUESTIONS

1. Where does the term *U-matic* come from?
2. What is magnetized during recording?
3. In helical scan recording, how is the video signal placed on the tape?
4. What is the purpose of the control track?
5. What is the video local production format?
6. What are the three components of a video system?
7. What is the name of the tube in a video camera that is sensitive to light?
8. What does 6:1 mean when describing a zoom lens?
9. Why is color temperature important in video recording?
10. Define *RF.*
11. What is the primary function of a VTR?
12. What is the difference between a TV monitor and a TV receiver?
13. What is the maximum distance a student should sit from a TV receiver while viewing a program?
14. What is the aspect ratio of a television screen?
15. How much of the video image falls onto the "bleed area"?
16. Define *resolution,* when discussing a television image.
17. What are the names of the four lights normally used in a studio setting?
18. What is insert editing?
19. Is it possible to play a Beta cassette on a VHS machine? Why or why not?

ANSWERS TO REVIEW QUESTIONS

1. From the way the tape cassette is threaded by the VTR, in a U-like pattern, automatically
2. The oxide on the videotape
3. In a slanted, diagonal pattern or track
4. The control track aligns the video playback machine to the videotape in a way similar to the way in which sprocket holes are used to align motion picture film correctly in a movie projector
5. The ½-inch format, because it is least expensive and very widely found in schools and training centers
6. Inputs, storage/retrieval, outputs
7. The vidicon tube
8. An object can be made 6 times (6x) larger, or closer, by manipulating the lens. A 6:1 zoom lens has a 1x to 6x magnification range
9. Various light sources have different color temperatures. Color is determined by the color temperature of a light source. To obtain good color in a video recording, the color temperature of the light source used to illuminate a scene must be matched to the settings on the camera used to record the scene
10. Radio frequency, a modulated radio wave
11. To convert electronic information to magnetic signals that are stored on videotape. The VTR's recording head accomplishes this
12. A TV receiver can receive RF signals only from a wire (RF cable) or from the air via its antenna. A TV monitor can input or output (send or receive) only nonmodulated, non-RF (often called raw-video) signals. A TV receiver/monitor can do both
13. 16–24 H (height of TV tube picture), or about 30–35 feet for a 21-inch TV
14. 3×4
15. ⅙ of the graphic's length or width, all the way around the margin
16. The quality of the TV image seen on the screen, as described by potential (or actual) CRT sweep lines projected on the face of the picture tube
17. Key, fill, accent, background
18. When a new scene is electronically inserted on a videotape between two scenes, either as new information or to replace another scene
19. No, the machines' formats are not compatible. A duplication from the Beta machine to the VHS machine to make a VHS copy would be necessary.

REFERENCES

Ackerman, J., and Lipsitz, L., eds. *Instructional Television: Status and Directions* (Englewood Cliffs, N.J.: Educational Technology Publications, 1977).

Atkinson, S., and Locatis, C. *Media and Technology for Education and Training* (Columbus, Ohio: Charles E. Merrill, 1984).

Bensinger, C. *The Home Video Handbook,* 2d ed. (Santa Barbara, Calif.: Video-Info Publications, 1979).

Bensinger, C. *The Video Guide,* 2d ed. (Sante Fe, N.Mex.: Video-Info Publications, 1981).

Combes, P., and Tiffin, J. *Television Production for Education* (New York: Hastings House, 1978).

Kybett, H. *The Complete Handbook of Videocassette Recorders* (Blue Ridge Summit, Pa.: Tab Books, 1977).

Marsh, K. *Independent Video* (San Francisco: Straight Arrow Press, 1973).

Mattingly, G., and Smith, W. *Introducing the Single-Camera VTR System: A Layman's Guide to Video-Recording* (New York: Scribner's, 1973).

Murphy, J., and Gross, R. *Learning by Television* (New York: Fund for the Advancement of Education, 1966).

Robinson, R. *The Video Primer* (New York: Quick Fox, 1974).

Stoddard, E. *Television* (New York: Franklin Watts, 1970).

Utz, P. *Video User's Handbook* (Englewood Cliffs, N.J.: Prentice-Hall, 1980).

Williams, R. L. *Television Production: A Vocational Approach* (Salt Lake City: Vision, 1976).

Zettl, H. *Television Production Handbook,* 3d ed. (Belmont, Calif.: Wadsworth, 1976).

VIDEO EQUIPMENT MANUFACTURERS

Akai America, Ltd.
2139 East Del Amo Blvd.
Compton, CA 90220

Ampex Corp.
401 Broadway
Redwood City, CA 94063

Hitachi Shibaden Corp. of America
58-25 Brooklyn Queens Expressway
Woodside, NY 11377

JVC Industries, Inc.
58-75 Queens Midtown Expressway
Maspeth, NY 11378

Magnavox Video Systems
1700 Magnavox Way
Fort Wayne, IN 46804

Panasonic Video Systems
Matsushita Electronic Corp. of America
One Panasonic Way
Secaucus, NJ 07094

Sanyo Electronics Inc.
1200 West Artesia Blvd.
Compton, CA 90220

Sony Corp. of America
9 West 57th St.
New York, NY 10019

3M Company
3M Center
St. Paul, MN 55101

PRODUCING INSTRUCTIONAL SYSTEMS: AN INTRODUCTION

D

The media discussed in Part C—audio, still photography, motion photography, and television—have several concepts in common. They all are produced from a script. They all can be used as part of a longer instructional program, or as stand-alone presentations, and they all have categories of components that make up equipment systems.

For many media users, the easiest way to remember how to use audiovisual equipment to produce materials is to think of all hardware components generally since they are fairly similar for many media types. The equipment systems in Part C have the components shown in the chart on the right.

Components

Medium	Inputs	Storage/Retrieval	Outputs
Audio	Microphone Record player	Cassette recorder Reel-to-reel recorder	Loudspeaker
Still photography	Viewfinder camera Twin-lens reflex camera View camera Single-lens reflex camera		Slide projector Filmstrip projector Display
Motion photography	Super 8mm camera 16mm camera Microphone (for sound films)		Super-8 projector 16mm projector
Videorecording/Television	Video camera Microphone Tuner (for off-air recording)	Video player/recorder	Television receiver Television monitor

Complex media equipment are organized into systems. Similarly, complex instruction can be organized into a system. In Part A the idea of instructional development and instructional systems was introduced. In Part D these concepts are explained in more detail. Additionally, the relationship between the media presented in Parts B and C and instructional systems is explained. Chapter 15 wraps up *Media Planning and Production* by presenting techniques for preparing for teaching with media, so the result is professional looking and delivered effectively. These last two chapters should tie together the various goals of this book. They relate the topics of how people learn, how instruction should be organized, and how media are produced, to the teaching process. While the primary purpose of this text is to discuss media production and equipment operation, these specific skills are important only when they are used as a part of an instructional system.

GOALS

□ Become aware of the variety of media available for use in teaching.
□ Understand the strengths and weaknesses of selected media, so that each medium can be chosen for the task for which it is best suited.
□ Compare the systematic planning processes that are inherent in the design of effective instruction.
□ Learn several of the factors that must be considered more or less at the same time during the conceptualization of a teaching strategy.

OBJECTIVES

□ List the strengths of each medium as related to the cognitive, affective, and psychomotor domains.
□ Outline the steps necessary in planning an instructional strategy.
□ State accepted psychological principles that support the need for a variety of teaching strategies.

The title of this chapter implies that a classroom can be compared to a concert stage. The teacher might be the conductor, and the class might be the audience. The music or program compares to the lesson, and the instruments creating the music are the media used in teaching the lesson.

This analogy, like others, begins to break down if you press it to extremes. But let us stay with it for a moment. Think of an orchestra comprised entirely of violins. There would be no variety in sound, a limited dynamic range, and pitches spanning only about three octaves. Certainly the rich variety that we expect of music would be impossible. Furthermore, certain types of music, like rock, or folksongs, would be difficult to achieve with the violin orchestra. There would be no rhythm section—no drums—and deep bass notes as well as unusual melodic effects would be lacking. A violin concert would not be effective for certain types of music because of the limits of the instruments.

Compare the situation to a lesson that is being taught in a classroom. Imagine a lecture, nothing more. The teacher is talking without the use of visual materials, chalkboard, or other media. All of the information is being transmitted with words. As you think about this case, consider the subject area and content you are preparing to teach. Review the concepts, the examples, the diagrams, the problems you plan to pose to the class. Undoubtedly you can identify some difficult times ahead if you plan to lecture exclusively when you teach.

On the other hand, suppose you choose the overhead projector as your favorite medium for use in teaching. Everything you plan to discuss, explain, or teach about would be committed to transparencies. Armed with a stack of ten, or twenty, or fifty transparencies you would face a class day after day. A teaching style based exclusively on the use of transparencies would be as dull—for both teacher and students—as one that relied on any other single medium all of the time.

14

ORCHESTRATING INSTRUCTION/ DESIGNING INSTRUCTIONAL SYSTEMS

Fortunately, no teacher needs to feel trapped by a single technique. Even the most conservative instructor may show a film once in a while, or use the chalkboard, or make a spirit-duplicated handout from time to time. Yet, without conscious effort to maintain versatility it is possible to gravitate toward one's favorite things either because they are more familiar, or more appealing, or the production techniques required to use them are easier and faster.

This chapter is about the need to assemble a wide range of media into unique combinations so that variety, originality, and an "air of the unexpected" can be maintained in the classroom.

THE NEED FOR VARIETY

The importance of variety is an appropriate starting point for this section of the chapter. You might wonder why several media might be needed, when one—books, for example—could do. After all, books have been effective teaching tools for a very long time. It should be possible to provide an information base with print and pictures that would fit nearly any subject. Books have the advantage of being relatively inexpensive, quite portable, and always dependable. How could books possibly be improved?

The question is inappropriate; the emphasis is misplaced. It is more meaningful to ask how instruction or learning could be improved. If the end result is the more efficient acquisition of information, under more amenable circumstances, the motive shifts from the improvement of the medium to the improvement of teaching and learning. Large numbers of people may not enjoy obtaining the majority of their information from print. Consider the number of hours people watch television, or listen to the radio, or simply converse with others. One recent study indicated that we watch television, on the national average, nearly four hours a day, close to 1500 hours a year. This compares to an average of ten hours a year reading books.

As a variety of stimuli have been added to our world, it has become possible to reach each other with computers, television, newspapers, recordings, live broadcasts, and many other ways. The basis for this expansion lies in two fundamental areas:

1. The unique nature of each medium to perform its task.
2. The psychological need that people have for variety.

Variety is a fundamental theme in much of our living. In choice of food, variety is important. Clothing and shelter are widely varied within or across cultures. Languages, ideas, technology, goods, and services manifest variations that human beings feel are important. Because the need to be similar but different is a fundamental human trait, it has become an important consideration in teaching and learning.

The connection between variety and learning may be apparent to anyone who has been a student. Some of the conditions that prevail during learning include:

☐ *Attention span* A wide variety exists between people, ranging from a few minutes to several hours, depending on the person and the activity.
☐ *Activity* Intellectual endeavors, such as reading, writing, or discussion—as well as psychomotor or manipulative activities.
☐ *Location* Some people like to knit under a tree, others prefer a living room chair. Certain types of learning are more appealing in certain locations, based on individual needs.
☐ *Timing* Each person may prefer to work or study at certain times when efficiency seems highest. It is important to determine when these peaks of productivity occur so that full advantage may be taken of them.

In addition to the psychological needs for variety, there is another aspect related to variety which is inherent in the media. Each medium is unlike others in form, ability to convey certain types of information, and mechanical or logistical features such as production difficulty, cost, and availability. Of these elements, the most significant is the ability to convey certain types of information, skill, or attitudes.

The three domains of learning were discussed in Chapter 4, as outlined by Bloom and others. The domains are cognitive (facts), affective (feelings and attitudes), and psychomotor (performance). There are accepted methods for accomplishing very efficient teaching in each of the three domains by using certain types of media. For example, if sheer cognitive acquisition of the multiplication tables is required, one of the most satisfactory methods is drill and practice, using printed material or simple computer-assisted instruction. Attitudes and opinions are best formed through the use of films, discussion, or reading. Physical ability to perform a skill may be demonstrated with film or videotape—or in a live situation—but the actual acquisition of the skill is almost entirely dependent on practice and hands-on experience.

You can see that certain general guidelines have been formulated to help the teacher choose the most appropriate medium for a given teaching situation. In the next section of the chapter the relationship between the media and the content to be learned will be examined.

CHOOSING A MEDIUM

For just another moment, please consider another comparison. Think about choosing food. You may enjoy ice cream, or hamburgers—not particularly because you want a balanced diet, but because you enjoy them. If your criterion is enjoyment rather than proper nutrition, your choice is dictated by feelings instead of science. A natural tendency or craving for food is somewhat comparable to a tendency toward a learning style.

There has been considerable research in the last two decades on the learning styles of individuals. Students process information differently. Some seem to prefer visual rather than verbal information. It has even been proposed that these visual learners translate verbal information into visual symbols. Verbal processors translate pictures into words.

Another learning style is field dependence/field independence. Some learners require a great deal of direction from their environment to assist them in processing information. They are described as *field-dependent.* Others, called *field-independent,* provide the internal framework for new information as they receive it. How a teacher presents information to students who have tendencies toward field dependence, or field independence, can influence the extent to which learning can occur.

An additional area of research is concerned with the hemispheres of the brain. It has been shown that one hemisphere is usually dominant, and that left-brain individuals tend to process information analytically and in terms of separate concepts. Right-brained people are often holistic and intuitive. Only a handful of people are totally right- or left-brain dominant, or exclusively field-dependent or -independent. Most of us lean toward one learning style or another, however, and prefer a particular learning style because of this acclimation. The implication is that teaching methods should be developed with regard to learning styles.

These ideas may hold true when choosing media for instruction. You might enjoy using transparencies, or audiotape, or film—even though the inherent capabilities of the media and the requirements of the subject matter might make another choice more suitable. It may be impractical to attempt to talk someone out of a choice which he or she prefers, but nevertheless it is important to know something about the characteristics of each medium and to understand its strengths and weaknesses.

The media selection charts on the following two pages represent a helpful approach to the problem of choosing the correct mode of delivery and software to communicate a message.

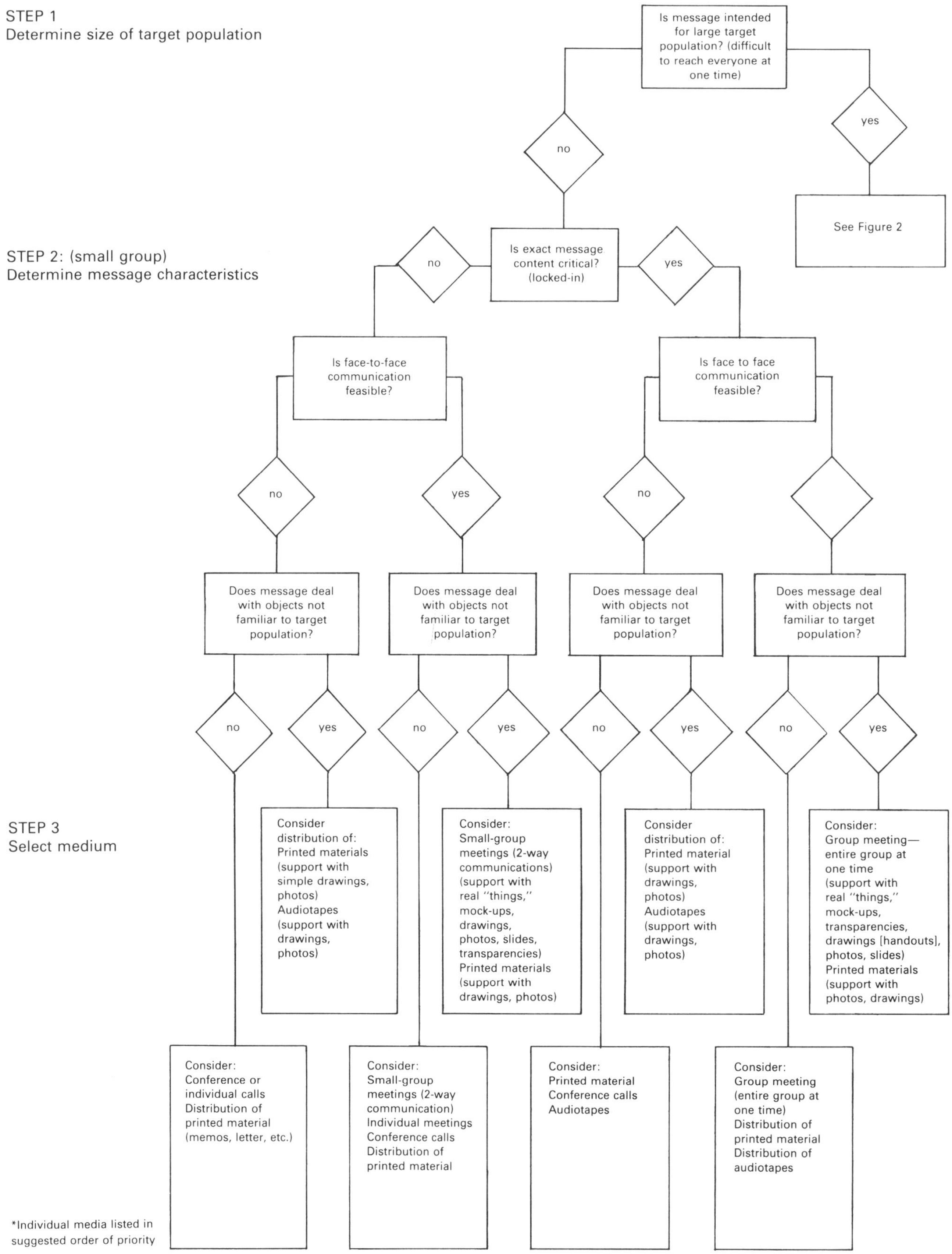

Chart 1: Media selection information.

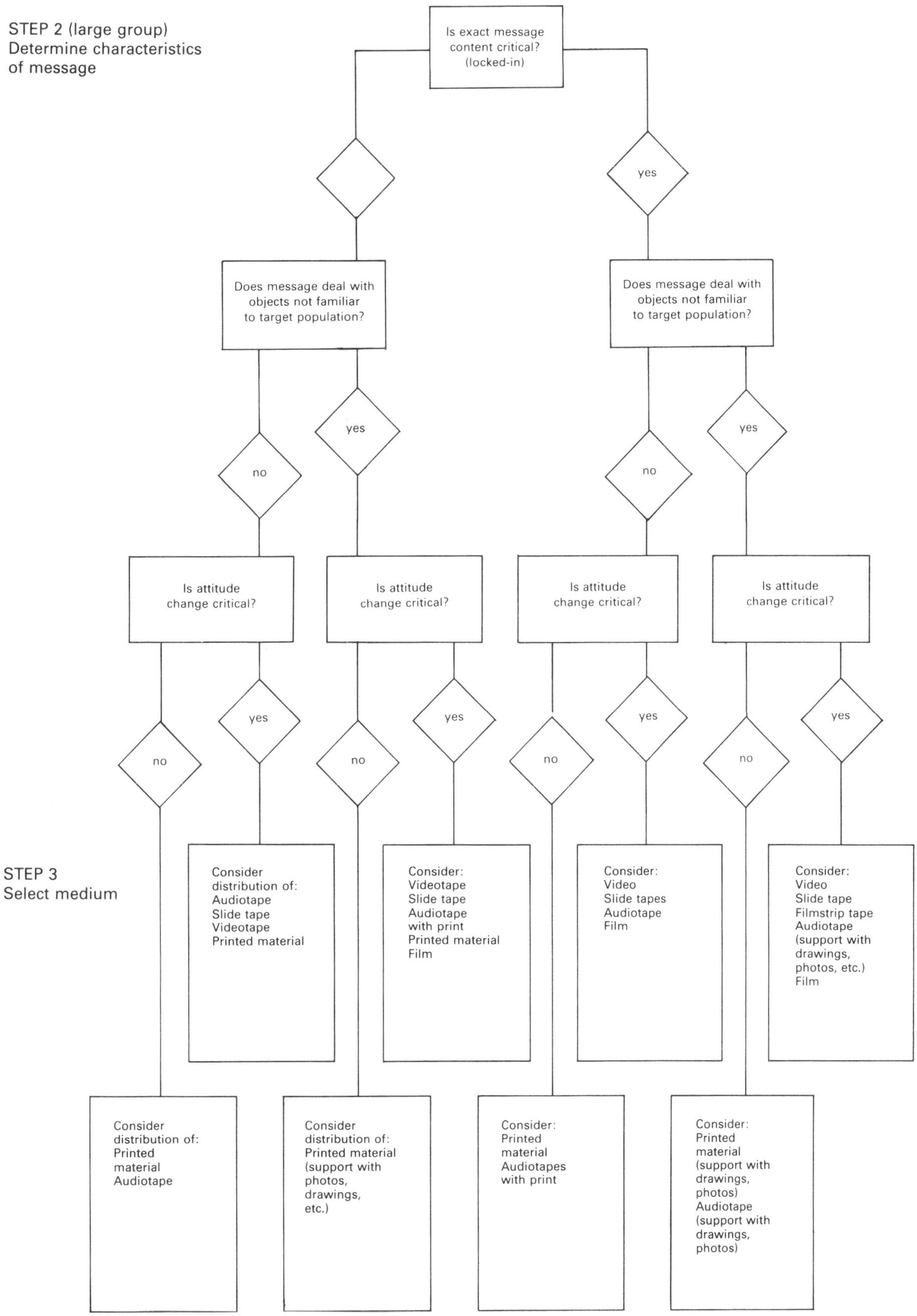

Chart 2: Media selection information.

PRODUCE

A Treatment for a Specific Subject

Suppose that you want to change attitudes, as well as provide information, about the effect of industrial waste on the environment. Using the Media Selection Information charts make the appropriate decisions leading to your choice of methods and materials. Follow these steps:

1. Start with chart 1, step 1, to determine size of the target audience (target population).
2. If the message is intended for a large group, see chart 2, step 2. If the message is more suitable for a small group, go to step 2 in chart 1.
3. Move down through the appropriate media selection information, making decisions as called for, until you reach step 3.

PRODUCE

A Lesson Based on a Selected Treatment

Suppose that you want to work with the medium of television, or audiotape/slide sets, or transparencies. Select one of these media, then examine your subject area for a suitable topic that would fit the medium. Follow these steps:

1. Refer to the Media Selection Information charts, but begin at the bottom of either one, on step 3. Note that the media you have chosen (tape/slides or transparencies) are found in more than one place in step 3.
2. Move up to step 2, progressing from bottom to top through the branches in the chart. As you pass each decision point, note the characteristics of the media that indicate audience response.
3. Continue upward in the chart until you reach step 1, noting further characteristics of the media in each succeeding set of circumstances and conditions.

GETTING DOWN TO BUSINESS

So far the chapter has brought up the need for variety in learning materials, teaching style, and media—and has provided some rather specific guidelines about the capabilities of selected media. With this background it is possible to begin consideration of the planning process that is needed to develop a full-fledged version of a teaching plan or lesson for a specific subject.

It appears as if there is no single best starting point. Student objectives might be considered early in the planning. On the other hand, the type of medium could also be considered early. In addition, it is necessary to determine if the subject matter will require motion for understanding, or other visual amplification—and in which of the three domains the majority of the learning is to occur. In short, several considerations occupy the teacher's attention as a lesson is pondered in the early stages. For this reason it may be unwise to list in 1, 2, 3 order the "steps" necessary in planning instruction. It turns out that nearly half a dozen facets surface, and they all must be considered simultaneously as the teaching approach takes shape.

A practical way of getting down to business and carrying out the planning is to think of the process as a series of clusters rather than steps. They include:

☐ *Cluster 1* Objectives, topic, content, learning experiences, type of media, cost, personal interest the teacher has in the subject and the media, and characteristics of learners, including age, IQ, experiences, and learning style.
☐ *Cluster 2* Length of time to be spent on the subject, student activities, cost, time needed for production, long-term benefit or payoff of learning related to difficulty of production.
☐ *Cluster 3* Method of delivery, evaluation, follow-up.

There is a certain amount of overlap from one cluster to another, which further emphasizes that planning is not a definite, step-by-step process. Rather, it is somewhat circular or spiral in nature, gradually increasing and widening to take into account more and more information and conditions. It is in these early stages that intuition, talent, skill, and creativity play a major role. However, don't become discouraged because you don't see yourself as particularly intuitive, talented, or skillful. Remember that outside help is available. Some approaches are described in the next section.

HELP FROM OTHERS

One explanation for the nest-building ability that birds seem to be "born with" is that they learn by modeling, or imprinting. The habits and training they acquire from others in the environment set an example that can, in turn, be imitated. Learning by watching others, playing by ear, imitating, mimicking—all are useful methods of acquiring behaviors.

Teaching skill can be acquired by modeling or imprinting. It can also be strengthened by using ideas that other teachers have made available to us. Some of the sources of teaching ideas, techniques, use of media, and other suggestions include:

☐ *Texts* Examine texts, laboratory manuals, teacher's guides, and other appropriate materials that students will be using so that you can glean ideas for preparing media for teaching. Often a chart in a text will suggest a transparency you can make, sometimes by copying directly from the original. A series of illustrations might be designed into a slide set. Descriptions of certain teaching situations could be the basis for the production of a film or videotape. And the overall content of a chapter might be used to make a bulletin board or worksheet that summarizes the unit.
☐ *Professional literature* The methods books that describe techniques and strategies for teaching are very helpful in the lesson-planning stage because of the wide range of approaches they suggest. Many of these books have chapters that are concerned with objectives, evaluation of students, media, and teaching style as well as ideas for teaching specific content and concepts. The bibliographies in the methods books often list sources of supplies and software that can be used during the planning stages.

A number of publishers have teaching-methods books that provide hints for presenting information.

☐ *Contemporary media* Both the mass media (newspapers, magazines, radio, television) and educational media serve as ongoing examples of accepted communication principles. For example, if you are planning to create a videotape that is designed to change attitudes, study techniques used by television or film producers. As you view a program on television or a film in a motion picture theater, sensitize yourself to the methods and devices used. Pay attention to the music, the camera angles, the dialog, and the overall treatment or story line. Note how the characters use selected nonverbal mannerisms and body language to communicate emotions and attitudes. It is helpful to point out these production techniques to students, since they watch a great deal of television and attend movies regularly. Given an inside glimpse of acting styles and a little coaching on dialog delivery, students often are able to develop meaningful approaches in media production. If you are a teacher, seek the assistance of the school's theater arts people in producing dramatized materials.

☐ *Current experiences* Examples in museums, display centers, and exhibits in larger metropolitan areas offer opportunities to take advantage of contemporary design techniques used to merchandise ideas. Science centers, planetariums, botanical exhibits, and natural history museums are particularly helpful in suggesting ways of teaching. Using a camera and high-speed film you can frequently photograph exhibits, with permission of the staff in the facility. Historical collections, dioramas depicting events in natural history, and displays with moving parts to simulate complex processes can all be photographed or videotaped for later use in teaching. Most of these facilities have a shop where materials are on sale; filmstrips, slides, postcards, models, and other items can be purchased and then used in the classroom.

PRODUCE

A Slide Set

Choose a public exhibit in your area (in a local museum, gallery) and create media that might be used to substitute for a field trip to the exhibit. Follow these steps:

1. Visit the site and take detailed notes to indicate the nature of the displays, and the sequence in which they should be viewed. Use the notes as a basis for the script.
2. Visit with the officials at the site to obtain background information, printed materials that might be furnished to accompany the exhibit—and don't forget to obtain permission to take pictures in the area.
3. Write the narrative to accompany the pictures. Inquire about the availability of any recordings that would aid in enhancing your production.
4. Carefully analyze the visual elements of the exhibit so that you can plan the photography in terms of long shots and close-ups. Note the type of film and lenses that will be required.
5. Complete the script, photograph the scenes, make final selection of the finished pictures, and narrate the audiotape.
6. Consider writing a pre- and posttest to further contribute to the teaching and learning effectiveness of the production.

When important events are scheduled on television or radio, plan to record them or alert students to tune in on the program. If you attend a lecture, it is sometimes possible to obtain an audiotape of the event. Frequently the speaker is available for informal conversation after a presentation, providing the opportunity to discuss ideas that were included in the talk, or to probe additional areas of interest.

VARIETY IN YOUR VOICE: CHARACTERISTICS AND CONTENT

It is natural to assess people's enthusiasm, outlook, demeanor, and attitude merely by listening for clues as they speak. Emphasis of certain words, inflection, and dynamic range of the voice provide an important index as to how one feels about an idea or situation. The connection between a person's voice and personality is a real one, as shown by studies conducted by psychologists. Using audiotapes of selected speeches or presentations, it has been found that an audience can detect fear, rage, kindness, patience, and many other emotional characteristics.

TRY THIS

Select a short paragraph from a novel, the newspaper, this text, or a script you have written. Make a tape recording of your reading of the selection, paying special attention to the mechanics of speaking. Provide variety in voice inflection, choose the correct rate of speech, make sure you pronounce the words correctly, and use correct grammar. Do not listen to the tape for a few days, and then critique it after you have partially forgotten how it might sound. Attempt to be objective in your analysis, reacting to it as if you were a student listening to an instructional tape.

These "content-free" aspects of the voice cannot be left to chance because of the prominent role they play in teacher/student relationships. In order to refresh your memory about the role a teacher's voice plays in the instructional process, review Chapter 1. In addition, plan to make an audiotape of yourself in a teaching situation, and then analyze it for these characteristics:

1. *Voice inflection* Check for variety in pitch (rise and fall of the voice) so that you are not labeled as a monotone.
2. *Rate of speech* Determine if you seem to be speaking too fast or too slowly. It may be necessary to ask others to judge this characteristic.
3. *Pronunciation* Watch for certain words that may be used improperly, such as "grammarically" instead of *grammatically,* "nucular" instead of *nuclear,* "irrevelant" instead of *irrelevant.*
4. *Grammar* Listen to your speech for mismatched subjects and verbs, as in "There's all kinds of reasons. . . ." (There is all kinds of reasons is incorrect.) Avoid verbal mannerisms too, such as "kind of," "OK," and "you know. . . ."

In addition to the mechanics of speaking, pay attention to the content of what you are saying. Review Chapter 1, in which analogies, comparisons, and mnemonic devices were discussed. Continually look for unusual, unique, and meaningful ways to explain ideas and concepts.

FINALLY, A FORMULA

If you recall the title of this chapter, you may be wondering when the actual process of orchestrating instruction will be described. We purposely delayed providing a formula in order to emphasize that many considerations must be made more or less simultaneously. After you have let the ideas germinate and you have a pretty good idea of the subject area, the teaching approach and strategy, and the type of media with which you would like to work, you are ready to apply these formal steps to the design of instruction:

1. *Narrow the topic* Do not plan to design a major instructional unit on the Second World War. Instead, choose a major battle or a specific set of conditions. Begin your orchestration process on a small scale, creating components that can be fitted together into a larger framework that emerges over a long period of time.
2. *Determine objectives* Decide what students should be able to do when the instruction is finished. A block of instruction that includes about half a dozen objectives is often appropriate.
3. *Choose learning activities* Consult your objectives and invent activities that will provide students the opportunity to become involved. The choice of media is inherent in this process, since viewing, discussing, working in a laboratory, writing, and listening are necessary activities.
4. *Write lesson plan/script* Outline your teaching strategy step by step, with instructions complete enough that you can preteach the material mentally. Imagine yourself in the classroom; anticipate student questions, your answers, their reactions, your explanations. If your instruction includes more complex forms of media such as slide sets, videotapes, or individualized instruction, develop scripts for those aspects. Include instructions for making the visuals and notes on where certain prepared illustrations can be located.
5. *Produce the lesson* This is the point at which you must create all of the material you have specified. If time is short or some elements of your plan must be delayed, go as far as you can to complete the lesson. You can always continue to add as you gain experience and expertise.
6. *Teach and evaluate* It should be obvious that these are the remaining steps in the process, although they may not occur simultaneously. Some types of evaluation can be informal, by sensing the expressions and comments from students. On the other hand, you may wish to develop a formal evaluation sheet to obtain student reactions.

TRY THIS

Select a lesson plan, unit plan, script, or similar project that you have designed as an assignment for this course and evaluate it on the basis of the formula for orchestrating instruction (the six steps) presented in this chapter. Write a paragraph or two for each of the six steps, commenting on strengths and weaknesses of your teaching plan.

CHOOSING A DELIVERY SYSTEM

We have spoken at length about the need for variety in teaching by using a range of materials and techniques, and by varying your verbal delivery of information. On a larger scale, variety can be the basis for the types of classes you teach. One convenient method for categorizing instruction is to describe systems for delivering it. Three modes are generally recognized:

1. *Large-group instruction* Refers to teaching groups of more than twenty people. The size of the group is only one consideration; the method of handling the instruction is another criterion. For example, when you watch a television show you may be part of an audience of several million. Certainly that is a large number, but the instructional style upon which the program is based is equally important. There is no opportunity for individual participation and interaction. Large-group teaching generally makes little provision for audience interaction, either with each other or with the instructor.

2. *Small-group instruction* Employs a freer, more interactive atmosphere, where questions and interchange of ideas are encouraged. The group usually numbers about five to ten people. A more relaxed climate is possible, and various techniques such as brainstorming and role playing may be used. Since it is sometimes necessary to break a large group into smaller segments, it will be important to designate group leaders, or peer leaders. That way each small group will have a specific contact person.

3. *Individualized instruction* Designed for use by individual students, this method of delivering instruction has many advantages. Generally the materials are portable, relatively simple, and interactive. Examples of appropriate items include computer programs, tape/slide sets, study prints, and reading material—all tied together with an audiotape to serve as a tutor. The scope of individualized instruction can range from a learning package on a single concept to large and extensive audio-tutorial systems designed to deliver an entire one-semester course.

PRODUCE

A Flowchart of a Delivery System

Using a series of rectangles (see illustration below), make a map of the instructional sequence you might use in teaching a selected subject.

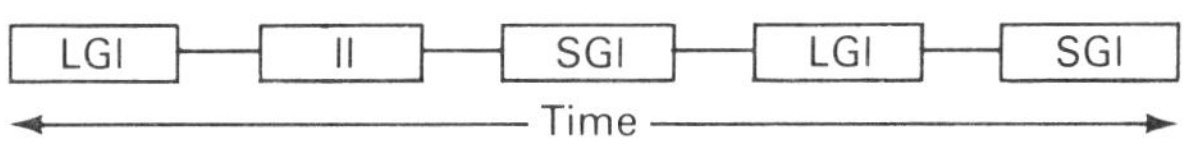

1. Consider the advantages of each method of delivery, as related to your intended sequence of topics.
2. Using the stated characteristics of large-group, small-group, and individualized instruction, analyze each topic to determine the best delivery method.
3. Make a map of the sequence, using the example in the illustration. In each rectangle specify the type of delivery. Under each rectangle make a brief statement about the content. Beneath the entire sequence draw a time line to indicate the amount of time to be spent on each mode of delivery.

A FINAL WORD

While some of the contents of this chapter may have repeated information found earlier in the text, it is important enough to require additional emphasis. You may have felt somewhat unprepared to handle questions of instructional design earlier but, with the background acquired in media production, the options and opportunities for producing strategies of your own design should now be much clearer.

Teaching with media requires a type of double vision; both a macro- and microview. Sizing up the entire scope of a teaching task is essential. At the same time, it is necessary to be thoroughly competent to deal with the smaller production and utilization that comprise the broader picture.

LOOKING AHEAD

Think of the entire preceding portion of this text as preparation for a performance. The planning, production, and careful attention to detail that a conscientious teacher carries out parallel the activities that precede the opening night of a new theater production.

All that remains is the teaching act. Arrangements similar to those necessary for any public gathering must be kept in mind. Since media hardware will be involved, the arrangements take on an added dimension of complexity. In the final chapter the critical aspects of teaching performance will be discussed.

REVIEW QUESTIONS

1. List some disadvantages of teaching exclusively with a single medium.
2. List some advantages of orchestrating teaching with a variety of media.
3. List and provide examples of some of the conditions that prevail during learning.
4. How might a knowledge of field dependence/field independence help a teacher who wishes to choose and use media in teaching?
5. Choose one of the elements in Cluster 1 and discuss its significance in planning instruction.
6. How can contemporary mass media be gleaned for suggestions of teaching style and ideas?
7. List the characteristics of your voice that contribute to effective use.
8. List the steps in orchestrating instruction.
9. List and describe three types of delivery systems.

ANSWERS TO REVIEW QUESTIONS

1. Uninteresting, routine, dull instruction. Furthermore, no single medium can represent all ideas or concepts.
2. Capturing and holding attention and interest, as well as the ability to portray selected concepts in a particularly distinctive and effective way.
3. Attention span, type of activity, location of learning, and timing. (Examples will vary)

4. Knowing that some students perform more effectively if a higher degree of structure is provided, a teacher can plan instruction for that contingency. For students requiring less structure, the teaching can be more open and variable.
5. Answers will vary.
6. Production techniques used in television, film, or newspapers and magazines can provide current examples of the process of developing scenarios, story lines, photo essays, and other forms of instruction.
7. Inflection, rate, pronunciation, and grammar use.
8. Narrow the topic, determine objectives, choose learning activities, write lesson, produce it, teach, and evaluate.
9. Large-group, small-group, and individual instruction. See explanations in chapter for descriptive characteristics.

REFERENCES

Cavert, C. E. *An Approach to the Design of Mediated Instruction* (Washington, D.C.: Association for Educational Communication and Technology, 1974).

Fleming, M., and Levie, W. H. *Instructional Message Design* (Englewood Cliffs, N.J.: Educational Technology Publications, 1978).

Gagné, R. M. *The Conditions of Learning,* 3d ed. (New York: Holt, Rinehart and Winston, 1977).

Gagné, R. M., and Briggs, L. J. *Principles of Instructional Design* (New York: Holt, Rinehart and Winston, 1974).

Gerlach, V. S., and Ely, D. P. *Teaching and Media: A Systematic Approach,* 2d ed. (Englewood Cliffs, N.J.: Prentice-Hall, 1980).

Kaufman, R. A. *Educational System Planning* (Englewood Cliffs, N.J.: Prentice-Hall, 1972).

Kemp, J. E. *Instructional Design,* 2d ed. (Belmont, Calif.: Fearon, 1976).

Romiszowski, A. J. *The Selection and Use of Instructional Media* (London: Kogon Page, 1974).

Russell, J. D., and Johanningsmeier, K. A. *Improving Competence through Modular Instruction* (Dubuque, Iowa.: Kendall/Hunt, 1981).

GOALS

☐ Become aware of the elements that comprise teaching performance and teaching effectiveness.

☐ Understand matters that need attention before teaching begins, skills to employ during the teaching, and details to keep in mind when the teaching is finished.

OBJECTIVES

☐ Describe teaching performance and teaching effectiveness, and list methods for evaluating each.

☐ List considerations to be kept in mind before teaching, during teaching, and after teaching.

In the last few sentences of the last chapter we compared all of the planning and production that precede teaching to the arrangements necessary for readying a play for opening night. The original idea, changes that occur to it as a treatment develops, the media needed to make it take shape and come alive, and delivery systems to bring it to an audience—all are a part of the important prior arrangements.

The situation is much the same with classroom teaching. The subject has been chosen, the objectives have been written, the media have been selected and made, and the assignments have been determined. Now it is time to stage the lesson.

THE NEED TO MERCHANDISE IDEAS

Teaching has been compared to salesmanship because both require that a client become convinced to invest in a product or an idea. Even though teachers have a captive audience and salespersons may not, the task is still the same. In both cases it is necessary to prove that a new product or new information will have value.

Because we live in a world that uses modern technology to sell ideas and goods, the population is accustomed to such sales treatments. Advertising, the promise of a "good deal," guarantees, and follow-up service are forms of leverage that evoke response from customers. Similar techniques enable a teacher to build confidence with students, and create an enthusiasm for learning. Evidently the public responds favorably to education, since public opinion polls consistently show that educators and our educational system are held in very high esteem.

15

TEACHING WITH MEDIA

It would seem foolish to attempt to sell anything without proper merchandising techniques. The same should be true for teaching. Knowing how to portray a subject in the most favorable circumstances makes its acceptance more probable. Much depends on the atmosphere and instructional climate created by the teacher. The performance of the teacher, including preparation before entering the classroom, is an important key in gaining acceptance and creating enthusiasm among students for learning. Media make the task easier and more effective.

TEACHER PERFORMANCE AND TEACHER EFFECTIVENESS

These two aspects of teaching are often confused with one another. *Teacher performance* refers to the methods and techniques used during instruction. It can be measured, although the evaluation is subjective. Questions that might be used on a performance rating scale are shown on page 302.

On the other hand, *teacher effectiveness* refers to the cognitive, affective, and psychomotor skills that are taught. Measurement of student acquisition of the skills is more objective, and shows up as scores on tests, quizzes, and examinations.

The relationship between performance and effectiveness is sometimes confusing and unclear. For example, it is possible to have a teacher whose performance is superior, as measured by rating sheets that students fill out. The teacher may be well liked, and conduct interesting and enjoyable classes. The pace and variety provided by such a teacher have great appeal to students. They may feel they learn a great deal. A study of their test scores, however, shows that they actually may not be learning too much. They may be getting more entertainment than education.

On the other hand, some entertaining and interesting teachers are able to create classroom situations that do often lead to above-average learning. Both their performance and effectiveness are at work to complement each other, resulting in what may be called "good teaching."

You can see that good performance does not necessarily correlate highly with effective learning. However, if performance is poor, the chance is that learning levels will be low, also. The teacher who is dull, who provides little variety in the classroom, and who uses few or no media stands a poor chance of encouraging students to become motivated.

Apparently the key ingredient is the connection between performance and effectiveness. Merchandising and selling (performance) must be kept at a high level, as well as the teacher's knowledge (effectiveness). In other words, teachers must have thorough knowledge of the subject as well as a knowledge of teaching methods.

A variety of approaches could explain how to increase skills in teaching performance. The approach used here examines classroom teaching from three points of view: before, during, and after. The necessary considerations in each segment are discussed within this chronological framework.

Some Questions to Use in Evaluating Teacher Performance

Question	Yes		Neutral		No
1. Was the instructor well organized?					
2. Was class time used efficiently?					
3. Was the instructor interested in the class and enthusiastic about the content?					
4. Did the instructor stimulate student interest?					
5. Was the instructor respectful, tolerant, and fair?					
6. Did the instructor explain the course material clearly?					
7. Were the objectives clearly stated and understood?					
8. Was the lesson plan format followed as outlined in class?					
9. Was the lesson well organized and prepared?					
10. Was the lesson appropriate for the grade level?					
11. Did the lesson pertain to a definite curriculum or developmental area?					
12. Was there an attention getting introduction to the lesson?					
13. Was there a good summary to the lesson?					
14. Was the lesson material well covered in the amount of time given?					
15. Was there good use of instructional materials?					
16. Was there good instructor-student interaction and communication, i.e., were all the students involved in the discussions and activities?					
17. Did the teacher react well to situations that arose during the lesson?					
18. Did the teacher demonstrate good classroom management?					
19. Did the teacher use questioning and answering effectively?					
20. Did the teacher use an appropriate speaking voice for the classroom situation?					
21. Was there a pleasant overall atmosphere created during the lesson, i.e., was the learning situation comfortable, cheery, open so everyone felt free to talk and was it inviting to both the students and teacher?					

PRODUCE

An Evaluation Sheet to Assess Teaching Performance

Using suggestions in this chapter and particularly the sample questions on the evaluation sheet included in the chapter, develop an assessment instrument of your own. Choose from this list of teaching characteristics:

- ☐ Verbal mannerisms
- ☐ Nonverbal mannerisms
- ☐ Attitude toward students
- ☐ Method of handling discipline problems
- ☐ Techniques of dealing with students who have special needs

1. Choose one of the five areas (above) and think of teacher behaviors that would characterize that area.
2. Phrase one statement for each of the behaviors so that it can be evaluated on an agree/disagree scale.
3. Design a format for the instrument, placing the statements on the left and the scale on the right.
4. Field-test the instrument to determine how useful and accurate it may be.

PARTITIONING THE PERFORMANCE

The portions of the performance to be discussed in this section include events *before, during,* and *after* teaching. In our examples here, the before and after are less than an hour or two away from the actual classroom teaching.

Before Teaching Begins

This is the critical preparation time in which the last-minute arrangements and details must be checked and taken care of. It might be appropriate to think of the preparation in terms of a checklist, similar to that which a pilot uses before take-off. The following list of concerns includes many of the important items you should keep in mind. You may wish to add more ideas, based on your experience as a student. Presentations sometimes include awkward moments because of improper preparation on the part of the teacher.

Practice Operating Equipment If you are unfamiliar with any of the hardware that is to be used, it is necessary to learn how to operate it. Find the operating instructions, which often are attached to the cover of the machine. Review all of the steps in the procedure, and learn how to replace lamps, reset the circuit breaker (both in the machine, and on the electrical panel that supplies power to your classroom), focus, rewind, attach patch cords, or perform any of the other operations necessary to use the machine.

Appoint an Assistant For last-minute errands, controlling the lighting at the proper moment, turning projectors on and off, and serving as a general assistant for troubleshooting, it is important to have a student assistant on hand to help. The school may have an organized club or group of students to assist the teachers. Otherwise, appoint a student who can take this responsibility. Make arrangements for the assistant to practice with the equipment in order to become thoroughly familiar with all of it.

PRODUCE

Practice Operating Equipment

Identify an item of equipment that you plan to use in teaching, such as a videotape recorder, projector, computer, or other item of media hardware. Follow these steps in reviewing your ability to operate:

1. Check this text or other laboratory manuals for operating instructions.
2. Locate operating manual or instructions furnished with the hardware.
3. Using both sources (above) read the instructions and follow directions to operate the machine.
4. After practicing, repeat the operating procedure from memory two or three times without referring to the instructions.

If the RF signal is coming from the antenna, or from a pigtail connector attached to the antenna post, this switch should be in the 300-ohm position. If the RF signal is coming directly from an RF cable, the switch should be in the 75-ohm position.

A 3-prong to 2-prong adapter can be used with older wall sockets.

Locate the circuit-breaker panel and the switch for your classroom in case you overload a circuit and have to reset the breaker during class.

"Cue up" the Software If you are planning to show a film, have it threaded into the projector and ready to use. If audio- or videotapes are included, cue them up to the starting point. A stack of transparencies should be numbered and then organized in the sequence that is planned. Flannelboard and hook and loop items should be stacked in order, with the first addition to the board in the topmost position on the pile. Some teachers prefer to lay out the items on a tabletop in the pattern they will form on the board.

Learn to Trouble-Shoot In the preparation time, when you are becoming familiar with the equipment, learn to identify what may go wrong. Film projectors, for example, have *two* lamps; one for the film, and the other (an "exciter" lamp) for the soundtrack. If external speakers are used, check the speaker plug and receptacle on the output of the player. Wires sometimes work loose in plugs, resulting in no sound coming from the external speaker. Television monitors and receivers may have a switch to send a signal *either* from the antenna or from the RF input, but not both at the same time. If the television set is intended to show a picture from a player, see that the input switch is in the correct position.

Extension cords (power cords) frequently create difficulty because of faulty plugs, or faulty receptacles. If you suspect the cord is not conducting electricity, hook it up to a table lamp or other electrical device. Sometimes the outlet on the wall does not grip the plug firmly, so that no power is able to get into the extension cord.

Locate Power In older classrooms the wall outlets may not be numerous enough to permit many appliances to be operated. Furthermore, the outlets may be the older, 2-prong type rather than those that can receive a 3-prong plug. Acquire a supply of 3-prong adapters with the grounded third prong to use between an electrical appliance plug and the wall outlet.

It is wise to identify the location of the circuit-breaker panel for the classroom, since older circuits may not accommodate the heavy electrical needs of modern equipment. It is possible to blow a fuse or trip a circuit breaker if more than one projector is powered from the same outlet, or same circuit.

Determine Placement of Equipment It may be necessary to rearrange some of the seats in the classroom to accommodate a projector or television set. If a projector has a zoom lens the placement is not too critical since the size of the image on the screen can be made larger or smaller with the lens. Zoom lenses are available for all types of projectors, but they are more often found on slide projectors.

If recorded sound (from film, audio-, or videotape) is a part of the presentation, place an extension speaker in the front of the room. It is important, psychologically, to have sound and picture emanate from the same location; do not allow the sound source to be placed behind the audience. Once the classroom is set up, preview the setting to determine if it is possible to see and hear from all angles. Adjust volume levels and then mark the volume controls with a piece of tape. A smoother effect is achieved if sound is faded up at the beginning and then faded out at the end, thus eliminating the abrupt on/off shock and contributing to a more pleasing esthetic atmosphere.

Plan Alternatives While all of this planning is good insurance against unfortunate circumstances occurring during a presentation, the possibility of equipment failure remains. When that happens, avoid the temptation to announce that you are sorry, or that you sympathize with the unfortunate audience. Calling attention to the problem does not make it go away. A more effective alternative is to have a substitute strategy in mind; some notes to write on the chalkboard, a transparency or two that represent an overview of the same lesson that failed, or a duplicated handout that can be used in continuing the lesson.

Alternatives are also important in the case of software that is expected to arrive just prior to use or does not arrive. When films are rented, when computers are scheduled, when kits and learning packages are reserved, or when other media are required, there is the chance that the materials will not be available as planned. In these cases an alternative teaching strategy should be on hand.

During Teaching

Even though preparations have been made for effectively presenting a lesson, certain accommodations must take place concurrent with the instruction. Some of the important concerns are listed below.

Assess the Audience It is always a good idea to take an up-to-date inventory of the level of understanding (or confusion) possessed by the class. As alternatives to the common "Are there any questions . . ." it is effective to hand out small cards or half-sheets of paper, asking participants to list any concerns, questions, or points of confusion. Advantages of this method of assessment include the protection of anonymity for those who do not want to identify themselves by asking questions that might be considered simplistic. The teacher is also protected by not having to attempt an answer that might not be correct. Surprises are eliminated if questions are submitted on paper. Furthermore, the teacher can sort the questions, answering only those that seem to be relevant.

Another method for assessing classroom climate is through the use of a simple cardboard device consisting of a disk with colored, pie-shaped segments. Students can vote "yes," "no," or "uncertain" by displaying the appropriate color (green, red, or yellow) on the disk.

Provide a Preview In an earlier chapter the utility of establishing set, introducing a topic, and providing an overview was pointed out. A few notes on the chalkboard or a quickly produced transparency is sufficient to indicate what is to be done during the class. Objectives and goals should be included also, as well as assignments. If possible, indicate the time to be spent on each segment of the lesson so that students can accustom themselves to the instruction, and pace themselves. Some media, such as tapes or films, should be introduced with a thorough overview, based on your own prior examination of the material. A duplicated guidesheet or brief notes may serve as an effective advance organizer.

Zoom lenses make it possible to change the size of the projected image without moving the projector.

The projector speaker should normally be placed in the front of the room under the screen.

PRODUCE

A Simple Feedback Device

These instructions are provided so that you can make a class quantity of the feedback device.

1. Using light-weight cardboard or a $8\frac{1}{2} \times 11$ manila folder, form an A frame. Tape the bottom edges of the frame so that it stands upright on a desk.

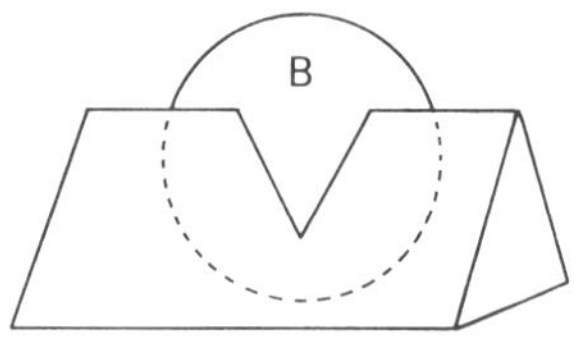

2. Using a 3-pound coffee can, trace and cut out a circle from another piece of cardboard.
3. Consider the circular cutout as a pie and draw seven wedges on it. Label A, B, C, D and color each of the three remaining wedges red, yellow, and green. Do the same on the reverse side of the circular piece.

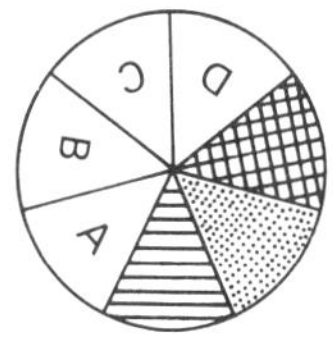

4. Mount the finished disk into a slot in the A frame with a brass fastener.

Students can use the feedback device to respond to A-B-C-D multiple choice questions that are asked in class, or to respond positively (green), neutral (yellow), or negatively (red) on questions requiring an opinion.

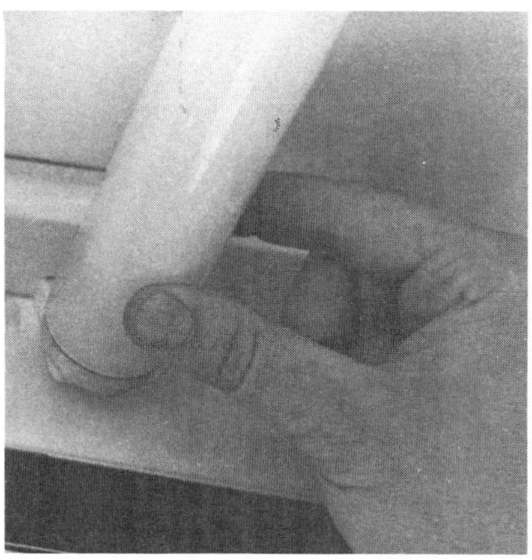

It is often a good idea to loosen the fluorescent bulb directly over the screen so that it does not light. This keeps the projected image from being washed out by this light.

Monitor Classroom Environment Light, temperature, sound level, external noise, the teacher's voice, and student demeanor contribute to the atmosphere of a class. All are important, but some are often overlooked. The light level in the classroom, for example, is sometimes too high or too low when projected media are being used. Remember, it is not necessary to darken a room during the use of an overhead projector. But it may be essential to keep an above-average amount of light from hitting the screen, either originating from windows or artificial lighting in the room. A ceiling light fixture near the screen can reduce the contrast of many projected materials.

Some suggestions for getting and keeping attention of a class include the technique of revealing one line at a time when transparencies are used. In the case of both an overhead projector and a slide projector, it is poor technique to allow the screen to be illuminated unless an image is being projected. The large, lighted screen is distracting. Thus, do not turn on the overhead projector unless a transparency is already in place on the stage. And remember to insert a 2×2 piece of cardboard at the beginning and at the end of a tray of slides. In this way the slide set begins and ends with a dark screen.

An effective means of calling attention to material that is being projected is to use a pointer. A wooden dowel, available at any lumber store, makes an effective pointer—particularly if painted white for increased visibility. A homemade pointer, such as this, should be wrapped with tape at the end so that the screen is not damaged when touched by the pointer.

Review the Lesson Set aside a few minutes at the end of each teaching session to review, recap, summarize, and assess the level of understanding of the students. If some of the media used in the lesson are available for further study, it is effective to place them in the school media center so that interested students can work with them for additional time. A helpful technique involving transparencies is to pass each transparency around the class after use to enable individuals to study further.

After Teaching Ends

According to some teachers, teaching never ends. The windup of one lesson is preparation for another. The processes of evaluation, revision, and recycling of materials are continual and ongoing. In this last section of the chapter, however, some attention is paid to those more important elements of teaching that might take place after a class period is over.

Evaluating Whether a formal assessment of the teaching is made, or informal verbal comments are sought from students, it is important and necessary to monitor classroom climate and the feelings of students. Solicit comments from students in an informal setting, talking with them before or after class to find out how things are going. Unthreatened by having to speak out in front of the entire group, individuals will frequently speak quite honestly and openly about your teaching.

TRY THIS

An Evaluation of a Media Presentation

Locate a videotape of a presentation or make a videotape, either of your own teaching or of another person teaching. Then evaluate the presentation using the criteria listed in the form on page 308.

Revising Based on the evaluation from students as well as the inherent feeling that comes about from the teaching performance, certain changes and modifications will present themselves for inclusion in the next lesson to be taught. It is possible that the sequence of topics needs reorganizing, or certain types of media need to be changed, added, or deleted. Explanations, analogies, examples, sample problems, and other elements of the presentation should be analyzed for effectiveness. Remember—teaching contains a portion of trial-and-error experimentation, and there will always be another time to try an idea again.

Putting Things Away The urge to walk out of the classroom when teaching is finished may leave a number of details undone. If any equipment and its associated software have been used, it is important to put things away. Films should be rewound, as well as audio- and videotapes. Projectors should be allowed to cool, then lids replaced, and cords wound up. If any equipment malfunctioned or lamps burned out, a note should be attached with the needed repair noted. If the equipment or software was checked out from the media center it should be returned, or the center should be notified that the use of the medium has been finished. If you have a student assistant, he or she can handle these jobs.

Evaluation of a Presentation

____
(Name)

____________ ____________
(Section) (Media)

	Poor		Average		Excellent
1. *Scope of Presentation* Topic too broad or narrow? Concept too complicated or simple? Comments:	1	2	3 (circle one)	4	5
2. *Organization and Continuity* Did material presented early help understanding of later information? Were terms defined? Was an overview provided? Comments:	1	2	3	4	5
3. *Creativity and Productivity* Uniqueness of presentation. How much time, effort, creativity demonstrated? Comments:	1	2	3	4	5
4. *Choice and Amount of Media* Were proper materials used? Were media over/under used? Comments:	1	2	3	4	5
5. *Quality of Materials* Care used in preparation? Comments:	1	2	3	4	5
6. *Technical Use of Media* Did student know operating procedures? Was equipment properly located? Was image easily seen? Were images projected correctly? Comments:	1	2	3	4	5
7. *Objectives* Were objectives obvious? Were objectives met? Comments:	1	2	3	4	5
8. *Overall Effectiveness* Was student effective in teaching? Was audio (voice) appropriate? Comments:	1	2	3	4	5

A FINAL WORD

While this section of each chapter has served the purpose of review of the chapter, in this case it may be appropriate to glance back over the entire book. As stated in the preface, the emphasis has been on production of media for teaching. However, the use of media has also been included where appropriate because it becomes nearly impossible to separate production and utilization. The planning and thinking that go into the overall design of any lesson involving media may not fit neatly into a production or utilization category.

Please try to remember specific techniques, and refer to them when you need to refresh your memory. The index to this text should make the job of locating the techniques much easier. While pigeonholing and classifying the production technique are necessary, it is also important to remember that the Gestalt approach might turn out to be the most practical. The broader consideration of a large number of elements that comprise that endeavor we call "teaching" might not lend itself to the neat classification of individual techniques. Many ideas and skills must be available so they can be used more or less simultaneously. Rather than a single best approach, it is more sensible to realize that several suitable approaches are at hand for the solution to an instructional problem. Superimposed on all of the methods for making and using media are the spirit and creativity of the teacher who adds an undefined, slightly mysterious ingredient. When all of this is synthesized into a lesson plan or a script for a production, the product is more effective instruction and more effective learning.

REVIEW QUESTIONS

1. Why do ideas in the classroom require merchandising, much like products sold in stores?
2. What is the difference between teacher performance and teacher effectiveness?
3. What is the relationship between performance and effectiveness?
4. List important considerations to be responsible for before teaching begins.
5. List considerations to be responsible for during teaching.
6. List matters that need attention after teaching is completed.

ANSWERS TO REVIEW QUESTIONS

1. Selling ideas and concepts is a process that closely resembles selling products and services. The student or customer must be convinced that acquiring the item will be important to welfare or happiness.
2. Performance is related to the physical movements, manner, explanations, and other characteristics normally associated with articulating ideas. Effectiveness is related to the skill, attitudes, and cognitive information that students acquire.
3. There can be a high degree of relationship between performance and effectiveness, but apparently a number of additional factors are also involved. Therefore, it becomes difficult to state a definite relationship with certainty.
4. Practice, appoint an assistant, organize the software, learn to trouble-shoot, locate power sources, determine placement of equipment, plan alternatives.
5. Assess the audience, provide an overview, monitor classroom climate and esthetics, and provide a review of the lesson.
6. Evaluate, revise the lesson, put things away.

REFERENCES

Brown, J. W.; Lewis, R. B.; and Harcleroad, F. F. *AV Instruction,* 5th ed. (New York: McGraw-Hill, 1977).

Heinich, R.; Molenda, M.; and Russell, J. *Instructional Media* (New York: Wiley & Sons, 1982).

Locatis, C., and Atkinson, C. *Media and Technology for Education and Training* (Columbus, Ohio: Charles E. Merrill, 1984).

Sleeman, P. J.; Cobun, T. C.; and Rockwell, D. M. *Instructional Media and Technology* (New York: Longmans, 1979).

Wittich, W. A., and Schuler, C. F. *Instructional Technology,* 6th ed. (New York: Harper and Row, 1979).

GLOSSARY

AC Alternating current. Standard 120-V, 60-cycle (-Hz) USA household current. Also called line voltage. VTRs and cameras often use the 60 Hz as a horizontal sync reference.

accent light A spotlight that accentuates and highlights an object in a scene.

acetate A plastic sheet permitting a high degree of light transmission, resulting in a transparent appearance.

acoustic feedback The howling caused when a microphone picks up vibrations from its own speaker system.

advance organizers The use of clues and previews before instruction to promote learning, as proposed by David Asubel.

affective domain Category of instructional objectives relating to attitudes, values, and appreciations within human behaviors.

AGC See *automatic gain control.*

agitation The process of.stirring, swirling, or otherwise causing a liquid to move freely over the surfaces of film or paper during processing.

ALC Automatic level control (same as AGC).

amplify To increase levels, as with a volume control.

ANSI. American National Standards Institute. The present title of the organization which used to be called the American Standards Association (ASA).

aperture The lens opening, F/stop, or iris which determines the amount of light that enters the camera. Expressed as a fraction of the focal length. The F number.

ASA The American Standards Association, now the American National Standard Institute. A system of film-speed rating now replaced by the ISO designation.

aspect ratio Proportion or format of an audiovisual material, such as 3 × 4 for a filmstrip or motion picture frame.

assembly editing "Add editing." A new video or audio sequence that is consecutively added to a previously edited scene.

attenuate To decrease levels, as with a volume control.

audio Sound portion.

audio dub The recording of sound only, without disturbing the picture.

audio head A magnetic recording head that records or plays back sound.

audio in Audio input jack.

audio mixer A device that allows the simultaneous combining and blending of several sound inputs into one or two outputs.

audio out Audio output jack.

audio track The portion of the videotape that stores the sound.

audio cassette recorder A recorder that uses magnetic tape on spools that are enclosed in a plastic case.

automatic gain control An automatic circuit that controls the audio or video intensity during recording.

automatic shut-off A special switch on a tape recorder which automatically stops a machine when the tape breaks or runs out.

auxiliary in Audio line input (nonmicrophone).

auxiliary out Audio line output (nonmicrophone).

available light The light condition which the photographer finds existing at the subject position. The term usually implies an indoor or nighttime light condition of low intensity requiring fast film, large lens aperture openings, and/or slow shutter speeds.

B&W Black and white.

back light Illumination from a source beyond the subject, as seen from the camera position, which tends to silhouette the subject.

background light The illumination thrown on the background to lighten it, giving the scene depth and separating the subject from the background.

beam The flow of electrons in the camera or TV tube.

behavioral objective A planning technique that specifies in performance terms what a person should do as a result of instruction. The three components of behavioral objectives are the conditions under which learning and evaluation take place, action to be demonstrated by the learner, and evaluation criteria to indicate the level of learning considered essential.

between-the-lens shutter A shutter designed to operate in a space between the elements of the lens.

boom (v) Camera moves up, down, or sideways through space.

boom (n) A long arm or device used to suspend a microphone or camera over the action.

bracket To make a number of exposures (some greater and some less than one considered to be "normal") in addition to the "normal" one, with the intent of getting one near-perfect exposure.

broadcast TV A type of programming and bureaucracy that produces TV programs specifically for over-the-air broadcasting.

bromide Any salt of hydrobromic acid but commonly used as a shortened form of potassium bromide.

bulb A marked setting (B) of most shutters which permits the shutter to be held open for an indefinite period by continued pressure on the shutter release.

bulk eraser A strong alternating electromagnetic device used to erase the magnetic patterns on tape while still wound on a reel, or in bulk form.

C-mount Standard screw-in lens mount found on most video cameras.

cable release A long, flexible cloth or metal braid-covered plunger which screws into a special threaded socket on the shutter or camera body. Compressing the plunger with thumb and finger pressure will release the shutter with little danger of camera movement or vibration.

cable TV Television signals transmitted primarily by cable instead of antennas.

camera Literally, "room" in Latin. The instrument with which photographs are taken, consisting of a light-tight box, a hole, and a shutter.

camera obscura Latin for "dark room." Ancestor of the photographic camera; probably originally an actual room in which observers could watch the images of outside subjects formed by light rays entering the room through a small aperture and passing directly to the opposite wall. The camera obscura eventually evolved into a portable box with lens and viewing screen, and finally for actual photography.

capstan A roller and rotating shaft in the VTR that is motor-driven and determines the speed at which the tape is moved through the VTR.

CATV Cable TV.

CCTV Closed-circuit television. Any self-contained, nonbroadcast TV system.

channel A specific band or frequency assigned to a radio or TV station.

chassis The metal base or frame which carries the electrical or mechanical assemblies.

chroma (chrominance) The hue or saturation (intensity) of color.

chromakey An electronic matte where one image is matted on top of another without image bleed-through (an electronic blindfold).

close shot Only the face of a person is shown, filling the screen completely.

close-up A concentration of the camera on the subject, or on a part of it, excluding everything else from view.

close-up lens A supplementary lens which, when placed over a camera lens, shortens its focal length and thereby permits closer than normal focusing.

closure The summarizing of newly learned information, recognition of general achievement through recall of learning, and the development of foundations for using acquired knowledge in future learning activities.

cocking the shutter Winding or tensioning the shutter mainspring prior to making the exposure.

cognitive domain Category of instructional objectives relating to knowledge, information, and other intellectual skills within human behavior.

color lock A circuit that stabilizes the color hues during playback.

color temperature Measured on the Kelvin scale, and is appropriate only for the description of continuous-spectrum light such as is emitted by the sun or tungsten filament bulbs.

colored adhesive A translucent or transparent color printed on a thin acetate sheet having adhesive backing for adherence to cardboard, paper, acetate, or film.

compatibility The ability of one piece of equipment to interface with another (interchangeability).

concept Broad, general statement that summarizes fundamental ideas.

condenser microphone A high-quality microphone that transmits sounds when a plate that receives direct current is caused to vibrate with respect to an adjacent fixed plate. Condenser mikes contain built-in amplification and require batteries.

conditions of learning Proposed in 1965 by Robert Gagné as a way to sequence and organize learning. The eight levels are: signal learning, stimulus-response learning, chaining, verbal association, multiple discrimination, concept learning, principle learning, and problem solving.

contact paper 1. A printing paper, usually sensitized with silver chloride, which is intended for use in contact printing. 2. A pressure-sensitive plastic-coated paper usually used for shelf covering, but which may be applied to instructional materials.

contact printing A method of printing in which the negative is placed in contact with the printing paper, emulsion-to-emulsion, and held in that position in a printing frame. The exposure is made by exposing the frame to raw light so that the paper emulsion is exposed by light through the negative.

continuous tone Describes an image containing a gradation of grays as well as black and white extremes. See dropout.

contrast Density difference, usually of adjacent areas of the image.

contrast grade A number or descriptive term assigned to a particular printing emulsion which identifies its contrast characteristic. In the range from 1 through 6, a normal contrast paper is usually considered to be 2, with the lower numbers indicating a tendency toward lower contrast and higher numbers indicating higher contrast. Printing filters for use with variable contrast papers are numbered similarly.

control track The portion of the tape containing the speed-control pulses.

control-track head The head which records or plays back the control-track pulses.

control-track pulses Speed-control pulses created by the VTR.

copy stand A vertical or horizontal stand for accurately positioning a camera when photographing flat subjects very close to the lens.

credits The titles at the beginning or end of a film listing the names of people who made or acted in the production.

crop The cutting off of picture elements by the camera framing.

cross talk Signal leakage between two channels. Undesirable interference created by one track on another.

CRT Cathode ray tube. The vacuum tube used in all TV sets.

CU Close-up.

cue A warning or indication of an event that is about to take place.

cut The most common type of transition in which one scene ends and a new one begins immediately.

cutaway A shot used to introduce variety to a given scene by showing a simultaneous action.

cut film Another name for sheet film.

cut-in shot A close-up feature of a subject being filmed as a motion picture and usually placed between two scenes that have discontinuity of action.

daylight Sunlight or skylight or any mixture of the two. For the purposes of color photography, daylight is considered to have a color temperature of from about 5500 K to 6000 K.

DC Direct current. Electrical current that maintains a steady flow and does not alternate.

decibel A relative measure of sound intensity. One decibel is the smallest change in sound volume that the human ear can detect.

definition The degree of detail or sharpness in a TV picture.

depth of field The region of acceptably sharp focus around the subject position, extending toward the camera and away from it, from the plane of sharpest focus. The boundaries of the depth of field are referred to as the near limit and the far limit. Varies with the distance of the camera from the subject, the focal length of the lens, and the F/stop.

depth of focus The little zone including the focal plane of the lens through which the film can be moved, toward and away from the lens, and still record an acceptably sharp image.

developer The solution which produces the silver image in the normal photographic process.

diaphragm The assembly of thin metal leaves, usually incorporated into the lens barrel or shutter assembly, which can be adjusted to control the size of the lens aperture. Same as iris diaphragm.

diazo process A method for preparing overhead transparencies requiring film containing one of a possible number of dye colors, which is exposed, in contact with a translucent original, to ultraviolet light, and then developed in ammonia vapor.

DIN Deutsche Industrie Normen. The German system of film-speed determination.

dissolve An optical effect in motion pictures involving two superimposed scenes in which the second gradually appears as the first gradually disappears.

dissolve unit A device that activates one or more projectors to fade out an image on the screen while an image projected by another projector begins to fade in as a superimposed image over the first one.

distortion Any difference between the original sound and the recorded and reproduced sound.

documentary A film of fact as distinguished from fiction; it uses actual or restaged material from real life.

dolly The movement of a camera toward or away from an object. Also the wheeled apparatus on which the camera is mounted.

dolly shot A motion picture scene filmed as the camera is moved away from, toward, or across the subject.

dropout A loss of picture signal during tape playback—displays as a black or white streak in the monitor.

dry-mounting A method of mounting prints on cardboard or similar sheet materials. Dry-mounting tissue placed between the print and mount board is softened by the heat of a dry-mounting press to effect the bond.

dry-mounting press A machine for dry-mounting prints. It has a large, flat, metal pressure plate which can apply uniform pressure and thermostatically controlled heat on the prints to be mounted.

dry-mounting tissue A thin tissue paper impregnated with shellac or some similar material, which, when heated sufficiently, softens to become an effective adhesive for paper.

dry-transfer letters Transferring symbols from a sheet to the working surface by rubbing with a blunt tool.

dual-track recorder Type of monophonic recorder which records or plays back half of a standard ¼-inch tape in one direction and the other half in the opposite direction.

dubbing The duplication of a videotape or the addition of new audio information to the tape.

dynamic microphone An electromagnetic type which employs a moving coil in the magnetic field.

dynamic range The ratio between the softest and loudest sounds a tape recorder can produce without distortion.

ECU Extreme close-up.

editing Selection of certain sections of tape recordings and the deletion of unwanted portions and then splicing them together in the desired sequence. The same process applies to editing films.

editing in the camera Filming motion picture scenes according to their listing on the script.

EIA The Electronic Industries Association. The American TV standards Committee.

EIAJ The Electronic Industries Association of Japan. The Japanese standards committee which set standards for ½-inch helical scan videotape recorders.

EIAJ Type I The standard for ½-inch helical scan reel-to-reel VTRs manufactured after 1969.

electronic editing The process whereby a videotape is edited.

electronic flash A photographic light source which produces a brilliant flash of light by the discharge of electricity through a gas-filled glass or quartz flash tube. The flash duration is very short, usually less than 1/500 second and there is no firing delay.

emulsion The light-sensitive coating on photographic film or printing paper.

ENG Electronic news gathering.

enlarger A printing machine designed to project the image of an illuminated negative onto a sheet of sensitized paper.

erase head The magnetic assembly on a tape recorder over which the tape passes to remove previously recorded signals.

establishing shot A medium or long shot that establishes the whereabouts of a scene and serves as orientation to action photographed at close range. Often coming at the beginning of a film or sequence, the establishing shot sets the time and place of the action.

ETV Educational Television.

exposure 1. The act of subjecting a photosensitive material to the action of light. 2. The accumulated effect of the action of light on a sensitized material.

exposure index A number assigned to a film by the manufacturer which indicates the relative emulsion speed of the film for determining camera settings (F number and shutter speed).

exposure meter An instrument used to measure the illumination level at the subject position or the brightness of the subject, and to equate this information with the film speed to indicate appropriate camera-aperture and shutter-speed settings.

extended-play tapes Tapes that are longer than usual.

extension tubes A set of three or more (usually) rings or tubes of varying lengths, intended to be interposed between the body and lens of a small camera, so as to permit focusing the lens on subjects very close to the camera.

F number The numerical expression of the aperture diameter of a lens as fraction of the focal length. The lens setting selected from a series of numbers consisting partially of . . . 2, 2.8, 4, 5.6, 8, 11, 16, ... These numbers refer to the size of the lens aperture. They are the denominators of fractions.

fade A gradual change in the picture or sound intensity.

fade in An optical effect in motion pictures in which a scene gradually appears out of blackness.

fade out An optical effect in motion pictures in which a scene gradually disappears into blackness.

fade out, fade in This transition is often used to indicate passage of time. One scene gradually goes dark and the new one gradually emerges from the darkness.

fade to black The picture is faded out until the screen is dark.

fast A term used to describe lenses of large relative aperture or films of high sensitivity.

fast, or undercranked The camera is run at a speed lower than the normal 24 frames per second. When the film is projected at normal speed, the action appears to move rapidly.

fast forward Provision in a tape recorder to run tape rapidly forward through the machine for quick selection of desired position.

feed reel The reel on a tape recorder which supplies the tape.

fill light Light directed into the shadow areas of the subject to reduce the lighting contrast.

film Generally, the familiar light-sensitive material used in cameras in the practice of photography.

film chain 1. The part of a television system which includes the slide and motion picture projectors connected optically to a television camera. 2. An optical system whereby an image from a film or slide projector is transferred to a video camera for use in a television system.

film clip A brief filmed sequence used generally on television without having titles or special effects.

film gate The part of a camera or projector that presses the film on the open frame of the film mask.

film speed 1. The number of frames that pass through a camera or projector film gate per second (e.g., 18 fps). 2. The degree of sensitivity of a given emulsion (e.g., ASA 40).

film treatment A description of the film in narrative form written as if the writer were seeing the film being described.

first generation The original recording or master tape.

fish-eye lens A type of super wide-angle lens, or lens attachment, capable of covering a field of about 180°. Fish-eye images are circular, if the whole image appears on the film, and are notable for their barrel distortion.

fix To make the film insensitive to further exposure to light, usually by bathing the emulsion with a solution containing hypo or some other effective silver-halide solvent. Such a solution is called a fixing bath.

fixer A solution in which the chemicals harden the developed film image and change all undeveloped silver salts so they can be removed by washing.

flannel- (felt-) board A presentation board consisting of a flannel or felt surface to which objects backed with flannel, felt, or sandpaper will adhere.

flash General name for any photographic light source which produces a very brilliant, very brief pulse of light.

flip wipe The ending scene appears to run over on its vertical or horizontal axis, being replaced by the new scene.

focal length Loosely, the focal distance when the lens is focused on infinity; more accurately, the distance from the rear point of the lens to the image plane when the lens is focused on infinity.

focal-plane shutter A curtain shutter operating in the camera body just ahead of the film (or focal) plane.

focus To adjust a camera, for example, so that an image is formed precisely on the film plane.

frame 1. To adjust the position and angle of the camera with respect to the subject for the purpose of containing or composing the image within the boundaries of the viewfinder. 2. The useful area and shape of the film image. 3. A complete TV picture of 525 horizontal lines which is composed of two scanned fields of 262½ lines each. One complete TV frame is scanned in 1/30th of a second. 4. An individual picture in a filmstrip or motion picture.

freeze frame A single frame is selected and repeated for as long as needed.

frequency The rate of repetition in cycles per second (cps) of musical pitch, as well as of electrical signals. Low frequencies refer to bass tone, high frequencies to treble tone.

frequency response The ability of an electronic device to reproduce a range of frequencies.

gain The increase in signal provided by an amplifier between input level and output level.

genlock The locking up of the sync generators of one or several sources to a main sync source; i.e., a VTR locked up to a special-effects generator.

Gestalt learning Learning based on analysis of the unified whole, that the understanding of an entire process is better than the study of individual parts or sequences of the whole.

glitch Any picture distortion.

glossy Describes a printing paper with a smooth surface suitable for ferrotyping.

goal A desired outcome of instruction that is broad in scope, generally timeless, and nonbehavioral.

grain The visible granular texture of the silver image, caused by apparent clumping of the individual silver particles, not usually apparent until the image is enlarged or viewed under magnification.

gray card A card of known reflectance, usually 18 percent, intended to be placed in the subject area and used as a meter target in the determination of exposure.

gray scale The various shades of gray in a TV picture which correspond to color.

ground A point in any electrical system that has zero voltage, usually the chassis of any electrical component.

ground glass A general name for the focusing screen in a reflex or view camera.

guide number A number assigned to a film for the purpose of calculating exposure when flash bulbs or electronic flash units are used; it is based upon the film speed, the type of flash bulb or electronic unit, and the shutter speed.

halides Metallic compounds of the halogens, namely the elements fluorine, chlorine, bromine, and iodine. The chlorides, bromides, and iodides of silver are the light-sensitive materials upon which most photographic processes are based.

head An electromagnetic device across which the tape is drawn and which magnetizes the iron oxide coating of tape. Records or retrieves information from magnetic tape.

head alignment In tape recorders, the correct position of the tape head and gap, with respect to the magnetic field.

headphones Small sound reproducers in a suitable frame for wearing about the head. Close-coupled to the ears for private listening.

helical scan recording A helix-like method of recording video information on ape in which the signal is recorded diagonally in adjacent stripes. Also called slant track recording.

hertz (Hz) Cycles per second.

high angle The camera looks down at what is being photographed.

high-angle shot A scene photographed with the camera placed high, looking down at the subject.

high-contrast subject An illustration consisting solely of black lines or marks on white paper.

hook and loop board A presentation board consisting of a cloth surface textured with minute nylon loops to which display materials backed with strips of tape having nylon hooks will intermesh and hold firmly.

hum Low-frequency noise in an audio component usually induced from the power line or stray magnetic fields.

hypo The common nickname for the chemical sodium thiosulfate, which was originally called sodium hyposulfite.

Hz Hertz; cycles per second.

impedance Measured in ohms, it is the AC resistance of any electrical system. Generally referred to as either "high" or "low" impedance. For best results in connecting two components, output and input impedances must match.

incident light The light reaching the subject from any and all sources.

incident-light method The measurement of light falling on a scene by the use of an incident-light meter held in the hand and aimed at the camera.

incident meter An exposure meter designed, or adapted, for the measurement of incident-light intensity, or illuminance.

input A receptacle or other connection through which an electronic signal is fed into an amplifier.

insert edit The addition of new video and/or audio information into any point of a preexisting video program.

instructional development Process of designing an instructional program employing an objective, systematic procedure, such as an instructional-design plan.

instructional system A coordination of all aspects of a problem toward answering a specific objective and following a logical series of steps to reach a solution. The logical, systematic planning of instruction with these fundamental components: objectives, sequence, strategy, and evaluation.

instructional technology Includes two components: 1. the equipment and media resources for instruction and 2. the process of systematic instructional planning termed instructional development.

interlace The process of scanning whereby the alternate lines of both scanned fields fall evenly between each other.

ips Inches per second. Tape speed or writing speed.

iris The adjustable opening on the lens which controls the amount of light entering the camera.

jack Receptacle or plug connector leading to the input or output circuit of a tape recorder or other component.

jump cut Shots are put together in such a way that there is little continuity, giving a feeling of strife, confusion, or violent action.

kelvin (K) The unit of measurement which denotes the temperature of light.

key light The light used for primary illumination of a scene.

keying The matting of one picture over another.

kilohertz (kHz) Thousand cycles.

kinescope "Kine," the process of recording a TV picture by photographing a TV set with a film camera.

lamination Applying a thin adhesive-backed clear acetate coating over a picture or other graphic material.

large format Video equipment that uses very large videotape such as 2-inch.

latent image The invisible impression on the sensitized emulsion produced by exposure to light. Development converts the latent image to a visible one.

lavalier mike A small microphone worn around the neck.

leaf shutter A type of shutter, usually operating in the space between the major lens components or immediately behind the lens, which consists of a number of thin metal leaves or blades, arranged concentrically around the lens axis, pivoted so that they can either form an opening for the passage of light or overlap to block it.

LED Light-emitting diode.

lens diaphragm The opening through which light enters a camera; its size is controlled by an adjustable diaphragm.

lens speed The ability of the lens to collect light. A "fast" lens collects more light than a "slow" one.

level The average intensity of an audio or video source.

level indicator Indicates the level at which the recording is being made and serves as a warning against underrecording or overrecording. It may be in the form of a neon bulb, a "magic eye," or a VU meter.

light level The intensity of the light as measured in footcandles.

light meter An instrument which measures light intensity.

line in An audio input for other VTRs or an audio mixer.

line out Audio output.

lip synchronization (lip sync) A recording of the speech of a person appearing in a motion picture so that the sound is heard at the same time the corresponding lip movements are observed.

live The process of transmitting a program the instant it is taking place.

log sheet A written record of all pictures taken, including scene numbers, takes, camera settings, and special remarks.

long lens A telephoto lens.

long shot Although a relative term, a long shot is generally made from a sufficient distance to show a landscape, a building, a large interior, etc. Often an establishing shot is a long shot.

low angle The camera looks up at what is being photographed.

low light Subdued illumination.

luminance The photometric brightness or radiance of a light source.

macrolens A term used to describe lenses especially corrected for use at short subject distances.

magnetic-striped film 8mm or 16mm motion picture film to which a narrow stripe of magnetizable material is added which will accept sound impulses in the form of magnetic variations.

main light The light in a studio setup, usually the brightest one, which establishes the light and shadow pattern on the subject and thus describes the forms.

matching action Shots are put together in such a way that they give the impression of natural continuity. For example, if a man is shown crossing a room toward a door and the next shot shows him opening the door, the actions are matched and continuity is unbroken.

medium shot In basic film terms, the medium shot shows two men fighting and is taken from far enough back to reveal their full figures from head to foot. The camera view between a close-up and a long shot.

megahertz (MHz) Million hertz.

microteaching A form of teacher education where a student is videotaped while he or she practice-teaches in front of a small group. Later, the video recording is played back and critiqued in front of a small group. The five components of the microteaching model are set induction, closure, questioning, reinforcement, and nonverbal communication.

mil A unit of measurement used for denoting video- audiotape thickness—1 mil = 0.001 inch.

miniplug A small audio plug, sometimes called a Sony plug.

mixer A device by which signals from two or more sources can be combined and fed simultaneously into a tape recorder at the proper level and balance.

mixing The blending of two or more signals for special effects.

mnemonic Visual and verbal techniques used to remember things, such as ROY G. BIV (colors of the rainbow: red, orange, . . .).

modulation The process of adding video and audio signals to a predetermined carrier frequency.

monitor A TV set without receiving circuitry that is used primarily to display video signals.

monochrome Black and white.

monophonic recorder A tape recorder with a single recording head, capable of recording on only one channel at a time.

montage A series of short scenes in a motion picture used to condense time or distance or to show a series of related activities or places.

M.O.S. A sequence filmed without any sound being recorded.

multimedia Sequential use of a variety of instructional material for presentation or for self-study programs.

narration The verbal comments made to accompany the visuals.

negative Any photographic image in which the subject tones have been reversed. Specifically, the reversed-tone image resulting from the simple development of the film exposed in the camera in the conventional process of taking a picture.

neutral density filter A filter that reduces the amount of light entering a camera without affecting the color.

ni-cad Nickel cadmium battery.

noise Random undesirable picture or sound interference.

objective scene A scene recorded with the camera aimed toward the subject, from a theater audience point of view.

one-shot A shot with one person in it.

opaque projector A projector that can enlarge information from paper, pages from a book, or other nontranslucent or nontransparent materials.

open-reel recorder A tape recorder that uses magnetic tape on separate reels.

optical sound Sound recorded on film by varying the density or areas of a black and white track.

optical soundtrack A narrow band of light and dark areas or lines along one side of motion picture film which through the action of light is converted into the film's sound.

output A connection to an electronic device at which the signal leaves the unit.

out-takes Film footage shot but not used in the final, edited film.

overexposed Too much light reaches the film, making the picture appear "washed out."

overhead projector A projector that accepts transparent and translucent film or other plastic and projects the information prepared on them onto a screen.

overlay One or more additional transparent sheets with lettering or other matter that can be placed over a base transparency or an opaque background.

overprint The superimposition of one scene over another; generally titles, captions, or labels over a background scene or a specially prepared background.

oxide The magnetic particles that are bonded on the tape.

pan The movement of the camera horizontally.

panchromatic Describes an emulsion sensitive to blue and green and some, or all, of the red region of the spectrum.

paper The sensitized paper used in making photographic prints.

parallax The difference between the position of an object in a filmed scene as viewed through a viewfinder and that recorded on film through the camera lens.

paste-up The combination of illustrations and lettering, each unit of which is rubber-cemented in position on paper or cardboard by a temporary method.

patch cord An electrical wire used to connect two pieces of sound equipment (e.g., tape recorders and record players) so that electrical impulses can be transferred between the two units in order to make a recording.

phone plug An audio plug with a ¼-inch-wide connector shaft.

phono plug RCA audio plug.

photoflood A type of incandescent light bulb of high efficiency but limited life, designed to burn at 3400 K.

photographic light meter A device for measuring light levels, either incident upon or reflected from a scene.

pick-up tube The electron tube in the camera or TV set.

pinch roller (pressure roller) A rubber roller which engages the capstan and pulls the tape with constant speed and prevents slippage.

planning board A 3 × 4½-inch board with strips of acetate channels, designed at Eastman Kodak Company, for holding storyboard cards.

playback Reproduction of the sound previously recorded on the tape.

playback head The magnetic head which picks up signals from tape for playback.

plug A form of mechanical interconnector used for quick and easy connection of components, such as phone plug, phono plug, and AC plug.

porta-pak A portable video system.

positive An image in which the tones or colors are similar to those of the subject.

power amplifier An amplifier designed to operate a speaker system.

power cord Cable used to connect a tape recorder to AC current.

preamplifier An amplifier that boosts extremely weak signal voltages, such as those from microphones, magnetic playback heads, or phonograph pick-ups, to a level that is usable by power amplifiers, and at the same time accomplishes the necessary equalization for industry standards.

prerecorded tapes Recordings on tape that are commercially available.

pressure pads Felt pads mounted on arms which hold the magnetic tape in close contact with the heads. Almost always used in "one-motor" tape recorders.

production script A shot-by-shot description of the film with action and camera directions on one side of the paper and sound direction on the other.

psychomotor domain Category of learning that involves the development of physical manipulative skills and physical performance.

pumping in (pumping out) The change of camera position from one motion-picture scene to the next without change of camera angle with relation to the same subject.

quadruplex A video recording system that uses four video heads and scans the tape at a 90° angle.

quartz light Also quartz-iodine light or quartz-halogen light. An incandescent electric light of small size and high efficiency employing a tungsten filament burning in an atmosphere of iodine or bromine vapor and enclosed in a quartz envelope.

quartz lighting Very bright and efficient lighting.

questioning The component of the microteaching model that involves asking and answering questions.

rack focus The focus is shifted so that part of the picture becomes blurred and another part becomes sharp.

rack mounting The method of mounting equipment in a 19-inch-wide standard metal rack.

rangefinder A camera attachment that, upon proper setting, indicates the distance from camera to subject or sets the lens in focus at that range.

rangefinder camera A camera featuring a built-in, optical rangefinder, usually incorporated into the viewfinder and linked mechanically with the focusing mount of the lens so that bringing the rangefinder images into coincidence also focuses the lens.

raster The scanned or illuminated area of the TV picture tube.

rc papers Resin-coated papers.

rear-screen projection Projecting on the back side of a translucent screen, the projector being behind the screen or off to one side, projecting on a mirror.

reflected-light method The measurement of light reflected from a scene by the use of a reflected-light meter held near the camera and aimed at the subject.

reflection The rebounding of light from a surface.

reflex camera A type of camera in which the viewfinder image is formed by a lens and reflected by an inclined mirror onto a ground-glass screen mounted in the top of the camera body. When the exposure is made, the mirror swings up out of the way, and the film is exposed; then the mirror drops back into position.

reinforcement In the microteaching model, involves asking and answering of questions.

release form The form used to obtain written permission for use of pictures taken of persons or of their copyrighted materials.

resin-coated papers Printing papers employing a special base material, treated during manufacture with a "resin coating" which, by limiting water absorption, allows for very rapid processing, and reduces washing and drying time.

resolution The amount of resolvable detail in a picture.

reversal film Film which, after exposure, is processed to produce a positive image on the same film.

RF Radio frequency; the range of frequencies used to transmit electric waves.

RF adaptor A device that allows video and audio signals from a VTR to play back on a standard TV set.

RF amplifier A device used to amplify RF signals.

roll film Film supplied in rolls rather than sheets.

rough cut The second step in editing a motion picture, on completion of which the length of scenes has been shortened somewhat but not necessarily to final length, and action between adjacent scenes is not yet exactly matched.

safelight Illumination, used in various darkroom processes, which is of a color and intensity which will not appreciably affect the emulsions being handled.

saturation The intensity of the color.

scanning The horizontal and vertical movement of the electron beam in the camera or TV tube.

scene (shot) The basic element that makes up the visuals of audiovisual material; each separate picture or amount of motion-picture footage exposed when the release button is pushed and then released. Section of a film in which location and time remain unchanged.

script The specific directions for picture taking or artwork in the form of a listing of scenes with accompanying narration or captions.

SEG Special-effects generator.

sensitivity In photography, the susceptibility of an emulsion to light.

sequence A group of scenes unified by time, place, or action.

set induction Techniques used to open a class or a presentation. It is the setting of the stage for what is to follow in the lesson.

sharpness The subjective impression of clarity of definition and crispness of a photographic image.

sheet film Film supplied in individual pieces; also called cut film.

shooting ratio The amount of tape recorded relative to the amount of tape actually used.

short lens A lens of less than normal focal length; a wide-angle lens.

shotgun microphone A unidirectional type that accepts only a narrow angle of sound and rejects most other noises.

shutter The mechanism, sometimes electronically controlled, which opens and closes to admit light to the film chamber of a camera and control the length of the exposure interval.

shutter release The lever or plunger which, when pressed, allows the shutter mainspring to operate the shutter mechanism and make the exposure.

shutter speed The interval between opening and closing of the shutter of a camera, measured in fractions of a second. Duration of the interval of exposure.

signal Information transposed into electrical impulses.

signal-to-noise ratio The amount of visible "snow" or "noise" in a given picture relative to the actual picture signal. The higher the signal-to-noise ratio, the better the picture.

single-frame filmstrip A series of pictures, each of which is approximately 18×24 mm, oriented horizontally across the width of 35mm film.

single-lens reflex A reflex camera in which the viewfinder image is formed by the camera lens and reflected to a top-mounted viewing screen by a hinged mirror normally inclined behind the camera lens. During exposure of the film, the mirror flips up to seal the ground-glass opening, allowing the image light to pass through to the film chamber. In most designs, a focal-plane shutter is employed.

skew The tape tension.

slant track recording See *helical scan recording.*

slide A transparency, mounted in cardboard, metal, plastic, or glass, for projection onto a screen for viewing.

slow A term used to describe the longer exposure intervals provided by the shutter, e.g., "One-half second is a slow speed." Also applied to relatively insensitive emulsions, as a slow film.

slow- or **overcranked** The camera is run faster than 24 frames per second. When the film is projected at normal speed it gives the effect of slow motion.

SLR Abbreviation for single-lens reflex.

small format The small-width videotape equipment such as ¼-inch, ½-inch, ¾-inch, and 1-inch.

SMPTE The Society of Motion Picture and Television Engineers. A standards-setting committee.

solid state Electronic circuitry that contains no vacuum tubes.

sound effects Sounds such as a police siren, creaking door, wind, etc. Added in order to provide realism to the picture.

sound-on-sound A method in which previously recorded material on one track may be rerecorded on another track while simultaneously adding new material.

special effects generator (SEG) A video mixing device that allows switching between several cameras and a variety of special effects such as dissolves, fades, inserts, wipes, etc.

splicer Device for joining separate pieces of film or tape.

stereophonic sound "Dimensional" sound reproduction achieved through the use of two or more soundtracks recorded through microphones so placed as to provide separation of sounds and heard simultaneously through speakers arranged somewhat apart from each other, according to the size of the room in which it is played.

stop down To reduce the size of the aperture of a lens.

stop motion Technique of exposing film one frame at a time.

storyboard A series of sketches or pictures which visualize each topic or sequence in an audiovisual material to be produced; generally prepared after the treatment and before the script.

subject The thing or view photographed.

subtitles Words, phrases, or sentences that appear over a visual to explain, emphasize, or clarify a point.

superimposition The old and new scene both remain on the screen.

switcher A video switching and mixing device.

sync Synchronization; the timing pulses that drive the TV scanning system.

tacking iron A small thermostatically controlled tool used to tack or attach dry-mounting tissue to the back of a print or to the mount board, so as to hold it in place while the print is being trimmed and heated in the dry-mount press.

take A shot.

take-up reel The reel located on the right side of the tape recorder which accumulates the tape as it is recorded or played.

tally light An indicator lamp on the camera which shows when recording from the camera is being done.

tape deck A tape recorder designed for use in built-in high-fidelity music systems. It is encased in a metal cage rather than a carrying case. The electronics usually consist of record amplifiers and a playback preamplifier. A separate power amplifier is needed to drive speakers.

tape guides Grooved metal posts located on either side of the head assembly to keep the tape tracking properly across the heads.

tape index counter A digital counter used mostly to aid in referring to a particular portion of tape.

tape path The path of the videotape past the heads.

tape splicer A semiautomatic or automatic device used for splicing tape.

tape speed The speed at which tape moves past the heads; measured in inches per second.

tape transport The mechanical portion of the tape recorder mounted with motors, reel spindles, heads, and controls. It does not include pre-amplifiers, power amplifiers, speakers, or carrying case.

taxonomy 1. A device to organize things. 2. A technique of classification.

telephoto lens A camera lens which permits a closer view of a subject than would be obtained by a normal lens from the same position.

test strip A piece of paper or film which is subjected to a sequence of regular and cumulative exposures and controlled development, so as to sample the estimated range of useful exposures in an effort to determine the optimum one.

third generation. Two copies removed from the original tape.

tight shot Close-up.

tilt The movement of a motion picture camera in a vertical plane.

time base corrector A device that electronically corrects the mechanical and electronic errors created by a VTR.

time-lapse photography The exposing of individual motion picture frames at a much slower rate than normal speed; the method accelerates action that normally takes place too slowly for motion to be perceived.

TLR Abbreviation for twin-lens reflex.

tone control Used to vary base and treble response to achieve individually desired balance of tone.

tracking The position of the video head over the recorded signal on the videotape.

transitional devices The use of such techniques as fade out, fade in and dissolves to bridge space and time in motion pictures.

treatment A descriptive synopsis of how the content of an audiovisual material may be organized and presented.

tripod A three-legged stand, usually adjustable in height and provided with a tilting and swiveling head on which a camera can be fastened for support and stability during use.

tungsten light Generally, the light emitted by a heated tungsten filament such as is contained in conventional electric light bulbs. Sometimes used to refer specifically to the light of special photographic tungsten-filament bulbs which are designed to burn at either 3200 K or 3400K. Also often used loosely to apply to artificial light in general, as distinguished from daylight.

twin-lens reflex A type of reflex camera which uses separate but similar lenses in separate compartments of the camera body for the individual functions of viewing and recording the image. See single-lens reflex.

type A Refers to those color films which have been balanced or specially sensitized for use in tungsten light of 3400 K.

type B The group of color films intended for use in tungsten light of 3200 K.

U-matic Sony trademark name which refers to the ¾-inch video cassette format VTRs.

UHF Ultrahigh frequency; a connector used for video cables.

underexposed Insufficient light reaches the film, making the picture darker than normal.

unidirectional A selective microphone pick-up pattern.

VCR Video cassette recorder.

view camera A type of camera in which the image is viewed and composed on a ground-glass screen placed precisely at the film plane. The viewed image is therefore identical to the one presented to the film during exposure.

VHF Very high frequency.

video cassette One reel of videotape and one empty reel in a closed plastic container.

videodisk Video information recorded on a large, round disk.

video gain The amplitude of the video signal.

video in A video input.

video out A video output.

vidicon tube An expensive and versatile electronic pick-up tube used in TV cameras.

voice over A commentary or narration accompanying the picture.

VTR Videotape recorder.

VU meter A volume unit meter which indicates the relative levels of sounds being recorded.

warm tones In photography, shades of red or orange (browns) in the black and white silver image.

white balance The video camera's ability to reproduce white correctly.

white light Generally used (as distinct from safelight) to refer to any light which is capable of exposing film. Also, in color photography, it describes light of a Kelvin temperature suitable for use with a particular emulsion type. Thus, white light for daylight film is considered to be of about 5600 K, while for "type A" film, white light is 3400 K, etc.

wide-angle lens A camera lens which permits a wider view of a subject and its surroundings than would be obtained by a normal lens from the same position.

wild sound Sound recorded without being synchronized with the picture, or with no picture being filmed.

wipe An optical effect in motion pictures in which a new scene seems to push the previous scene off the screen.

workprint An inexpensive copy of original motion picture footage used during editing, and to which the original film is matched before duplicate copies are made.

wow Repetitive slow variations in tape speed.

zoom The focal length of the lens is increased or decreased giving the impression that the camera is approaching or moving away from the object or scene being photographed. A zoom is sometimes mistaken for a tracking shot. However, in the latter, the perspective changes, giving a more dynamic effect. A zoom shot involves no change of perspective.

zoom lens A type of lens of very complex structure which can be adjusted in use to provide a continuous range of focal lengths within its design limits. A camera lens of variable magnification which permits a smooth change of subject coverage between distance and close-up without changing the camera position.

INDEX

W

Z